Hollywood Economics

Just how risky is the movie industry? Is screenwriter William Goldman's claim that "nobody knows anything" really true? Can a star and a big opening change a movie's risks and return? Do studio executives really earn their huge paychecks?

These and many other questions are answered in *Hollywood Economics*. The book uses powerful analytical models to uncover the wild uncertainty that shapes the industry. The centerpiece of the analysis is the unpredictable and often chaotic dynamic behavior of motion picture audiences.

This unique and important book will be of interest to students and researchers involved in the economics of movies, industrial economics and business studies. The book will also be a real eye-opener for film writers, movie executives, finance and risk management professionals as well as more general movie fans.

Arthur De Vany is Professor Emeritus of Economics at the University of California, Irvine and President of Ars Analytica, a consulting company specializing in energy, motion pictures and risk-return analysis.

Contemporary Political Economy Series

Edited by Jonathan Michie

Birkbeck College, University of London, UK

This series presents a fresh, broad perspective on the key issues in the modern world economy, drawing in perspectives from management and business, politics and sociology, economic history and law.

Written in a lively and accessible style, it will present focused and comprehensive introductions to key topics, demonstrating the relevance of political economy to major debates in economics and to an understanding of the contemporary world.

Global Instability
The political economy of world economic governance
Edited By Jonathan Michie and John Grieve Smith

Reconstructing Political Economy
The great divide in economic thought
William K. Tabb

The Political Economy of Competitiveness
Employment, public policy and corporate performance
Michael Kitson and Jonathan Michie

Global Economy, Global Justice
Theoretical objections and policy alternatives to neoliberalism
George F. DeMartino

Social Capital versus Social Theory
Political economy and social science at the turn of the millennium
Ben Fine

A New Guide to Post Keynesian Economics
Steven Pressman and Richard Holt

Placing the Social Economy
Ash Amin, Angus Cameron and Ray Hudson

Systems of Production
Markets, organisations and performance
Edited By Jonathan Michie, Jill Rubery, Brendan Burchill and Simon Deakin

New Economy, New Myth
Jean Gadrey

Hollywood Economics
How extreme uncertainty shapes the film industry
Arthur De Vany

"With this book, the agnostic mantra of the motion picture industry 'nobody knows anything' finally loses its legs. Somebody does know something, and it's Art De Vany. When the history of motion picture industry thought is written, this author and this book will have its unique place and special recognition in that story."

Bruce Mallen, PhD, *Director, The DeSantis Center,*
Dean, College of Business, Florida Atlantic University

"Arthur De Vany has revolutionized the way in which Hollywood can be thought about. He has brought a rigour to the study of film that simply cannot be ignored, even where the reader does not always agree with his conclusions. These articles are essential reading for anyone wanting to understand the manner in which Hollywood works."

John Sedgwick, *Principal Research Fellow in Economics,*
London Metropolitan University, UK

"Screenwriter William Goldman is famous for having said that 'nobody knows anything' about the economics of the entertainment industries. His claim is now proved wrong for the movie industry. Arthur De Vany's book makes it transparent. Even those who are not very familiar with the sophisticated statistical and econometric tools that he uses will understand the arguments, which are explained with great clarity and lucidity. De Vany's book will soon be among the very few classics on the entertainment industries."

Victor Ginsburgh, *Professor of Economics at Université*
libre de Bruxelles, Belgium

Hollywood Economics

How extreme uncertainty shapes
the film industry

Arthur De Vany

LONDON AND NEW YORK

First published 2004
by Routledge
2 Park Square, Milton Park, Abingdon, Oxon, OX14 4RN

Simultaneously published in the USA and Canada
by Routledge
270 Madison Ave, New York, NY 10016

Routledge is an imprint of the Taylor & Francis Group

Transferred to Digital Printing 2006

© 2004 Arthur De Vany

Typeset in Galliard by
Newgen Imaging Systems (P) Ltd, Chennai, India

British Library Cataloguing in Publication Data
A catalogue record for this book is available
from the British Library

Library of Congress Cataloging in Publication Data
A catalog record for this book has been requested

ISBN 0–415–31260–4 (hbk)
ISBN 0–415–31261–2 (pbk)

Contents

Figures

Tables

Acknowledgments

Many of the chapters of this book were published in scientific journals or presented at scientific meetings. I gratefully acknowledge the permission by the publishers of these journals to republish the articles in this collection. I am also grateful to my coauthors and collaborators Ross Eckert, David Walls, Cassey Lee and Hank McMillan for permission to publish our joint work here. The papers and the journals where they were published or the professional meetings where they were presented are as follows:

Arthur De Vany and David Walls, "The market for motion pictures: rank, revenue and survival," *Economic Inquiry*, XXXV (4) (October 1997) 783–797 by permission of Oxford University Press.

Arthur De Vany and David Walls, "Bose–Einstein dynamics and adaptive contracting in the motion picture industry," *Economic Journal*, 106 (439) (November 1996) 1493–1514 by permission Blackwell Publishing.

Arthur De Vany and David Walls, "Quality evaluations and the breakdown of statistical herding in the dynamics of box-office revenue." Presented at the American Economic Association Meetings 2003.

Arthur De Vany and David Walls, "Uncertainty in the movies: can star power reduce the terror of the box office?" *Journal of Cultural Economics*, 23(4) (November 1999) 285–318 by permission Kluwer Academic/Plenum Publisher.

Arthur De Vany and David Walls, "Does Hollywood make too many R-rated movies? Risk, stochastic dominance, and the illusion of expectation." *The Journal of Business* 75(3)(April 2002) 425–451 by permission University of Chicago. All rights reserved.

Arthur De Vany and David Walls, "Big budgets, big openings and legs: analysis of the blockbuster strategy." Presented at the Econometric Society Meetings 2002.

Arthur De Vany and Ross Eckert, "Motion picture antitrust: the *Paramount* cases revisited," *Research in Law and Economics*, 14 (November, 1991) 51–112, JAI Press.

Arthur De Vany and Hank McMillan, "Was the antitrust action that broke up the studios good for the movies? Evidence from the stock market," *American Law and Economics Review*, Oxford University Press.

Arthur De Vany and Cassey Lee, "Stochastic market structure: concentration measures and motion picture antitrust," presented at the Private Enterprise Economic Association Meetings 2000.

Arthur De Vany and David Walls, "Motion picture profit, the stable Paretian hypothesis, and the curse of the superstar," *Journal of Economic Dynamics and Control*, Elsevier.

Arthur De Vany, "Contracting with stars when 'nobody knows anything'," portions presented at Rotterdam Conference of the Society for the Economics of Art and Culture.

These papers were written over a period of nearly two decades. For a good part of that time my research was supported by the Private Enterprise Research Center at Texas A&M University. Its Director, Tom Saving, has been a lifelong supporter and colleague. I'm grateful for his support and friendship. My wife, Bonnie, has been my constant supporter for longer than I can say; everything in this book has passed the Bonnie Test for common sense, so it must be right.

Prologue

I have been thinking about the movie business and writing papers about it for about twenty years. This volume is a collection of those papers. In these papers I have sought to understand the industry at a fundamental level as I puzzled over many questions. Why haven't the corporate studios been more successful? Why was the studio system so successful? What are the sources of risk in the business and why is risk hard to deal with? Is it possible to measure how risky the movie business is? Why are returns so low when uncertainty and risk are so high? Do stars and marketing really alter a movie's risk or revenue? Is a star really worth $20 million? Does corporate management get in the way of creativity? How do you price movies and pay talent when you can't know how much a movie will make? Can you predict how successful a movie will be? Why do studio executive earn so much? Are they really good at choosing which movies to produce? And, most important to me as a movie fan, why do there seem to be so few good movies?

The chapters in this book are the results of my work on these questions. I worked with a number of collaborators in this research: Ross Eckert, David Walls, Cassey Lee and Henry McMillan. David and Cassey are my former graduate students who became valued collaborators. Hank is a former colleague. Ross and I go way back to when we argued over the movie business more than thirty years ago as graduate students at UCLA. As a result of those discussions, Ross and I wrote the first paper in what was to become a long line of some fifteen or so papers that I have written on the movies over the past two decades. Most of these papers have been published in scientific journals, though a couple are original to this book. What distinguishes these papers from the research that had been done for years on the movie industry is their point of view and their scientific rigor.

I think of motion pictures as an information industry, arguably the first major information industry of the twentieth century. Many things flow from that observation, as you will see. To express this insight, I had to create mathematical and statistical models of the dynamics of information which I see as a mixture of signals about quality (word-of-mouth) and quantity (box office numbers and rankings). I wanted to see what possibilities there were for the evolution of a movie's box-office revenues and this had to be explored mathematically. We weren't really prepared when we found the answer: movie revenue dynamics are so complex that they are nearly chaotic. We have been able to show that movie revenues follow a nonlinear dynamics that bifurcates into two separate paths, one leading to long lives and high revenues and the other leading to brief lives and low grosses. Bifurcations are a signature of chaos.

Eventually, these complex dynamics converge to a group of statistical distributions that govern risk and returns, these are the so-called stable distributions. But, don't think that "stable" means that movie revenues are well behaved. They are not, they are wild. "Stable" refers to the fact that the process retains its structure on all scales. If structure is the same on all scales, then things become similar and they look the same from many different angles. You will see that this property—self similarity on all scales—is a recurring theme in this book. It is natural for movie grosses to converge to stable distributions because they are the limiting distributions of all dynamical processes that retain a self-similar structure under choice, mixture and addition. These are the natural operations through which box-office revenues aggregate during the course of a movie's run. For reasons that will become apparent, we call the model of this process the stable Paretian model.

The stable Paretian model opens our understanding to just how wildly uncertain this business is and how that uncertainty shapes its organization and operation. I call the uncertainty "wild" because the stable Paretian model implies that the probabilities of outcomes look nothing like the Gaussian distribution with its comforting mean and standard deviation. Neither the mean nor the standard deviation need exist for a stable distribution, even though they do exist for a Gaussian distribution.

If the movies followed the Gaussian distribution (the bell-shaped curve your teacher probably used to assign grades), most movies would be similar to one another and close to the mean. The "typical" movie would be represented by the average. A movie that earned box-office revenues three standard deviations above the mean would be extremely rare and virtually no movies would gross an amount more than four or five standard deviations above the mean.

In a Gaussian world, there would be no movies like *Titanic, Gone With the Wind* or *Star Wars.* Most movies would be comfortably mediocre and neither earn nor lose too much. But, the movies, we have learned in our research and you will see in what follows, are not like that. There is no typical movie and averages signify nothing. An average differs from an expectation. Averages of key variables are unstable over time. Expected values may not even exist. And the variance is infinite. There is room for extraordinary movies like *Titanic* and even higher grossing movies in the "heavy" upper tail of the stable Paretian probability distribution where the extraordinary events lie.

The movie business is completely and utterly non-Gaussian because it is a business of the extraordinary. Because extreme events drive the movies, it is high on kurtosis. Kurtosis refers to the shape of the probability distribution; it is peaked and skewed, with a heavy tail to the far right. A kurtotic distribution has the property that small events are plentiful and of little importance while extraordinary events are rare, but powerful. Thus, a Gaussian world where most movies are similar and close to the average is a mediocracy. There is no room in a mediocracy for extraordinary movies. In the movies, kurtosis is large and extraordinary movies and artists abound.

The high kurtosis of the movie business makes it a what Nassim Taleb calls a kurtocracy. A kurtocracy is dominated by a handful of extraordinary movies and artists that account for nearly all the industry's impact and revenue. The kurtocrats are Hollywood's elite, the actors, directors, writers and producers who are associated with the major movies. Hollywood's kurtocracy would not exist if the movies were Gaussian and, without its high kurtosis, the movies would be a pretty dull business.

You will see much more about this in the chapters to come, for the "wild" uncertainty that comes of the kurtotic and non-Gaussian nature of this industry trumps everything else. For now, this shouldn't surprise you because these are the key features of an information industry. Information products and producers can vary in the extreme because information variables are essentially unbounded and small differences can grow explosively through nonlinear feedback. Nonlinear processes and unbounded variables can "go anywhere" and these processes produce features of the movies that are counterintuitive and surprising. They confound standard thinking and ordinary managers.

Given this wild behavior, the industry is never in equilibrium and is constantly renewing itself. In our research we found such wild statistical behavior that we came to feel that nobody could optimize anything. Shedding this belief meant jettisoning one of the key elements of conventional economics. We do not believe that people don't try to make the best decisions they can. We believe they are so deeply shrouded in clouds of uncertainty that optimizing is just not possible. Decisions are made by gathering information and then choosing the best actions in light of what is known. Information flows are delayed and erratic. So, many errors are made and many false beliefs are held.

The movies is full of surprises because it is an industry of innovation and discovery. Movie makers innovate and put a movie on the screen and then the audience discovers what it likes. Box-office revenues relay the audience's likes and dislikes to the movie makers in the coin of the box office. Information gathered during a film's run sets the stage for other decisions and triggers a wide range of contingent clauses in contracts involving prices, careers, bookings and compensation.

While many of the chapters included here demonstrate these issues, we also take up the problems inherent in trying to manage this business. It is a frightening business. There are many failures and just few successes. And the handful of great successes swamp all else. No wonder motion pictures is such an egocentric business, the uncertainty is so intense that decisions force one to be deeply self-conscious and self-involved.

The anguish of doubt and regret is largely hidden (with exception of the occasional drug charge or shoplifting incident that makes the news) but it underlies nearly everything that people in the business do. Every actor, writer, director and studio executive knows that their fame and success is fragile and they face the imminent threat of failure with each film they make. It can all go away in a flash. We see the successes, which are rare, but are seen so often that they might be taken as the norm. They are not. Most Hollywood careers are struggles from the unemployed actor doing casting calls up to the newly hired or fired studio head.

Careers can skyrocket and flame out in a wink of time. It is an industry of many people and wannabes and not a whole lot of work. Only a handful do most of the work and make the movies that generate the bulk of revenue and profit. The movies, we show, obey the de Solla–Price law which holds that half the work in a profession is done by the square root of the number of its participants. In his usual insightfullness, Robert Evans has his own form of this rule: At any time, most of the people in the business are not working.

Artist incomes and motion picture revenue follow laws similar to the Price–Evans law, as you will see in the chapters that follow. In fact, you will see that statistical self-similarity holds in all areas of Hollywood. Whether it is revenue, work, profit or productivity, every statistical distribution in the movies looks the same, with adjustment for their measurement scale. They are all skewed and kurtotic. Work, revenue and money are distributed like the movies are a third world country—the princes get almost everything and the paupers almost nothing. And everyone knows who the princes and the paupers are when the box-office reports come in. There is little privacy; your successes and failures are out there on the screen and in the box-office reports for everyone to see.

Yet the industry draws investors and would-be participants by the yard; far more than it could ever absorb if returns were Gaussian or careers were long-lived. The wild uncertainty makes turnover high, except for the kurtocrats who enjoy long-lived success, but it is renewing at the same time. There is always a new face or writer or director with something original to contribute. And there is always a green face in the executive suite to turn a greenlight on something that might be different.

The frightening thing about trying to manage this business is that there are no tangible means to reliably change the odds that a movie will succeed or fail. Marketing can't change the odds. There is no evidence to show that marketing has much to do with a film's success. Marketing is mostly defensive anyway; a studio has to market its films just to draw attention in a field where everyone is shouting. If you don't shout too, you will be drowned out and may not be noticed.

When you take all the marketing together, it is probably a wash. A studio can gain a slight edge through marketing, but there is no gain at the industry level. But a temporary advantage, gotten through advertising, quickly washes away if the film doesn't have the goods.

Nor can casting stars in movies increase the odds of success. Star movies cost a lot of money and have a tendency to run over budget. A single person just can't put that much production value on the screen in a story involving many characters and events. It is really a strong performance in an outstanding film that makes a star. Movies make stars, not the other way around.

What then, do stars do for a movie? Aside from earning a higher least revenue, a star movie has only a slightly higher chance of making a profit than a non-star movie. If the star's agent extracts the higher expected profit in the star's fee, then the movie almost surely will lose money. I call this the curse of the superstar.

Opening big and leading at the box office is a momentary success. A movie has to attain or sustain box-office dominance over many weeks to make major money. The size of the opening does not predict how the ensuing battles will evolve or how much money the film will take in. Why do executives compete so strongly for stars when they can assure no more than a higher expectation of a movie's least revenue? It seems to be based on a belief that the opening predicts how much a movie will make. That turns out to be false, as you will see.

A word about the organization of the book. I have grouped the articles into four parts: dynamics, wild uncertainty, judges and lawyers and extremes. There are three chapters in each of these parts. I have written a brief introduction to each part noting the main issues, techniques and results of the papers contained therein. I have done some editing of the originally published articles to remove similar tables and graphs and to make the book easier to read. But, I have not sacrificed rigor or completeness; these are refereed articles, published in scientific journals and their results have been independently confirmed and replicated by other authors many times over.

I have written a couple of new chapters for this book that were not published previously. One of these concerns artists, primarily actors and directors. It examines how productive they are and how they are paid. I establish the Price–Evans law of artists, estimate the half life of a star and see if we can separate luck from talent in career patterns. In another new work for this book, I try to put all this work into a more complete model, a model that begins to bridge the gap between standard management and economic models and the reality of the business.

In the Epilogue I muse on how one might manage a business where nobody knows anything. It is here that I take up the fundamental flaws my research reveals about the way the modern corporate studio manages the movie business.

Last, I think you will see why conventional models fail spectacularly to explain the movies and why we had to do a new kind of economics to come to grips with this endlessly fascinating business. I believe we have built a consistent and fundamental model of the industry that is of interest not just to scientists, but to movie fans and moviemakers. And I believe these models apply to other industries.

In science, as in the movies, creativity takes you to unexpected places. This research has been exciting because we made unexpected and wonderful discoveries. It would have been hard to imagine at the outset that by applying high-brow mathematical and statistical science we would end up proving Goldman's fundamental truth that, in the movies, "nobody knows anything."

None of our results is more surprising than finding that, hard-headed science puts the creative process at the very center of the motion picture universe. There is no formula. Outcomes cannot be predicted. There is no reason for management to get in the way of the creative process. Character, creativity and good story-telling trump everything else.

Part I

Box-office champions, chaotic dynamics and herding

The first three chapters set the stage for the dynamic point of view that pervades my research on motion pictures. They show in what sense the movies is an information industry and what that means. It means that audiences don't know what they like until they see it; every film is a discovery and audiences transmit their discoveries to others in a dynamic cascade of information. The process of many individuals choosing among movies and transmitting their knowledge to others amid a changing slate of competing movies induces a very complex dynamical behavior that leads to wildly diverging outcomes. This insight is fundamental to understanding the industry and how it is organized and functions. Because the process is stochastic and complex we have to be humble about what we can learn. Motion picture gross box-office revenue is an information variable so it has no natural limit in scale or size. Therefore, there is no typical or natural amount a movie might gross.

Movie tournaments

In the first chapter, David Walls and I model the motion picture market as an evolving dynamic tournament among movies. Motion pictures live unpredictable and brief lives. They are tested each week by the audience against a changing cast of competitors. The motion picture theatrical run is an accelerated stress test and when a movie breaks, its run is over. Fallen competitors are replaced by new contenders. Box-office reports on the news rank the movies by their grosses. The highest rankings pay the biggest prizes. When a movie can no longer command a prize among the top fifty, it dies to be replaced by a new contender. Films move up and down in the rankings as they play out their runs against old and new competitors. The distribution of motion picture revenue among movies and their survival rates evolve as a dynamic tournament among films contesting for audience.

The average rank attained by a newly released movie is right in the middle of the 50 film tournament: mean rank at birth is 25. The average ranking when a film dies is 36. Average run life is only 5.71 weeks. A film has less than a 25 percent chance of lasting 7 or more weeks and less than a 15 percent chance of lasting more than 10 weeks. The hazard of death rises the longer a film runs. A film has a lower hazard if it is re-released, if it opened on more screens, and if its revenues are high.

As in any tournament, the distribution of prizes is biased in favor of the winner and close contenders. Box-office revenue declines with falling box-office revenue, sharply from first to second, less dramatically from third to fourth and continues to decline at a more modest rate at the lesser ranks. A film loses $2.4 million in weekly revenues when it declines from rank 1 to rank 2; but it only loses $235 thousand when it falls from rank 4 to rank 5. The revenue distribution is convex in rank, so much so that the top 20 percent of the movies earn more than 80 percent of revenues. The champion gets the lion's share of the revenue.

In the movies, as in sports, the box-office champion stays longer at the top than contenders do in lower ranks. The number of entrants and exits each week are completely random and independent of the number of entrants and exits in previous weeks. The features of the exhibition contract—its declining rental rate and hold-over clause—are explained within the tournament framework.

In all, this chapter says that the theatrical market is remarkably like a sports tournament; prizes are unequal, champions outlast contenders to which they eventually succumb, the turnover among lesser ranks is random and independent, and each week is a new tournament in which last week's leaders are the top seeds. No wonder box-office numbers are reported on the evening news like the scores of contestants in a sports contest.

A turning point in research

Chapter 2 is a turning point in my research on motion pictures. My view of the industry has been based on the idea that movies are unique and that audiences have to discover what they like (cf. Chapter 7). The way they process and exchange information leads to complicated dynamics that create extreme differences among motion picture revenues. This chapter nails that down and opens the door to many of the other chapters in this book because it is our first look at the relationship between the dynamical behavior of audiences and box-office revenues and the sorts of statistical distributions they produce. David Walls and I modeled the distribution of box-office revenues over weeks of the run and found that this sequence followed a Bose–Einstein process that converged to a power law.

The end-of-run distribution we discovered in this chapter is our first hint that motion picture distributions are attracted to Pareto distributions. From that point there was no turning back and the stable-Paretian hypothesis becomes a centerpiece of our analysis in many subsequent chapters, including Cassey Lee's doctoral dissertation, which was done at UCI under my direction.

In trying to understand what dynamical process converged to a Paretian attractor we found that the Bose–Einstein statistical process was a good fit.

This was a rather remarkable finding that has received a lot of attention and has been verified in our own research and by others. This chapter was featured in John Cassidy's article "Chaos in Hollywood" in *The New Yorker* (March 31, 1997), pp. 36–44.

The Einstein of Bose–Einstein is that Einstein, the one of relativity. The Bose–Einstein theory is not as hard as the theory of relativity. In fact, it is completely natural if you think about it a bit. Movie fans imitate one another to some extent. They also share information with one another about their likes and dislikes. This means that a consensus about movies grows over time as the audience explores movies. The process of discovery and convergence to a consensus is part imitation, part communication. As the consensus begins to converge, so does the way the audience distributes itself over movies.

The Bose–Einstein process is a stochastic version of this part-imitation, part-communication process. As it evolves, the probability that *n* people will be drawn to a movie depends on the number who saw it before. The probability of a growth in revenue depends on the level of revenue already earned. This implies that the movies with the largest revenues have the largest expected growth which produces a nonlinear feedback in demand. But the process is not deterministic because it is only the probabilities that evolve. Given this nonlinear, stochastic feedback process, revenue can grow or shrink rapidly and random variations in demand can be magnified dramatically.

As the probability distribution evolves it becomes more stable against random perturbations. Eventually the distribution converges to a distribution that reflects fully informed preferences.

A turning point in motion picture revenues

Chapter 2 led David Walls and me to ask when and how the nonlinear dynamics of the Bose–Einstein process became a driving force. In Chapter 3, we found the answer. In this chapter we showed that audiences at first key in on box-office revenue rankings to choose movies. This imitative behavior is called statistical herding in the economics literature. But, things change about 4 weeks into the run. At that point revenues undergo a turning point where they bifurcate into two branches, a "hit" branch and a "non-hit" branch. At this turning point, the distance (in box-office revenues) between good and poor movies grows at an accelerating pace. At the turning point, the nonlinear mapping of current on past demand rapidly separates the hits from the misses.

By the fourth week of the run people have a fair amount of information about whether a movie is good or bad. Paradoxically, when they begin to act on word-of-mouth information they appear to be herding because so many go to the same movies. But, their apparent herding is not imitation. A consensus appears to emerge at the bifurcation point when the information from positive word of mouth and growing box-office revenues confirm one another.

Given how audiences choose films—part imitation, part information—and the powerful effects of nonlinearity, box-office revenues can really go anywhere.[1]

Contracts and adaptation

All three of the chapters in Part I investigate in some way the question: How does the business remain poised to adapt to such complex and unpredictable audience behavior? We suggest that the industry is ingeniously well-organized to successfully follow the hits wherever they may go and to shed the losers quickly.

In each of the chapters in this part of the book we discuss some aspect of contracts and practices employed by the industry. They show that the industry has evolved a way of letting the motion picture run go where the audience wants it to go. Then it uses clever contingency rich methods to respond to the information revealed during the run to extract the theatrical value of a movie. These methods are decentralized, adaptive and self-organizing rather than optimal. Semi-stationary admission prices, local information and contingent contracts flexibly match the film rental price and supply of theatrical engagements to demand as it is discovered by the audience and revealed in box-office revenues.

Just a few hits dominate the industry's revenues and profit. Nobody knows which films these will be until the audience tells them. Once the audience signals its demand, the industry is ready to flexibly adapt supply (theater screens) and prices (rental rates) to demand.

1 The market for motion pictures

Rank, revenue and survival

1.1 Introduction

On any given day in every major city there may be about 50–100 motion pictures playing on theater screens. Each is unique and its producer hopes it will catch that bit of magic that lights up the screen and the box-office. Yet, few achieve this feat and most lead lives that are unpredictable and brief. Each film competes for screens and audiences during its brief life against a changing array of imperfect and equally unique substitutes. How is one to understand this market? Is there any sense in which this market could be called competitive and how is it organized? What institutions and contracts are adapted to this market environment and how do they shape the results?

We seek to answer some of these questions by examining the relationship between contractual practices and motion picture hazard and survival distributions. Our modeling is based on a large sample of motion picture revenues and theater bookings during their theatrical runs. Our detailed sample of motion picture lifetimes is a rare look at this fascinating industry whose unusual and complex features challenge economic theory in interesting ways. As Smith and Smith (1986) observe, "Given the interesting characteristics of movies as ideal examples of differentiated products and of the institutional arrangements governing their production and distribution, such increased data availability would make this an exceptionally attractive area for applied micro-economic research" (p. 506).

Much of the empirical work on the motion picture industry has focused on the attributes of successful movies. Smith and Smith (1986) examined the impact of Oscar awards on the cumulative rentals of movies released from the 1950s through the 1970s. In more recent work, Nelson *et al.* (undated) have quantified the value of an Oscar award on movie revenues using a large panel of data and an event study methodology. In a somewhat broader empirical study, Prag and Cassavant (1994) examined the determinants of movie revenues and marketing expenditures and found that marketing expenditures and quality are important determinants of a film's success, and that production cost, star actors and awards are positively related to marketing expenses.

In this chapter, we attempt to account for the dynamic patterns in the data. Because each film is unique and plays in its own way, its life as a commercial product in the theatrical market is hazardous. Indeed most motion pictures have short and unpredictable lives because audiences must discover what they like and films compete against an ever-changing cast of competitors. For these reasons it is productive to use evolutionary models of survival and death to model the data.[1] We model competition among films in the theatrical market as an evolving rank tournament of survival.[2]

Motion pictures live and die in the box-office tournament as they are challenged during their run by a randomly evolving cast of new competitors. The challengers come from films previously released and from newly released films. The contending films are ranked by filmgoers and those with high rank survive and are carried over to the next week. Low ranked films fail and are replaced by new contenders.

These attributes of the motion picture market contribute to increasing returns and give the box-office revenue distribution the distinctive convex shape of a tournament prize distribution.[3] The leading products command a disproportionate share of the market and they have longer runs. Even then, a film's rank in the tournament is ephemeral and its life unpredictable. "In fact, of any 10 major theatrical films produced, on the average 6 or 7 are unprofitable, and 1 will break even" (Vogel 1990).

1.2   The motion picture exhibition market

Once a film is produced, it is distributed to theaters who "exhibit" it for audiences. The distributor chooses a release pattern—the number and location of theaters in which the film is "booked" or licensed to play. Distributors also choose a date at which to release their films for exhibition, looking for high demand periods and seeking to avoid playing against films that are strong substitutes. Distributors time some films for release during Easter and Christmas because they are high demand periods; films also vie for screens during the period preceding the Academy Awards. But, release timing is difficult because finishing a production and editing and preparing copies for release are highly uncertain processes.

The number of theaters and their locations for the initial release are based on the distributor's a priori estimate of demand. The size of the initial release determines the number of "prints" or copies that are needed for distribution to each of the theaters. The number of viewers who are able to see the film is limited by the availability of seats in the theaters booked in the release. The Motion Picture Antitrust Decrees of the late 1940s limit the distributor's release strategies significantly. All the signatories, the major theater owning studios during the 1940s, were made to divest themselves of their theaters. Until very recently, no distributors owned theaters. In addition, the decisions leading to the Decrees found illegal certain contracting practices that can limit a distributor's release strategies: generally, long term contracts, franchises, or repeated licensing between a distributor and a theater are illegal under the antitrust laws.

Most distributors use an auction process to license films to theaters, following the advice of their lawyers who interpret the court rulings to say that an auction of individual licenses, one theater at a time, is the approved method.[4] They send out bid letters announcing the tentative date the film is available to be played and the suggested terms the distributor is seeking. Theaters bid or respond indicating a willingness to negotiate a license and one or several of them are selected, depending on the distributor's release plans. If the film plays better than expected and fills all these seats, there is an economy of scale up to the capacity of the system in each week the seats are filled.

Supply adaptation is dynamic. The supply of seats for a successful motion picture is adjusted by expanding the number of weeks the film runs in theaters where it is booked; the exhibition license includes a "hold-over" clause which specifies a box-office revenue in the latter weeks of the contracted run that will cause the run to be extended another week. Lengthening the run conditional on each week's revenue is a source of increasing returns as the film's production, print and advertising costs are fixed with respect to changes in the length of run.

Distributors use the information they acquire from box-office reports to adjust the release pattern dynamically to match supply to demand. New exhibitors may seek to play the film if it is drawing large audiences and they can be added in accord with contractual commitments to the initial exhibitors and the availability of prints.[5] Strong demand may even lead the distributor to produce more prints. Supply adjusts dynamically to high demand by adding exhibitors and lengthening runs. The decisions which distributors and exhibitors make during each week of a film's run rely on information that is widely circulated in daily and weekly trade publications and overnight reports via proprietary channels.

A film's release pattern represents its distributor's strategy for acquiring the demand information that will guide subsequent decisions. The release also sets the initial supply of engagements. Because of the flexibility which the exhibition license affords to adapt the supply of seats to demand, the industry is able to capture bandwagon effects if word of mouth and other information starts one.[6] A wide release on many screens draws a large, simultaneous sample in many theaters and cities. A tailored release strategy samples sequentially, starting at a few theaters and using the information from that sample to adjust bookings if the film builds an audience. Following its initial release, decisions to expand engagements or lengthen its run are both centralized and decentralized. The exhibitor uses local information about how the film is playing in his theater to decide how long to play it. This decision is conditioned on the exhibition contract, but it uses local information and the exhibitor's assessment of how much other motion pictures might earn at his theater. Exhibitors in other locations monitor box-office information in making their booking decisions if they have not already committed all their screens to other motion pictures.

The contract to exhibit a film usually requires the theater to show it for a minimum number of weeks; 4 weeks is a common minimum, though 6 and 8 week minimums are sometimes used. In addition, the contract contains a hold-over clause that requires the theater to continue exhibiting the film another week if the previous week's box-office revenue exceeds a stipulated amount. The contract also clears an area near the theater where the distributor cannot license the film to other theaters.

Exhibitors pay the film's distributor a weekly rental for the privilege of playing it. The rental usually is some percentage of the exhibitor's box-office revenue. The typical arrangement might call for the exhibitor to pay 90 percent of his box-office revenue in excess of a fixed amount negotiated as part of the contract; this fixed amount is referred to as the "house nut" in industry parlance. In addition, the rental rate is subject to a minimum percentage of total box-office revenue. The minimum rental rate usually declines over the length of the run; a typical arrangement for a four week run might be minimum box-office percentages of 70, 70, 60, 60 in the first, second, third and fourth weeks, respectively. Additional weeks beyond the contract period often have a minimum rental rate of 40 percent of total box-office revenues. Throughout the run, the 90 percent rental rate will be triggered any time the weekly box-office gross exceeds the house nut.[7]

The declining rental rate is an incentive to the exhibitor to continue the film even as its box-office revenue declines over its run. After it has run for the contracted minimum number of weeks, the film must then gross more than its hold-over figure to continue playing. Or, it must yield the exhibitor higher earnings than the alternative motion pictures available. Thus, the decision to continue the exhibition is placed in the hands of the exhibitor after the contingent hold-over clause no longer is binding.[8] Whether it is the hold-over clause or the exhibitor's decision to hold a film over, the decision to extend the run at each theater uses local information about how the film is playing at that theater. The temporal and spatial distribution of a film's engagements is self-organized; it evolves after the opening release in response to the pattern of demand throughout the film's run, adapting to the audience through contingent contracting and exhibitor decisions that rely on a mixture of global and local information.

1.3   Evolutionary survival tournaments

Each film must earn a critical level of box-office revenue to survive. Otherwise, theatrical exhibitors will choose to exhibit another film. The critical level changes as audiences change, films play out their runs and other films begin their runs. The level of revenue that ensures survival is the amount the film must earn each week to be carried over in theaters to the following week. As stated earlier, the survival level partly is set by the hold-over amount in the exhibition contract. But, it depends also on what the film earns against the other films competing with it for screens.

Motion pictures can be modeled as a competition for the top fifty spots in a tournament.[9] Suppose there are $m(t) > 50$ motion pictures competing at time t for these 50 spots. Rank box-office revenues at time t from high to low in the m-dimensional vector $r = (r_1(t), r_2(t), \ldots, r_m(t))$, where low numerical ranks indicate high box-office revenues. Revenue is a random variable which may depend on the motion pictures in the tournament, which change each week. The cutoff revenue a film must earn to survive from week t to $t+1$ is a random variable $C(t) = r_{50}(t)$. A film i playing at time t survives if its revenue exceeds the critical level, $r_i(t) \geq C(t)$, and it dies otherwise. Every film in *Variety's Top-50* sample meets this condition by definition. When a motion picture does not meet the condition, it dies and falls out of the sample to be replaced by another contender. Generally, a film will move up or down in rank during each week of its run. The length of occupancy time in each rank is a random variable and we place no restrictions on this stochastic process, letting the data determine the statistics of survival in rank. In order to have a long life, a film can achieve high rank and occupy several ranks before death, or it must spend a long time at low rank or ranks. This latter path to a long run seems unlikely and the data confirm this.

A few films will achieve high rank in their first week, but most will not. If a film gains acceptance it may increase its rank over time, but its revenues eventually must fall as it exhausts its potential audience. Over its lifetime, then, a film's rank must eventually decline. As it ages and declines in rank, it becomes vulnerable to newly released pictures and we might expect that its occupancy time in rank becomes more brief; its hazard also should rise.

We should expect a film's rank at birth to be the mean rank, 25.5, and expect its rank at death to be between 26 and 50. These do hold for the sample (see Table 1.1); mean rank at birth is 25.51 and mean rank at death is 36.02. The median rank at birth is 26 and the median rank at death is 39. Similarly, longer lived films must achieve and sustain high ranks. In the sample, the mean life is 5.71 weeks and the median is 4 weeks. Of 31 films that ran 6 weeks, 28 were at sometime in their run ranked in the Top 10 or higher. Films that achieved a highest rank of 1 during their run had a mean run length of 17.67 weeks. The top ranked film holds its first place position an average of 2.86 weeks, nearly twice as long as films hold second place in the ranking. Mean occupancy time in rank declines as rank declines.

1.4 The *Variety* data

The data are from *Variety*'s domestic box-office sample. The *Variety* sample is a computerized weekly report of domestic film box-office performance in major and medium metropolitan market areas. The sample includes the box-office revenue, the numbers of theaters and screens showing a film, the number of weeks of the run on any prior release, the rank and the number of weeks in the current release on the chart of the top fifty grossing movies in the United States. These data are from a sample of theaters covering approximately 12–15 percent of the nation's screens. Summary statistics of the sample are contained in Table 1.1.

Each film in the sample was tracked from its birth to its death: birth is defined as the inception of a run and death is defined as falling off the *Top-50* chart.[10] The sample includes 350 unique motion pictures that were listed on *Variety's Top-50* chart during the interval from May 8, 1985 to January 29, 1986, inclusive. Of these movies, 251 entered for the first time and 99 re-entered after previously falling out of the *Top-50*; the two groups of films are referred to as *New-Entries* and *Re-Entries*, respectively.

The top four grossing movies in our sample accounted for 20.5 percent of the total box-office revenue, and the top eight grossing movies accounted for 28.1 percent of the total box-office revenue. In descending order the eight highest grossing movies in the sample were *Back to the Future, Rambo, Rocky IV, Cocoon, The Jewel, The Goonies, The Colorado* and *White Knights.* The four lowest grossing movies accounted for only 0.0036 percent of the total box-office revenue, and the eight lowest grossing movies accounted for 0.0082 percent of the total box-office revenue. The highest grossing movie, *Back to the Future,* earned cumulative revenues of $49.2 million, while the lowest grossing movie, *Matter Of Importance,* earned $4,615. The mean of box-office revenue is $1,918,261 and its standard deviation is $4,528,685, more than twice the mean.

Table 1.1 Summary statistics of *Variety's* box-office sample

Variable	Obs.	Min.	Max.	Median	Mean	Std dev.
Rank at birth	350	1	50	26	25.51	16.19
Rank at week 2	266	1	50	22	23.06	15.49
Rank at week 3	232	1	50	24.5	24.24	15.43
Rank at week 4	191	1	50	21	23.46	14.30
Rank at week 5	173	1	50	24	24.28	13.39
Rank at death	309	1	50	39	36.02	11.84
Weeks at rank 1	14	1	12	1.5	2.86	3.03
Weeks at rank 2	27	1	4	1	1.48	0.75
Weeks at rank 3	30	1	4	1	1.33	0.71
Weeks at rank 4	35	1	3	1	1.14	0.43
Weeks at rank 5	33	1	3	1	1.21	0.60
Weeks in *Top-50*	309	1	40	4	5.71	5.18
Showcases	2,000	1	347	13	53.12	73.25
Cities	2,000	1	191	4	7.44	8.08
Weekly revenue (1,000s)	2,000	2.5	10,000	57.5	335.70	685.26

Note
The sample contained 350 movies of which 309 were observed to fall from the *Top-50* chart. The remaining 41 movies were censored. Rank at death and weeks in *Top-50* were calculated using only the uncensored observations.

Table 1.2 Distribution of highest rank and length of run

Length of run	Number of films by highest rank achieved					
	All	*Rank 1*	*Top 5*	*Top 10*	*Bottom 10*	*Bottom 5*
1	84	0	1	2	39	22
2	34	0	1	2	7	6
3	41	0	3	6	6	· 1
4	18	0	3	8	1	0
5	19	0	4	8	2	0
6	31	3	10	15	0	0
7	31	2	12	18	0	0
8	13	0	5	7	0	0
9	14	1	5	8	0	0
10	11	2	7	8	0	0
11	7	0	3	3	0	0
12	10	0	1	1	0	0
13	10	1	4	5	0	0
14	9	1	5	8	0	0
15	3	0	0	2	0	0
16	3	0	1	0	0	0
17	2	0	0	1	0	0
18	2	0	0	0	0	0
19	1	1	1	1	0	0
20	2	0	0	0	0	0
21	1	0	1	1	0	0
22	1	1	1	1	0	0
27	1	0	1	1	0	0
30	1	1	1	1	0	0
40	1	0	0	0	0	0

In Table 1.2 we see that half the films that ran just one week opened in the bottom 10 ranks, and half of these were in the bottom 5. The highest rank achieved by 34 films that ran 2 weeks was in the top 5, but 6 of these were in the bottom 5. How can a film rank in the top 5 and last just 2 weeks? It could if it were heavily promoted and widely released but received unfavorable word-of-mouth information after it opened. The longest run by a film whose rank never exceeded the bottom 10 was 5 weeks (two made it).

The highest ranked films live longer and the lowest ranked die sooner. Films that made rank 1 ran at least 9 weeks and at most 30 weeks. The longest running film ran 40 weeks without ever getting into the top 10 ranks. Another feature of the market that fits the tournament characterization is that the market renews weekly with new contenders and old survivors. Each new entrant replaces a former contestant, so for each birth there is a death. Figure 1.1 shows how births and deaths varied over the sample. As few as 3 and as many as 14 films were replaced each week in the sequence of tournaments and the average rate of births/deaths is 7.69 per week. Deaths and births are Poisson distributed.[11] Most films showing in a given week are survivors from

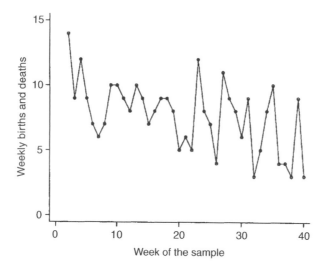

Figure 1.1 Births and deaths by week.

prior weeks. The average number of deaths is nearly 8 per week and between 35 and 47 films in the *Top-50* tournament are carry-overs.

The Gods must be Crazy was the longest running film at 40 weeks, but it generated only 0.74 percent ($4.98 million) of the total box-office revenues. *Back to the Future* was the second longest running film at 30 weeks, and it generated the highest revenue ($49.2 million) at 7.34 percent of the total revenue. The third longest running film at 27 weeks was *Kiss of the Spider Woman* with 0.89 percent (or only $5.95 million). The fourth longest running film was *Rambo* with about 5.25 percent (or $35.2 million). The fifth longest running film at 21 weeks was *Prizzi's Honor* with 1.56 percent ($10.5 million). Films with runs between 5 and 11 weeks generate 50 percent of the total revenue.

1.5   The survival model

The analysis of duration is important in motion pictures as increasing or shortening the length of the run is the most common way to adjust supply to demand. The other margin of supply adjustment is the number of screens playing a film. So, the time pattern of box-office revenue, the number of screens showing it and the duration from its birth to its death contain most of the information about the interplay of demand and supply.

The theatrical life cycle of a motion picture is a pure birth–death process in a system that is time-dependent. Define the *survival time* of a motion picture as the time interval from birth until death. This time interval is a random variable τ with distribution function $F(t) = \text{Prob}(\tau \leq t)$. The *survivor function* is defined as the probability that a movie is still alive at time t and is denoted by $R(t) = 1 - F(t)$.

The probability that a movie alive at time t will fail prior to time τ is the conditional distribution $F(\tau | \tau > t)$. Given that the movie is alive at time t, the probability that it will die between t and $t + dt$ is given by the product $f(\tau | \tau \geq t)\, dt$. At time $(\tau = t)$ the conditional density is a function of t alone and is known as the *hazard rate*, the instantaneous rate of failure for each time t: $h(t) = f(\tau | \tau \geq t) = -(r(t)/R(t))$, where $r(t) = R'(t)$ is the density of the survivor function. The hazard rate representation of the survival model has a natural interpretation: it is the probability that a movie will fall from the charts during a time interval given that it is on the charts at the beginning of the interval.

As a function of time the hazard rate may be increasing, decreasing or constant.[12] An increasing hazard would reflect the saturation of demand, while a decreasing hazard would reflect the increased attendance effected by an information cascade. De Vany and Walls (1996, figure 1) identify several patterns relating box-office revenue, a film's rank in the *Top-50*, and the length of a film's theatrical run that illustrate the various forms of the hazard rate. For example, some films, such as *Back to the Future* open at a high rank and revenue and gradually decline in rank and revenue until falling from the charts. Other films, such as *Pumping Iron*, open at a relatively low rank and build revenue and increase in rank. Other films enter the charts at at low rank and fall rapidly from the charts.

In addition to modeling the shape of the survivor function and the hazard function, it is of interest to model the effects of explanatory variables on these functions. Postulate the survivor function $R(t)$ to depend on a vector of variables z that represents the attributes of a movie's theatrical run. Suppose further that there is a function of the attributes $\psi(z)$ such that the survivor function can be expressed as $R(t; z) = R_0(t\psi(z))$ where $R_0(\cdot)$ refers to the baseline survivor function where $z = 0$ and $\psi(0) = 1$. In this formulation the z variables simply increase or decrease the argument of the survivor function.[13] Cox and Oates (1984) have shown that this survivor model can be rewritten as

$$\log T = E[\log T_0] - \log \psi(z) + \epsilon \qquad (1.1)$$

where ϵ is independent of z. This parameterization is referred to as the accelerated life model and it is estimated in the following section.[14]

1.6   Empirical results

1.6.1   *Empirical survivor functions*

We initially estimate the survivor function using a nonparametric estimator so that we need not make any assumptions about the distribution of survival times. We use the product-limit estimator which is a function of the data only; moreover, it is a maximum likelihood estimator of the survivor function in that it maximizes the general likelihood over the space of all distributions (Kaplan and Meier 1958). As a purely empirical approach to examining the survival and hazard functions the product-limit estimator can provide only a limited amount of information, but it is useful to employ this method since it is largely immune to heterogeneity and the restrictions imposed by parametric models.[15]

Each movie in our sample was observed for a single spell on the *Top-50* charts. However, some movies had previously appeared on the charts (Re-Entries) and other movies made their debut on the charts (New-Entries). We estimated the baseline survivor functions for the groups New-Entries and Re-Entries, and we tested them for equality using the Logrank and Mantel–Haenszel tests.[16] The null hypothesis that the survivor function for New-Entries into the *Top-50* is the same as the survivor function for Re-Entries could not be rejected at conventional significance levels.[17] Thus, we pooled observations on Re-Entries and New-Entries.

Figure 1.2 plots the product-limit estimate of the survivor function for the entire sample. The estimated survivor function is decreasing and convex, showing that most movies tend to drop out of the *Top-50* soon after birth. A movie has less than a 25 percent chance of lasting 7 weeks or more in the *Top-50* and less than a 15 percent chance of lasting 10 weeks or more. A film with "legs," surviving more than 15 weeks on the charts, is an aberration when compared to the population of motion pictures that breaks into the *Top-50*. In interpreting the empirical survivor function it is important to emphasize

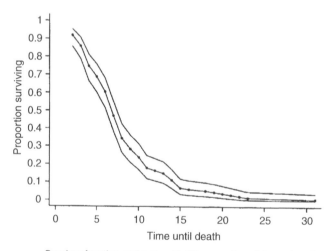

Survivor function estimate with 95 percent confidence bands

Figure 1.2 Motion picture survivor function.

that the *Variety* data are a national sample, so that movies will generally move between theaters within the nation during their theatrical runs. Rarely would all theaters book a film simultaneously so that a film would fall from the charts when its contracted run ends. Moreover, theaters in different towns will make their own agreements as to minimum length of run and hold-over clauses. Thus the theatrical run would rarely, if ever, wholly be determined by the minimum run specified in the first-run contract and is dependent on box-office revenues.

Figure 1.3 plots the maximum likelihood estimate of the hazard function.[18] The plot indicates increasing time dependence of the hazard rate. Thus it is appropriate to consider theoretical survival distributions which allow for a *nonconstant* hazard rate. One such distribution that is often used in practice is the Weibull distribution.

When survival times follow a Weibull distribution, the survivor function is

$$R(t) = \exp\left\{-\left(\frac{t}{\sigma}\right)^{\lambda}\right\} \tag{1.2}$$

and this can be written in the log expected time form as

$$\log\{-\log[R(t)]\} = -\lambda \log(\lambda) + \frac{1}{\sigma}\log(t). \tag{1.3}$$

In this specification the time-dependence is a parameter to be estimated: when $\sigma = 1$ the hazard rate is time-independent; when $\sigma < 1$ the hazard rate

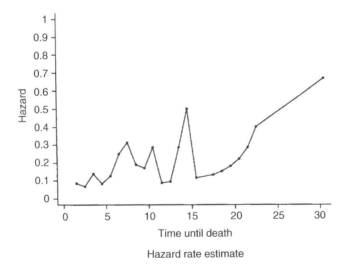

Figure 1.3 Motion picture hazard rate function.

is an increasing function of time; when $\sigma > 1$ the hazard rate is a decreasing function of time. The Weibull model also allows the survival model to be estimated more efficiently since it takes advantage of the actual survival times and not only their ordering.[19]

1.6.2   *Estimates of the survival model*

We now examine how weekly revenue, the release pattern, rank and time into the run affect expected survival. We are interested in discovering how these variables condition a film's expected life given its history at each point in the run.

We estimated the survival model with the vector of explanatory variables including the number of first run bookings, the week's revenue, the number of weeks the film previously had been in the *Variety's Top-50*, the film's rank and the number of showcases in which the film played in its debut on the charts.[20] Many of the explanatory variables are factors that would cause an exhibitor to continue showing a film either voluntarily or as the result of a contingency in the exhibition contract; for example, high revenues in a given week would trigger the hold-over clause and would be expected, *ceteris paribus*, to increase the life a film. Our inclusion of the number of weeks a film previously had been in the *Top-50* is intended to investigate how this previous exposure altered the play pattern of a film that was re-released. Tracking such films (*ET* is an example) permits inference to be made on the effect of previous exposure. Finally, we controlled for seasonality in movie attendance that might affect the hazard rate by including a set of dummy variables for the movie release dates.

Table 1.3 reports the estimation results for the censored Weibull regression model. We can statistically reject the hypothesis that the hazard rate is time-independent ($\sigma = 1$) in favor of the alternative that the hazard rate is an increasing function of time ($\sigma < 1$). For the explanatory variables we report the coefficients as $\exp(\beta)$ or time ratios for ease of interpretation: each reported coefficient indicates the multiplicative factor for expected survival time for a one unit change in the regressor. The number of first run screens, revenue and the number of weeks in which the film previously was in the *Top-50* each increase significantly the survival time of a movie. The effect of weekly revenue was an expected result of the model and is implied also by the "hold-over" clause of the exhibition contract. For a re-released film, the number of weeks it ran in its previous release is a measure of its entertainment value and this value carries over to the subsequent run; this is a strong confirmation of information transmission and memory. That the number of first run engagements might increase survival time is evidence that release strategies are reasonably well-designed, for this positive association indicates that films with good prospects are booked in more first run engagements.

Only the number of first runs appears to be economically significant, increasing the survival time by 11.3 percent. The number of showcases, our

Table 1.3 Accelerated life model for censored Weibull regression estimates[a] and controlling for the date of release[b]

Parameter	Estimate[c]	Standard error[d]	t-statistic[e]
First runs	1.113241	0.0365918	3.264
Revenue	1.000165	0.0000603	2.744
Release	1.019550	0.0052376	3.769
Showcase	0.996470	0.0011858	−2.972
Rank	0.973358	0.0067019	−3.922
σ	0.420	0.028	−20.714

Ancillary statistics
Log Likelihood $= -98.322$
Model $\chi^2(5) = 118.954$
Pseudo $R^2 = 0.4020$

Notes
a 41 out of the 350 motion pictures were not observed to fall from the *Top-50* chart during the sample. The Weibull regression model explicitly accounts for the censored data.
b A set of week-specific dummy variables was included in the regression equation. The marginal significance level for the set of dummy variables was 1.67 percent.
c Estimates of explanatory variables are reported as time ratios.
d Standard errors are conditional on σ.
e *t*-statistics are for the null hypotheses that the parameters equal unity.

proxy for the total number of screens playing the film, and the film's numerical rank decrease survival time. After its early weeks, a motion picture begins to lose rank and it becomes more vulnerable to replacement by the exhibitor with another motion picture. If a film is widely released initially, it may die quickly if it dilutes revenues among many theaters. If a film earns low revenue per screen exhibitors will replace it with another with better prospects. Thus, a wide release cannot guarantee high revenues after the early weeks and may lead to a short run. If many exhibitors are willing to show a film, then its revenues will be high in early weeks. But, only films with high revenue potential will be taken by a large number of theaters. Moreover, in order to induce many theaters to take a film, the distributor might have to accept lower terms for it.

1.6.3 *Rank and revenue*

The survival model has shown how each of the attributes of a movie's theatrical run affect the survival time conditional on the other attributes. In this section we quantify the cross-sectional relationship between rank and revenue across movies that were included in the *Top-50*. We examine this relationship in order to demonstrate the convexity of the prize distribution for each week in the survival tournament. It is through a highly convex prize distribution that some films with relatively short lives are able to earn high revenues and some films with relatively long lives earn low revenues.

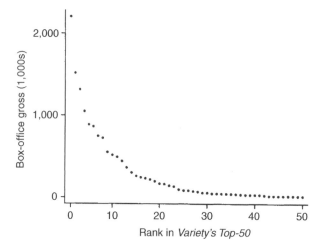

Figure 1.4 Box-office revenue and rank.

Figure 1.4 plots the rank and revenue for the movies in the *Top-50* for the final week of our sample. It is apparent that the prize distribution is highly convex in rank for this week. The plots of revenue versus rank for the other weeks of the sample also looked like rectangular hyperbolas. Thus, we quantified the relationship between weekly revenue and rank by estimating regressions of the following form: revenue $= \beta_1 + \beta_2\, 1/\text{rank} + \mu$. This functional form has the desirable property that the change in revenue with rank declines as rank declines, that is, $\partial(\text{revenue})/\partial(\text{rank})$ varies as rank varies. Moving from rank 2 to 1 may lead to a large decrease in revenue, while moving from 50 to 49 will not. Since $\partial(\text{revenue})/\partial(\text{rank}) = -\beta_2/\text{rank}^2$ and $\partial^2(\text{revenue})/\partial(\text{rank})^2 = \beta_2/r^3$, the payoff distribution is decreasing and convex in (the numerical value of) rank if $\beta_2 > 0$. We included week-specific dummy variables in the estimated equations to allow the distribution of revenue across ranks to shift between weeks; this controls for seasonality or other unobserved changes in movie attendance may occur on a week-to-week basis. We also controlled for heteroskedasticity and the length of run in some of the regressions.

We suspected that when a film earns high revenue (a numerically low rank) the variance in revenue earned would be high, and when a film has a low rank (numerically high) the variance would be low. To test for this form of heteroscedasticity we sorted the data by rank, separated the data into two halves and applied the Goldfeld–Quandt heteroskedasticity test. We rejected the null hypothesis of homoskedasticity at the 1 percent marginal significance level. We corrected for the heteroscedasticity by weighting the observations by $\sqrt{Rank}$. Table 1.4 reports the unweighted OLS regression results and the heteroskedasticity-corrected GLS results of the revenue—revenue relationship.

We found that β_2 was significantly different from zero at the 1 percent marginal significance level in all of the estimated regressions. Thus, the distribution of prizes is highly convex in the rank. To illustrate this, we calculated the change in the prize from a one step fall in rank from rank 1 through 5 using the estimates shown in column (3) of Table 1.4. The results of the calculations are shown in Table 1.5 and they illustrate the implications of the convex prize distribution for a change in rank: falling from rank 1 to

Table 1.4 Distribution of box-office revenue and rank

Independent variables	Coefficients			
	(1)	*(2)*	*(3)*	*(4)*
1/(Rank in *Top-50*)	3,644.86[a] (53.37)	3,601.98[a] (53.55)	4,703.30[a] (64.12)	4,692.78[a] (63.96)
Weeks in current run		−9.40[a] (1.72)		−2.24[a] (0.59)
Constant	−89.61 (52.95)	−76.35 (52.62)	−87.87[a] (18.27)	−85.22[a] (18.22)
Week-specific variables[b]	yes	yes	yes	yes
Observations	2000	2000	2000	2000
R^2	0.70	0.71	0.74	0.74
Estimation method	OLS	OLS	GLS	GLS
Weighting factor	—	—	$\sqrt{\text{Rank}}$	$\sqrt{\text{Rank}}$

Notes
a Indicates marginal significance level ≤1 percent.
b Dummy variables were included for each week in the sample except the first week. The set of weekly dummy variables was significant at the 5 percent level in each regression equation.
Dependent variable is weekly box-office revenue in thousands of dollars. Standard errors are in parentheses.

Table 1.5 Decline in revenue with one unit decline in rank

Rank	$\partial Revenue/\partial Rank$
1	−\$4,703,300
2	−\$2,351,650
3	−\$1,567,770
4	−\$1,175,820
5	−\$940,660

Note
These effects of rank on revenue were calculated from estimated equation (3) in Table 1.4.

rank 2 decreases weekly box-office revenue by about \$2.4 million, while falling from rank 4 to rank 5 decreases box-office revenue by only about \$235,000.

This highly convex revenue function is consistent with Rosen's (1981) superstar phenomenon in which small differences in talent are magnified into enormous differences in success.[21]

1.7   Conclusions

Motion pictures lead hazardous lives. The hazard that a film's run will end is high and rising over the run and its expected life is brief. The variance of film revenue is high. Just a few highly ranked films earn nearly all the revenue. The hazard rate varies with time; it is low early in a run and high late in the run. A time-varying hazard implies that audiences have a taste for variety.

Long runs do not guarantee success because revenue is highly convex in rank. Many of the longest-lived films in our sample earned a small box-office revenue and some of the top-grossing films had relatively short lives. A specialty film may play on only a few screens and it may run a long time if its audience develops slowly and its revenue is not diluted over many screens. A mass appeal film usually will be licensed to many theaters. Because there are many screens, it may play off rapidly because demand is saturated quickly. Its revenue per screen will fall rapidly and exhibitors drop it. Only a few widely released films are so appealing that they have "legs" and run for many weeks. When they do, they become the superstars. A long and broad run is typical of the handful of films that lie at the upper tail of the revenue distribution. Durability is unpredictable because each film meets many challenges from existing and new releases during its theatrical run.

Time-dependent survivor functions incorporate information about revenues, rank and other characteristics of the film's play from the opening to its last week. Information flows through word-of-mouth transmission among viewers, through advertising, and is disseminated through trade channels to exhibitors and distributors. The adaptive run is the primary mechanism for capturing motion picture value and casting out failures. It depends on a mixture of flexible contracting and the use of global and local information that decentralizes decisions to expand or contract exhibitions in response to demand.

Current antitrust decisions and policy with respect to the licensing of films and ownership of theaters by distributors have implications for what sorts of films are produced and how they are exhibited. The courts have required films to be individually licensed, theater by theater, and solely on the merits of the film and theater. This requirement has stood in the way of ownership, franchising or other forms of long-term contracting between exhibitors and distributors. It also has been interpreted to restrict multiple-picture licensing. In practice, what this has meant is that it is not possible for a theater to agree with a distributor to exhibit more than one film at a time. No contracts can be made for the whole season of a distributor's releases, nor for any portion of them. This makes it impossible for a theater to be the outlet for a single distributor such as Twentieth Century or New Line Cinema or Buena Vista. Nor is it possible to license a series of films to theaters as a means of financing their production. Moreover, the inability to contract for portfolios of motion pictures restricts the means by which distributors, producers and theaters manage risk and uncertainty.

In general, the licensing restrictions of the *Paramount* decisions limit the ability of exhibitors to supply funds to finance motion pictures. At one time, when the studios owned their own theaters and when they could contract for many productions with theaters for a period of years, the theaters were among the primary means of financing production. This was labeled "block booking" and halted by the *Paramount Decrees*. The licensing and ownership restrictions of the *Paramount* antitrust litigation has made it impossible for exhibitors to finance the production of motion pictures for release in their theaters. As a result, motion picture financing falls almost entirely on the distributors who rely on internally generated funds or the capital market. This, in turn, has probably been a major factor in the emergence of the concept of a "bankable" star whose participation in a project can assure its financing because the star will get it on theater screens where it has an opportunity to earn revenue.

2 Bose–Einstein dynamics and adaptive contracting in the motion picture industry

A film is like no other product . . . it only goes around once. It is like a parachute jump. If it doesn't open, you are dead.

Producer Robert Evans (Litwak 1986, p. 84)

2.1 Introduction

The hard part about understanding the motion picture industry is coming to grips with the way demand and supply operate. Film audiences make hits or flops and they do it, not by revealing preferences they already have, but by discovering what they like. When they see a movie they like, they make a discovery and they tell their friends about it; reviewers do this too. This information is transmitted to other consumers and demand develops dynamically over time as the audience sequentially discovers and reveals its demand. Supply must adapt sequentially as well, which means there must be a great deal of flexibility in supply arrangements. Pricing must be equally flexible.

The crucial factor is just this: nobody knows what makes a hit or when it will happen. When one starts to roll, everything must be geared to adapt successfully to the opportunities it presents. A hit is generated by an information cascade. If supply can ride the cascade, a superstar might be the result. A flop is an information bandwagon too; in this case the cascade kills the film. The discovery of preferences, the transmission of information, and state-contingent adaptation are the key issues around which the motion picture market is organized. This organization is supported by adaptive contracts.

In this chapter we explore theoretically and (mostly) empirically how demand and supply dynamics and the path of the distribution of film revenues are related. What makes the sequential discovery of demand and adaptive supply difficult to model is the complicated distributional dynamics they can produce. If there are 50 films playing, how will information flows affect how well they do against one another? As demand unfolds, how will box office revenues change and how are these distributional dynamics linked to the opening? How do the dynamics shape the final distribution of total revenue at the end of the run? How must supply adapt to support these dynamics? What mechanisms support this adaptation? These are some of the issues considered here.[1]

We model the discovery and transmission of information and analyze the sequential dynamics of demand under various transmission mechanisms. We show that a Bayesian demand process leads to Bose–Einstein distributional dynamics for motion picture revenues. Other processes lead to distributions such as the uniform, the geometric, the Pareto and Zipf laws, and the log normal. By testing characteristics of the distribution of film revenues and its dynamic path during the course of the motion picture run, we are able to gain an understanding of the nature of the information transmission and adaptive supply dynamics in this industry. The evidence shows that there is positive information feedback among film audiences that is captured by the Bose–Einstein process and that the industry's flexible supply adaptation to this feedback produces increasing returns. The Bose–Einstein process can produce information cascades (Bikhchandani *et al.* 1992), path-dependent dynamics (Arthur 1989), and superstars (Rosen 1981), all of which are exhibited by motion pictures.

We begin this section with a discussion of some of the institutional features of the motion picture industry that are essential to understanding its demand and supply dynamics. The discovery and transmission of demand information is modeled in Section 2.2 and various processes and their associated revenue distributions are posed as candidates to be tested against the data. Section 2.3 tests the empirical implications of the models of information transmission with a large sample of motion picture runs in *Variety's Top-50*. In Section 2.4 we examine various aspects of the industry's contracts and practices in light of our empirical results and argue that the motion picture run is decentralized, adaptive and self-organizing—by using semi-stationary admission prices, local information and contingency-rich contracts the industry flexibly matches the film rental price and the supply of theatrical engagements to demand. Section 2.5 concludes the chapter.

2.2   The mechanics of supply

2.2.1   *Some examples of motion picture runs and revenues*

To fix some ideas about the motion picture theatrical "run" and how it responds adaptively to demand, consider some of the patterns relating box office revenue, the rank or position of a film in the weekly *Variety's Top-50* and the length of the run of a film in theaters. In Figure 2.1 we show how a few films in our sample played.[2] The figure shows the rank, weekly revenue (box-office gross), and week of run for four films: *Back to the Future, Krush Groove, Streetwise* and *Pumping Iron*. The films shown in the top two panels opened strongly and played well; they had long runs and high revenues. *Back to the Future* played 30 weeks in the Top-50; it achieved rank 1 in its first week and held that rank for over 10 weeks and then declined in rank and box-office

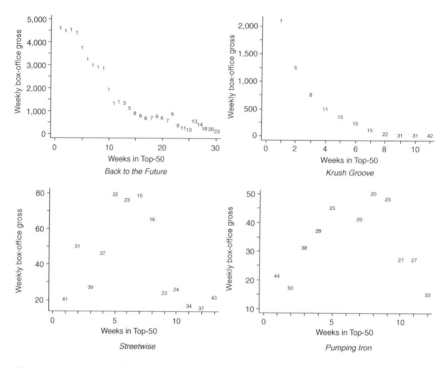

Figure 2.1 Patterns of film runs.

revenue to the end of our sample. *Krush Groove* showed a similar pattern of a high opening revenue and near top rank followed by a uniform decline of rank and revenue. This pattern contrasts with films that build revenues over their lives, examples of which are given in the lower two panels of Figure 2.1. These films opened moderately at low ranks and then began a climb in rank and revenue. *Pumping Iron*, Arnold Schwarzenegger's first film, opened at rank 44, fell to 50 and then began a steady climb to 18. It ranked in the 20s for a few weeks and then declined in rank and played off in a few more weeks. *Streetwise* followed a similar pattern—a climb in rank and revenue followed by a descent. Most movies follow neither of these patterns, but all successful ones do. The "bombs" or failures open poorly (at low rank and revenue) and decline quickly, ending their run in just a few weeks.[3]

2.2.2   *Booking and supply*

Once a film is produced, it is distributed to theaters who "exhibit" it for audiences.[4] The distributor chooses a release pattern—the number and location of theaters in which the film is licensed to play or "booked"—based on an a priori appraisal of demand. The size of the initial release determines the number of copies that are required for distribution to each of the theaters. Beyond that initial supply, the number of viewers who can see the film can grow to the capacity of theaters. Adjustments in supply are made by expanding the number of weeks the film runs in the theaters.

2.2.3 Contracting

The contract to exhibit a film usually requires the theater to show it for a minimum number of weeks; 4 weeks is a common minimum, though 6- and 8-week minimums are sometimes used. In addition, the contract contains a "hold-over" clause that requires the theater to continue the film another week if the previous week's box office revenue exceeds a stipulated amount. The contract also "clears" an area near the theater where the distributor cannot license the film to other theaters.

Theaters rent a motion picture when they exhibit it. The film rental is a percentage of the box-office revenue which the film attracts at the theater. The typical arrangement calls for the exhibitor to pay 90 percent of the theater's box-office revenue in excess of a fixed amount that is negotiated as part of the contract. In addition, the rental rate is subject to a minimum percentage on total box-office revenue (without the fixed deduction). The minimum rental rate declines over the run; a typical arrangement might be a minimum of 70, 70, 60 and 40 percent of the total box-office revenue in the first, second, third and fourth weeks, respectively. Additional weeks beyond the contract period often go at 40 percent, but the 90–10 split continues to hold throughout the run for reasons we will explain later.[5]

Given these arrangements, the cost to the distributor of expanding the seats available to accommodate customers is the incremental cost of extending the run at theaters another week. This cost is the opportunity cost of booking the film in another theater. If the prospective revenues from additional bookings at other theaters are sufficiently high, then the print may be moved (at essentially the same marginal cost of extending the engagement at the present theater). The exhibitor and the distributor usually prearrange in the contract to extend the run whenever the film earns a high revenue in the preceding week. This is done by specifying a "hold over" amount of revenue in the prior week that obligates both the distributor and the theater to continue the run another week.

2.2.4 Adaptation

Distributors and exhibitors use box-office reports to match supply to demand over time and geographically. New exhibitors try to book the film if it is drawing large audiences; they can be added in accord with contractual commitments to the initial exhibitors and the availability of prints. Distributors may produce more prints if demand is strong, though this may take some time. Supply adjusts dynamically to demand through changes in bookings at exhibitors, and by lengthening the runs at the theaters where it is booked. As noted, the decision to extend runs is decentralized to the local theater where the film is booked and are made weekly conditional on the revenue earned the preceding week.

A failure produces a different pattern. Poor box-office revenues in early weeks fail to attract new exhibitors. Poor reviews and negative word-of-mouth information discourages additional viewers. Exhibitors try to cut their losses by dropping the film, though the exhibition contract limits this adjustment. If the film is bad enough, distributors may even permit the theater to end the run before the contract term. Absent that drastic action, distributors may let exhibitors add a second feature on the marque; this is an implicit reduction in admission price, since the admission buys two films rather than one. In the worst cases, heavily promoted films like *Howard the Duck* or *Last Action Hero*, fail early leaving wounded exhibitors, distributor losses and broken contracts in their wake.

A third form of supply adaptation is noteworthy: films not considered to be suitable for a mass audience, like *Piano*, *Smoke* or *Hannah and her Sisters*, may be released in a few theaters to see if it will catch on. If filmgoers like it and spread the word, then the audience will grow over time as theater bookings and run lengths adjust to positive word-of-mouth communication.

2.2.5   *Release strategies*

Release patterns are strategies for acquiring demand information. A wide release places a film on many screens. This strategy draws a large, simultaneous sample in many theaters and cities. A tailored release strategy samples sequentially, starting at a few theaters and using the information from that sample to adjust bookings if the film builds a following. The risk of releasing a film widely to many theaters is that it may play off rapidly, say, in 4–8 weeks. If it is then dropped by theaters, many potential viewers will miss it and the information dynamic that might make it a hit may not get under way. A film released through the small sample, sequential strategy may capture good word-of-mouth information dynamics and play 10 or 15 or even 20 weeks in a varying number of theaters each week.

2.3   Demand discovery and distributional dynamics

Given the earlier discussion of the contracts and release strategies that support an adaptive supply response, we now turn to the demand discovery process. Our aim is to examine how information discovery and transmission determine the dynamics of revenues over the course of the run and how they shape the size distribution of revenue.

2.3.1 Search, communication and demand dynamics

Suppose for the moment that there is just one movie and only one person at a time can see it. Consumers choose in random sequence whether to go to the movie or not. If potential viewers share an evaluation of a film's quality, there is a common probability p that a randomly chosen person will see the film. If we let X be the number attending the film, then X is a binomial random variable such that

$$P\{X = k \mid p\} = \binom{x}{k} p^k (1 - p)^{n-k}, \quad k = 0, 1, \ldots, n. \tag{2.1}$$

In the simplest case, where consumers share a common, error-free evaluation of a film's quality, the film's attendance, run length and consequently, its revenues would follow a binomial distribution.

The more interesting case is where the quality is unknown and consumers differ in their evaluation of a film's quality. Because quality is unknown and evaluations differ among viewers, p is a random variable. By conditioning on p and integrating over the binomial distribution, we get:

$$\begin{aligned} P\{X = k\} &= \int_0^1 P\{X = k \mid p\} f(p) \mathrm{d}p \\ &= \int_0^1 \binom{n}{k} p^k (1 - p)^{n-k} \mathrm{d}p \\ &= 1/(n + 1), \quad k = 0, 1, \ldots, n. \end{aligned} \tag{2.2}$$

The last equation shows that each of the $n + 1$ possible outcomes is equally likely. Uncertainty in the distribution of the personal quality evaluations of the movie transforms the distribution of run length, attendance and revenue from the binomial to the uniform distribution.

To model the information dynamics that produce the uniform distribution, we consider a sequential Bayesian decision process. A Bayesian film buff can rely on information revealed during a film's run to refine her initial valuation of its quality. Suppose a potential filmgoer has information about the opinions of other viewers; this information is not exact and it may include word-of-mouth reports from friends, movie reviews, advertising and information about how the film is playing that can be extracted from box office reports and/or crowds at theaters. Jovanovic (1987) has modeled this kind of information-sharing and our model is a variation on his.

The return to filmgoer i is a function of the film's unknown quality which depends on viewer type and an unknown parameter θ which is common to all valuations. We assume the benefit to viewer i is

$$b_i = \theta + \epsilon_i, \quad i = 1, \ldots, N, \tag{2.3}$$

where the ϵ_i are the personal experiences, modeled as deviations from the common quality θ, and are assumed to be independently and uniformly distributed with zero mean. No consumer knows her valuation of the film until she sees it and experiences its quality $\theta + \epsilon_i$.

The information set available to potential viewer $n + 1$ in the sequence of potential customers is $s_n = (\theta + \epsilon_n, \theta + \epsilon_{n-1}, \ldots, \theta + \epsilon_0)$ where n is the number of previous trials and each element of the vector is the experienced quality of each previous trial. The expected return to i conditional on s_n is

$$E(b_i \mid s_n) = E(\theta \mid s_n). \tag{2.4}$$

If the cost of seeing the movie is common to all moviegoers and equal to c, then a randomly selected consumer on trial $n + 1$ will choose to see it if

$$E(b_{n+1} \mid s_n) > c, \tag{2.5}$$

and not otherwise.

Given an unknown θ and the uniform distribution of ϵ, the distribution of quality of the movie is unknown.[6] There is a common, but unknown, conditional probability that a randomly selected person will choose to see the film. This probability is equal among all consumers who inherit the same information set because they will all have the same estimate of θ, given s_n and the expectations of their ϵ's are all zero. Since p is equal for all trials, it can be shown (Ross 1993) that the conditional probability that the $(r + 1)$st trial is a success given k successes and $r - k$ failures in the first r trials is

$$\frac{P\{(r + 1)\text{st trial is a success, } k \text{ successes in first } r \text{ trials}\}}{P\{k \text{ successes in first } r \text{ trials}\}} = \frac{k + 1}{r + 2}. \tag{2.6}$$

If the first r trials go to the movie, then the next trial will, with probability $(k + 1)/(r + 2)$, go to the movie. The probability will evolve dynamically over the run as the discovery process unfolds. The revenue earned by the film in each week of its run (the trial period in our sample) will reveal information as to the "true" probability p and the underlying quality parameter θ. The unconditional distribution of revenue is uniform, but the sequential dynamics leading to the final outcome will exhibit conditioning of weekly revenues on prior outcomes; weekly revenues are not independent trials because they reveal information that will cause revenues to be autocorrelated.

2.3.2 *Demand dynamics and the Bose–Einstein distribution*

Now, consider the situation when there are m motion pictures playing on s theater screens. We allow for the distributor's booking strategy to incorporate multiple screen releases for motion pictures with high prior expectations. An equilibrium booking pattern will be such that the number of screens per motion picture equalizes a priori revenue expectations. Thus, by considering the distribution of moviegoers over screens to be multinomial uniform, we obtain a Dirichlet prior.[7] We have only to extend the previous model a little to let the outcomes be of s types. Let there be m movies, then the types are $1, \ldots, s, s + 1$, where type i is "went to theater i" and type $s + 1$ is "did not go to a theater and saw no movie." Denote by X_i the revenue at theater i. The number of type i outcomes in n independent trials, each resulting in one of the s outcomes is the vector $\mathbf{X} = (X_1, \ldots, X_s)$.

The Dirichlet is a multinomial uniform distribution whose dynamics have an urn model interpretation that is natural in the present setting. To develop the distributional dynamics associated with the conditional choice logic underlying the process with a Dirichlet prior, we follow Ross (1993) and compute the conditional probability that the $(n + 1)$st outcome is of type j if the first n trials result in x_i type i outcomes where $i = 1, \ldots, s$, and $\sum_{i=1}^{s} x_i = n$.[8] Ross (1993, pp. 118–19) shows this computation leads to

$$P\{(n + 1)\text{st is } j \mid x_i \text{ type } i \text{ in first } n, i = 1, \ldots, n\} = (x_j + 1)/(n + s).$$

(2.7)

The probability that the $(n + 1)$st outcome is type $j = 1$ depends directly on the proportion of previous n trials that are type $j = 1$. Because p is a random variable, the successive outcomes are not independent and a Bayesian movie buff gains information about the quality of the movie by observing the previous trials.

The unconditional distribution of the vector $\mathbf{X}$ corresponding to this dynamic can be shown to be

$$P\{X_1 = x_1, \ldots, X_s = x_s\} = \frac{n!(s - 1)!}{(n + s - 1)!} = \binom{n + s - 1}{s - 1}^{-1}.$$

(2.8)

This is the Bose–Einstein distribution. This distribution is uniform, but not in the way you would usually think of it. The usual way to think of a uniform distribution of outcomes over s theaters would be to say that the probability that a customer will go to any one of them is $1/s$. But, this is according to the multinomial distribution, not the Bose–Einstein distribution. In the dynamics leading to the Bose–Einstein distribution movie customers sequentially select movies and the probability that a given customer selects a particular movie is proportional to the fraction of all the previous moviegoers who selected that movie.[9] This is a natural result of the fact that the probabilities are not known and sampling reveals information that causes previous selections to attract new ones.

The Bose–Einstein distribution has the property that all of the possible outcome vectors $\mathbf{X}$ are equally likely. This means the vector $\mathbf{X}$ in which the attendance at every theater is equal to zero is as likely as one in which all n trials go to only one theater. Every other vector is equally likely.[10] The Bose–Einstein distribution has uniform mass over a space of $s + 1$-vectors of large dimension; the *s*-vectors are urns containing the revenues of the *s* theaters and one urn collecting those who go to no film.[11] What is important about the evolution of choice probabilities under the Bose–Einstein choice logic is the way past successes are leveraged into future successes; as soon as individual differences emerge among the films, these differences can grow at exponential speed. Differences among audiences that emerge at the opening can be compounded by information feedback through the run into very large differences by the end of the run. Path dependence from the opening depends on the stochastic evolution of demand at each theater. A broad opening at many theaters can produce high and rapidly growing audiences; but it also can lead to swift failure if the large early crowd relays negative information. Even though a film may have high initial expectations and appear on many screens, it can disappear rapidly if initial theater expectations are not met and negative information flows promote a declining audience.[12] The Bayesian choice model and its associated Bose–Einstein dynamics leads to genuine insights about the movie business.

2.3.3 *Demand dynamics and other distributions*

There are many other information and choice processes that one must consider as reasonable candidates to explain motion picture revenue dynamics. We develop a few reasonable alternatives here and indicate how they can be tested against the Bayesian model. Continuing to denote the revenue of film *i* as X_i, we now indicate revenue at time *t* as $X_i(t)$. We can generate another distributional dynamic if we assume that the revenue grows at a rate

$$r_i(t) = \frac{1}{X_i(t)} \frac{\mathrm{d}X_i(t)}{\mathrm{d}t}, \tag{2.9}$$

which may vary systematically or randomly because of challenges from other movies or external shocks to the market. This process leaves the rate of growth completely free to vary over the course of a film's run. Total revenue at time *t* may be obtained by integrating equation (2.9) to get

$$\log X_i(t) = \log X_i(0) + \int_0^t r_i(t')\mathrm{d}t'. \tag{2.10}$$

At week *t* of the run, each of the terms in equation (2.10) contributes to the total revenue: $\log X_i(0)$ is the contribution of the opening (week 0 of the run), and $\int_0^t r_i(t')\mathrm{d}t'$ is the contribution of random factors. If the first term dominates, then the opening is most important; if the second term dominates, the opening is unimportant, and stochastic factors that occur during the run predominantly determine the total revenue the film will bring.

The integral term is the log transformation of the product of independent random variables, $(r_i(t)r_j(t-1))$ and so is the sum of independent variables. If the film runs long enough, then according to the central limit theorem, the integral will be normally distributed. Hence, $X(t)$ is log normally distributed.

If the opening exerts influence, there is still a tendency for the revenue distribution to approach the log normal, but the final distribution will reflect the draw $X_i(0)$ at the opening, taken from some distribution that characterizes opening revenues, and the distribution that sums the stochastic factors over the run. If we consider the distribution of all movies, it will be a mixture of the distribution of opening revenue and the distribution reflecting the stochastic evolution of demand which tends to a log normal distribution.

If we raise the random variables to a power in the distribution of equation (2.10), the resulting distribution will be normal in the log of the log of revenue. Such a distribution would be evidence of strong interactions among filmgoers of the form $(r_i(t)r_j(t-1))^\alpha$.

Another distributional dynamic corresponds to a geometrical splitting of the audience among films. Suppose films interact in a complex way that expresses the idea that they competitively preempt one another in a hierarchical manner. They do this when they all seek to capture a large share of a static audience of a given size. If the most successful film takes fraction k of the audience and the next takes fraction k of the remainder, and so on, then the distribution of box-office revenue will be geometric.[13] In this case, the relation between sales (X_i) of film i and its rank with rank R_i is approximately

$$X_i = k(1-k)^{R_i - 1}. \tag{2.11}$$

Hence, geometric market shares, of which constant shares is a special case, will produce a power law distribution of revenue.[14] If the geometric market division process is not realized precisely, then the distribution will be a slightly modified geometric distribution called the log series distribution (May 1983).

Power laws commonly are found in economics. Steindl (1965) found that the relationship between a firm's size and it's rank could be approximated by a power law. Simon (1955) has shown that a power law is an implication of a model in which the growth rate of firms is independent of size (Gibrat's law), and the rate of birth/death of firms is constant.

2.4 Empirical box-office revenue distributions

2.4.1 Data

Our data are the box-office revenues of *Variety's Top-50* motion pictures by week. The *Variety* sample is a computerized weekly report of domestic film box-office performance in major and medium metropolitan market areas. These data follow the revenues of each picture for the duration of its run, starting from the picture's opening and progressing by week throughout its run, or until it drops off the Top-50 list. Each week, the pictures are ranked by order of their box-office revenue. The number of screens on which each movie played and other information are also given in the data. Our sample includes all the motion pictures to appear in *Variety's Top-50* list from May 8, 1985 to January 29, 1986; some 300 motion pictures are in the sample.

Descriptive statistics of the sample are given in Table 1.1. The top four grossing movies in our sample accounted for 20.5 percent of the total box-office revenue, and the top eight grossing movies accounted for 28.1 percent of the total box-office revenue. In descending order the eight highest grossing movies in our sample were *Back to the Future*, *Rambo*, *Rocky IV*, *Cocoon*, *The Jewel*, *The Goonies*, *The Colorado* and *White Knights*. The four lowest grossing movies accounted for only 0.0036 percent of the total box-office revenue, and the eight lowest grossing movies accounted for 0.0082 percent of the total box-office revenue. The highest grossing movie, *Back to the Future*, had a cumulative revenue of $49.2 million, while the lowest grossing movie, *Matter Of Importance*, had a cumulative revenue of $4,615. The median box-office revenue was $311,284; the mean was $1,918,261; the standard deviation at $4,528,685 was over twice the mean.

2.4.2   *Lorenz curves and Gini coefficients*

We initially quantify the revenue distributions through the use of Lorenz curves and Gini coefficients. These confirm that the distribution of total revenues among films and movie distributors is highly uneven. Revenues are concentrated on a few movies, and these hit movies are concentrated among a few distributors.

Figure 2.2 plots Lorenz curves for the distribution of box-office revenues across movies and distributors. If the revenues were distributed equally, the Lorenz curves in the figure would be straight lines from the lower left corner to the upper right corner. One can readily see that both curves are far below the diagonal, which would indicate a uniform distribution, and the curve of

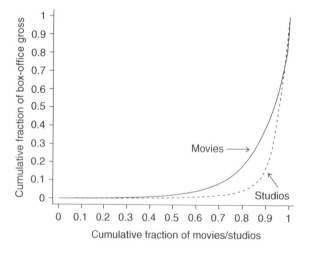

Figure 2.2 Lorenz curves: distribution of revenues over movies and studios.

distributor revenues is farther away from uniformity than is the Lorenz curve of film revenues.

The Gini coefficients for the distribution of revenues among films and distributors are 0.777 and 0.873, respectively; these values indicate a high degree of inequality in the distributions.[15] Just 20 percent of the films earned 80 percent of box-office revenues; this is such a durable relationship, it has been dubbed "Murphy's Law" after Art Murphy, a well-known industry reporter who has reported on the business for decades. The distribution of revenues of distributors is more concentrated than film revenues: 90 percent of all the revenue earned by distributors was earned by just 10 percent of the 79 distributors in our sample.[16] Distributors of successful films tended to have several successes and those that distributed failures tended to have multiple failures or only moderately successful films. There is no evidence of distributor diversification: hits received the bulk of the box-office revenue and a few distributors had most of the hits.

2.4.3 *The rank–revenue relationship*

Some of the distributional dynamics we consider as candidates imply the presence of power laws between rank and revenue; others do not. We can use the rank/revenue relationship to accept or reject some of the candidate dynamics. Ijiri and Simon (1974) show that firms have autocorrelated growth when the plot of $\log X_i$ against $\log R_i$ is concave downward.[17] This means that deviations in the concave direction from a log-linear rank/revenue relationship imply autocorrelated growth.

Figure 2.3 plots the empirical rank–revenue relationship and the fitted Pareto distribution for our sample of 300 movies. It is clear from the figure

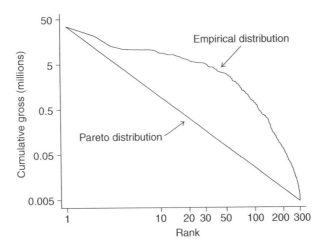

Figure 2.3 Pareto distribution of cumulative revenues over movies.

that the rank–revenue relationship is downward concave, which is the evidence of autocorrelated growth in movie revenues. To test the rank–revenue relationship, regressions of the following form were run:

$$\log X_i = \log \alpha + \beta \log R + \gamma (\log R_i)^2 + \mu \qquad (2.12)$$

where X_i is a movie's revenue, R_i is its rank and μ is a random disturbance with mean zero and finite variance.

If the geometric dynamic and its Pareto law describes the rank–revenue relationship, then the regression coefficient γ should be statistically indistinguishable from zero; finding that $\gamma = 0$ would be inconsistent with the hypothesis of increasing returns to motion pictures caused by information feedback. Finding that γ is less than zero would be evidence of positively autocorrelated growth in motion picture revenues. Conversely, if $\gamma > 0$ the rank–revenue distribution is convex downward and we have evidence of negatively autocorrelated growth.

Table 2.1 reports the estimates of the rank–revenue relationship for several levels of revenue aggregation. Columns (1) and (2) report the estimates when the rank–revenue relationship is estimated across the 50 movies in each week's sample. The data from the 40 weeks of the sample were pooled and an

Table 2.1 Estimates of the rank–revenue relationship

$$\log(\text{Revenue}) = \beta_1 + \beta_2 \log(\text{Rank}) + \beta_3 (\log(\text{Rank}))^2 + \mu$$

Aggregation	(1)	(2)	(3)	(4)	(5)	(6)
	Weekly by film		*Cumulative by film*		*Cumulative by distributor*	
Regressors						
Log rank	−1.7206	0.1859	−1.9826	2.1366	−2.7222	−0.3589
	(0.0146)	(0.0435)	(0.0515)	(0.1031)	(0.0787)	(0.1772)
(Log rank)^2	—	−0.4033	—	−0.5299	—	−0.4512
	—	(0.0089)	—	(0.0130)	—	(0.0329)
Constant	16.6333	14.8405	21.7375	14.5904	21.4484	18.9070
	(0.0923)	(0.0759)	(0.2480)	(0.2004)	(0.2676)	(0.2306)
Weekly dummy variables	Yes	Yes	—	—	—	—
$\bar{R}^2$	0.8772	0.9397	0.8319	0.9744	0.9462	0.9858
Observations	2,000	2,000	300	300	69	69

Note
Estimated standard errors are in parentheses.

intercept dummy was included for each week (except the base week), so there are 2,000 (40 × 50) observations. The hypothesis of a linear Pareto distribution can be rejected in favor of the alternative of downward concavity: the coefficient on (log rank)^2 is negative and statistically significant. Columns (3) and (4) report the estimates of the rank–revenue relationship where the unit of observation is the film. There were 300 films in our sample which were observed to fall from the Top-50 chart; the remaining films were censored and thus were excluded from the regression. The rank–revenue distribution clearly shows statistically significant evidence of downward concavity. The distributor is the unit of revenue aggregation in columns (5) and (6). As with the other units of aggregation, the distribution of distributor revenues by rank is also characterized by downward concavity, which is consistent with autocorrelated growth.

2.4.4 *Revenue distribution dynamics*

We showed that some dynamics can lead to log or log–log normal distributions. A formal test shows that the distribution of cumulative box-office revenues is not log normal. Table 2.2 presents several test statistics for log normality, and the null hypothesis of log normality could be rejected in each case; we could also reject the hypothesis of 3-parameter log normality.

To quantify the dynamics of the revenue generation process, we have calculated the distribution of revenues across movies by the week-of-run (Figure 2.4). The distribution of revenues at the film's opening (week = 1) is far from log normal though the distributions appear to converge to log normality. However, we can statistically reject the hypothesis of log normality for each week-of-run revenue distribution as well as the end-of-run distribution at the 1 percent significance level. Thus, the opening performance is statistically a dominant factor in revenue generation, although it appears to diminish in importance near the end of a film's run.

What we find is evidence of convergence to normality of the log of the log of weekly revenue among long-lived films. Table 2.3 shows the marginal significance level for the null hypothesis that the log of the log of weekly gross is normally distributed. We can reject normality of log(log(revenue)) for weekly revenues across movies lasting 6 weeks or less. But, we can't reject normality

Table 2.2 Testing log-normality of box-office revenues

Test name	Test statistic	Marginal significance level
Skewness–Kurtosis	$\chi^2 = 22.560$	0.00000
Shapiro–Wilk	$Z = 3.018$	0.00127
Shapiro–Francia	$Z = 3.468$	0.00026

Note
The hypothesis of 3-parameter log normality could also be rejected at a marginal significance level of 0.02.

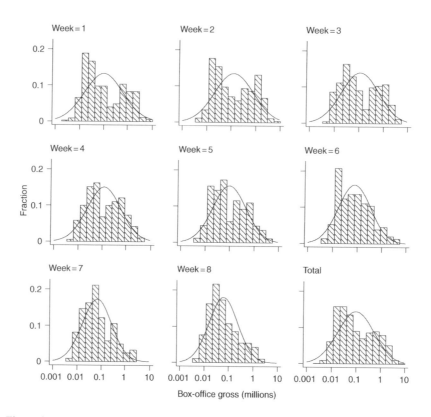

Figure 2.4 Distribution of revenues by week of run.

of log(log(revenue)) of films that last beyond six weeks. (Note that this test requires a minimum of eight observations.) Normality of log(revenue) is consistent with normality of the products of random variables, which is how one would expect information to spread through interaction among member of the population of filmgoers. But, normality of the log(log(revenue)) suggests a stronger form of interaction for those films that survive six or more weeks. The interactions must be products of random variables raised to a power. The extreme upper tail of the motion picture revenue distribution is populated by films whose revenues are generated by power laws. So, beyond a run life of 6 weeks, the power laws that Brock (1993) argues will usually hold in informationally rich economic environments do in fact hold.

2.4.5   *What information dynamic accounts for the distribution?*

From our formal tests we know the distribution of box-office revenue is not log normal; its departure from log normality is in a direction that indicates autocorrelated growth. There are several processes that could account for this.

Table 2.3 Testing log–log normality of box-office revenues

Week of run	Marginal significance level
1	0.0000
2	0.0000
3	0.0000
4	0.0000
5	0.0004
6	0.0082
7	0.1656
8	0.1841
9	0.0736
10	0.1983
11	0.1782
12	0.1710
13	0.3014
14	0.1330
15	0.4840
16	0.6070
17	0.4625
18	0.3060
19	0.4308

Movies may be hierarchically dividing up a fixed audience so that revenues grow geometrically. We rejected the geometric process and the Gibrat law by formally testing the hypothesis of a geometric revenue distribution and rejecting it. Movie revenues that grow at a constant or geometric rate would give a linear or nearly linear plot between the log of revenue and the log of rank and we formally rejected that too. The departure from power law behavior is evident in the rank and revenue plots and the formal test of log-concavity of the log(revenue) versus log(rank) relationship.

We are left with two possibilities that can account for both the dynamics of revenue and the shape of the end-of-run revenue distribution. One is that movie audiences are produced by independent stochastic growth factors and an opening that has a uniform log distribution. This process will exhibit the required from log normality in the form of a mass point in the upper tail for those films where the opening is more important throughout the film's run than the stochastic factors. But, for this process, the growth in sales should be random and independent of prior growth. We know this is not true because we can formally reject independent growth. The other possibility is the one we accept for now. The best account of the data is given by the Bayesian model and the Bose–Einstein dynamics.

Information transmission and Bayesian updating give rise to the revenue distributions and their dynamics. We can account for all their features if there is uncertainty over film quality and information sharing gives us Dirichlet priors. Then Bayesian updating will produce the Bose–Einstein dynamics.

If the unconditional distribution of movie revenues is Bose–Einstein, then they are maximally uncertain. This is true in three senses: One, the Bose–Einstein distribution is uniform and maximally uncertain if measured by entropy, mean-preserving spreads or stochastic dominance. Two, the space of outcomes is vast and consequently each outcome is highly unlikely. Three, even though the Bose–Einstein distribution is uniform but in a counterintuitive way that masks the real uncertainty it implies. A uniform distribution of outcomes suggests that the probability that a customer will go to any one of m movies is $1/m$. But, this is multinomial choice logic, not Bose–Einstein choice logic. In the information and choice dynamics that lead to the Bose–Einstein distribution, customers choose movies in proportion to the previous moviegoers who selected that movie. When revenues are Bose–Einstein distributed, outcomes which differ in the extreme are equally likely and similar outcomes are extremely unlikely—this is the quintessential characteristic of the movie business.

2.5   Explaining the industry

How do these results explain the way the industry operates? How do its contracts and practices adjust the supply of theatrical engagements to capture the demand processes that seem to be operating here? Since the choice probabilities evolve during the run, and they are not known before hand, dynamic optimization in this context must essentially be adaptive. In this section, we discuss industry practices and institutions from this perspective.

2.5.1   Launching the opening

Stars, and large production and advertising budgets can place a film on many exhibitor screens when it opens. This can generate high initial revenues in total, if not in each theater. If viewers like the film and spread the word, it will earn high revenue in the following weeks. But, if viewers do not like the film, the large opening audience transmits a large flow of negative information and revenue may decline at a rapid rate. A wide release is vulnerable to negative feedback. Because a wide release lowers the gross revenue per theater, exhibitors may drop the film sooner than they would drop a less widely released film. Thus, widely released films show more variance in the revenues and, on average, shorter run lives (De Vany and Walls 1994). A major read-stars film may earn advance distribution fees from foreign and the post-theatrical market. The willingness of exhibitors and downstream sources of revenue like cable television, VCR distributors, pay per view and network television as well as foreign distributors to pay advance guarantees for motion pictures before their theatrical run is a major inducement for distributors to produce big budget films and promote them heavily; for such films, the theatrical market (which we have studied here) can be less important than these other sources of revenues. There is a certain irony here. Some stars and movies earn so much from foreign distribution rights that third world motion picture tastes may play a large role in determining the kind of movies we see in US theaters.

2.5.2 *Adaptive contracting*

A difficult problem to solve contractually is how to keep a film on screens long enough for it to build an audience. If an exhibitor takes such a film, it is on the risk that it may build so slowly during his run that only exhibitors who show it later will benefit from information feedback. The *Paramount Consent Decrees* bar theater ownership by the major distributors; they also bar franchises, extended clearances in time and location, blind-selling (booking a film before the exhibitor has seen it) and long-term contracts. Because the decrees bar long-term exclusive showings, they make it difficult to guarantee that the first exhibitor who took the risk of introducing the film will benefit if the film later becomes a success (De Vany and Eckert 1991). When the audience grows recursively, as the evidence suggests, then the *Paramount* contracting restrictions prevent risk-taking exhibitors from capturing the demand externality which they helped to make. The effect of the *Paramount* prohibitions is to restrict the gateway to theater screens, allowing fewer films that would appeal to smaller and more sophisticated audiences to pass through the gate.[18]

2.5.3 *Admission pricing*

One industry practice, widely followed though not encoded in contracts, is to use relatively inflexible admission prices to reveal excess demand and to adjust to demand by lengthening the run. Prices rarely are raised even when excess demand is revealed.[19] A relatively stationary admission price combined with a count of admissions gives a reliable signal of demand, and this signal is transmitted throughout the industry by real time reporting of box-office revenues. This reporting is required in the exhibition contract and encouraged by other means as well.[20] Demand is accommodated by lengthening the run rather than raising the price and this is accounted for by the model of information transmission. If, at the first sign of excess demand, the admission price were to be increased, the number of people who would see the film in the opening weeks would fall and this would reduce the flow of information from this source to potential viewers. This lower rate of information transfer would lead to a shorter run and a lower total level of demand. The ability to extend the run makes an almost perfectly elastic supply response possible, so that price need not rise to choke off excess demand.[21]

The optimal admission price must trade off price and run length so as to maximize total revenue in the context of ignorance, not merely uncertainty, of demand. Charging a higher price will shorten the run and reduce the transfer of information; this strategy may capture only a fraction of the demand that eventually will be captured in the adaptive run so successfully used in the industry. Because motion pictures have short lives all adaptations must be made quickly and are of a short-run nature.[22] Inflexible admission prices lead to a pure quantity signal and an adaptive supply response, which seems ideally suited to the problem of discovering and responding to demand.

2.5.4   *Rentals*

The contingency-rich exhibition contracts used in the industry are highly adaptive: they rely on locally generated information; they set the rental fee in a precise and nonlinear way in response to demand; they share risk between exhibitors and distributors; and they create incentive for exhibitors to show films by granting a measure of exclusivity, although this is limited in scope by the *Paramount Decrees*. The rental price adapts precisely to the state of demand. It is usually calculated as a percentage of box office gross, as we discussed in Section 2.1, so the rental varies directly with demand. But, very high demand will trigger a switch from the minimum percentage, which may be 40 or 70 percent depending on the week of the run, to 90 percent on the amount in excess of the fixed amount. The rental schedule is nonlinear and has higher rates at higher grosses. Yet, in line with the declining revenues during the descent in rank and revenue that always occurs sometime in the run, the minimum percentage rental rate declines with the week of the run. This offsets the rising hazard the exhibitor faces as the run continues (De Vany and Walls 1997).

2.5.5   *Decentralization*

Each film's run through the market is sequential in order to exploit information dynamics. The run is self-organized because it decentralizes the decision to extend the run to each theater and uses only local information to extend or close the run at each location. The initial release is modified over time through this process and new engagements can be added subject to prior contractual obligations. These contractual features interact to adaptively capture revenue and generate strongly increasing returns from highly successful films. They help to create the Bose–Einstein distribution because they adapt sequentially to let exhibitors and the distributor ride the information cascade (up or down). When the cascade is headed in the right direction, the supply response is flexible enough for some films to ride the cascade to superstardom.

2.6   Conclusions

We have tied the information dynamics of motion picture audiences to the dynamics of the distribution of motion picture weekly box-office revenues for the top 50 films. Our results show that weekly revenues are autocorrelated: a movie which recently experienced increasing revenues is more likely to experience additional growth than a movie which experienced growth in the distant past. The end-of-run or total revenue distribution for motion pictures approximates a mixture of an opening distribution and a log normal distribution, suggesting that if the movie runs long enough it may acquire an independent life. Yet, the distribution never quite reaches log-normality; it has fatter tails than log normal and mass points at the far right, where the superstars are located. Films that show extraordinary staying power exhibit power law behavior in the extreme tail of the revenue distribution. The Bayesian information dynamics that gives the best account of the evolution of revenues is one that leads to the Bose–Einstein distribution.

The strong evidence supporting the Bose–Einstein distribution points to the stark uncertainty facing the motion picture industry; the space of outcomes in which the 50 films vying for audiences in our sample can evolve is of staggering dimension and the dynamics evolve unpredictably. Star vehicles may have higher initial expectations, but booking patterns and information dynamics produce highly unpredictable distributional dynamics and uneven revenues. The brief tenure of studio heads and the wide use of artist participation contracts is consistent with the profound uncertainty implied by our empirical findings.[23] Neither genre nor stars can guarantee success and a disaster can bankrupt a studio.[24]

Even though the motion picture industry exhibits profoundly the properties we have modeled, there are many other industries for which the model and statistical methods would seem to be appropriate: other entertainment industries such as television, music and book publishing might be usefully modeled in the framework we develop here. In addition, the sequential dynamics of innovation and demand discovery in many industries where information sharing and Bayesian demand discovery are at work may exhibit the Bose–Einstein process. The recursive generation of demand and adaptive supply and pricing that are the primary ingredients of our approach seem to be fundamental elements of many models that focus on distributional dynamics, information diffusion, superstars and path dependence.

The industry's organization and contracts make sense in the uncertain and dynamic environment of the motion picture industry. The adaptive run seizes opportunities to expand supply and adjust prices sequentially using local information and decentralized decision making to do it. This adaptive response is encoded in industry contracts and institutions; it is the mechanism that lets supply follow path-dependent demand sequences to capture box-office revenues and make the highly irregular distributions we reported here. The industry's market institutions and sophisticated contracts promote the information dynamics that discover extraordinary motion pictures and let supply and pricing adapt to capture the strongly increasing returns they create.

3 Quality evaluations and the breakdown of statistical herding in the dynamics of box-office revenue

3.1 Introduction

When *Star Wars* was released, it opened on nearly 3,000 theater screens in the United States, almost 10 percent of the available screens and closer to 20 percent of the screens in the most desirable theaters. *Godzilla* opened on nearly 5,000 screens. Yet, no movie that ever won an Academy Award for best movie or best screen play opened on more than 1,500 screens. Most award-winning movies—like *Amadeus*, *Life is Beautiful* (both best pictures) and *Thelma and Louise* (best screenplay)—opened on only a handful of screens. Some movies are engineered to be blockbusters. They are produced with big production budgets, they feature major stars, they are promoted on web sites and they open with heavy national advertising on thousands of theater screens. Studios use these strategies because they believe that a big opening will cause people to choose their movie over others. Many studios and producers believe the opening is the critical event in a film's commercial life. Producer Robert Evans stated the theory in a succinct and dramatic way when he compared a movie's opening to a parachute jump: "If it doesn't open, you are dead" (Litwak 1986, p. 84).

This theory seems to gain support each night on the evening news, where the box-office revenues of the top movies are reported as though they were sports statistics. If audiences rely on opening revenues to choose movies, then a movie's opening can determine the evolution of its revenues throughout its theatrical run. The statistical herding and information cascade models of Banerjee (1992) and Bikhchandani *et al.* (1992), and the contagion and path-dependence models of Arthur (1995) demonstrate the possibility of this behavior. Anderson and Holt (1997) produce information cascade in a laboratory setting. In statistical herding, cascade, or contagion models, agents rationally may ignore personal information to follow the public consensus that is summarized in box-office reports.

If motion picture audiences behave the way these models describe, the movie that wins the opening box-office contest could leverage that lead to capture the lion's share of revenue. Because private information is not revealed in the consensus choice, the consensus may be wrong and the best movies might lose out to movies that are not as good. An excellent movie with a small number of opening screens might get buried by public information about box-office revenues and never find its audience. This would be bad news for movie fans and people who try to produce quality movies. A contrary view is that moviegoers can and do communicate their private evaluations about the quality of a movie in the manner described by the evaluation models of Avery *et al.* (1999), De Vany and Walls (1996) and De Vany and Lee (2001).[1] At one time, this transmission of private information about quality was called word-of-mouth information, but the possibilities for the exchange of evaluations about movies go far beyond this now.[2] The internet has expanded the ability of movie consumers to exchange information, a fact the studios have not ignored. Other indicators of quality like reviews and awards are also widely available.

What is true about the transmission of information among moviegoers? Is there evidence of statistical herding? Does a movie that "opens big" dominate the market? Must a movie have a large opening to dominate? How important is information about a movie's box-office revenue in determining how many people will go to see it? Does *quality* information gain weight over *quantity* information as a movie's theatrical run progresses? What is the connection between the transmission of information and the size distribution of box-office revenues? These are broad questions about social learning and the exchange of information set in the context of one of the more visible and important markets where such influences are apt to hold. There are many models of social learning and demand, but few of them have been tested or even put in testable terms. We think the movie business provides us with a laboratory to examine these questions: The availability of weekly data in a market where each movie is a unique product whose quality is discovered only through the experience of seeing it makes the movie business an ideal testing ground for these competing theories.

3.2 Information and demand dynamics

Statistical herding, contagion and cascade models all have a similar structure. Each of these models asserts a proposition about the dependence of current demand on past demand. These models consider a sequential decision problem in which agents choose between alternatives under a particular information structure. Each agent is assumed to have a private signal about the object, known only to the agent. In addition, each agent receives a public signal, known to all agents, about the choices of the agents who preceded them in the choice sequence. Most of these models pose a binary choice problem in which the agent uses these sources to choose between two alternatives.[3] The sequence in which the agents choose is set and known to all and choices are made one at a time in the established sequence.

Thus, for an agent in position n in the choice sequence, the information set $\{\mathcal{I}\}$ consists of the private signal s_n and the sequence of all the observed previous choices $c_{n-1}, c_{n-2}, \ldots, c_{n-\lambda}$. When the sequence is predetermined, each agent's position in the sequence is unique so agents can be indexed by position in the sequence.[4] Since the private signals are not observable, each agent uses the choices of the previous agents to infer their private signals. It is important to note that these private signals are received *before* the agent consumes the good and are not based on their experience with the good. Thus, agents who follow in the choice sequence are not inferring the quality of the good. They are inferring the private signals the agents received, before any of them has any direct experience with the movie. We will return to this important distinction.

Now the agent's problem is to estimate the probability that the object is good or bad given the information structure and then to use the probability distribution to make the choice that maximizes expected payoff.[5] Since there is no cost of delay and the agent's position in the sequence is fixed, this sequential statistical decision problem reduces to a static decision with a fixed sample corresponding to the agent's position in the sequence.[6]

Let $V(\xi, \delta) = E\{V[\xi(c_1, \ldots, c_n; \omega)]\}$ be the expected value of the decision rule ξ when the agent is in position n in the sequence and observes all previous choices. The posterior probability distribution of the movie's quality is δ which depends on $(c_1, \ldots, c_n; \omega)$ ω represents quality information, which is silent in models where only the choices of other agents are observed. When both choices and quality are observed, then there is a competition between them and the weights attached to them in the decision rule.

If all the agents in the sequence are using a rule of the form ξ, then the dynamics of demand for a single movie take the form of a mapping F of choices on the history of previous choices:

$$q_{t+1} = F(q_t, q_{t-1}, q_{t-2}, \ldots, q_0). \tag{3.1}$$

In models of information cascades and path dependence this mapping has a fixed point where quality information may carry negligible weight. If the mapping is deterministic, then the opening revenue, q_0 wholly determines the evolution of demand. If the mapping contains a stochastic component, and it would in the Bayesian model, then the convergence to a fixed point may not be assured; this is why information cascades are considered to be fragile (Bikhchandani *et al.* 1992). If F is a contraction mapping with $|\Delta F| < 1$, then there is a fixed point q^* such that $q^* = F(q^*)$. This is the equilibrium to which the demand evolves and the movie ends its run. The requirements on $F(\cdot)$ to be a contraction mapping are weak; there must only be a kind of diminishing returns to further sampling of previous choices. On the other hand, $F(\cdot)$ may be an expansive mapping, $|\Delta F| > 1$, if there are increasing returns to information or if a consensus emerges that the movie is the best movie.

Because the decision rule ξ includes quantity and quality information, the mapping $F(\cdot)$ will reflect the timing of these signals, the degree to which they are relied upon, and the degree to which they confirm or disagree in their message. Arthur (1994) points to the possibility of a bifurcation of the mapping $F(\cdot)$ in the context of two or more choices, and this point is similar to the fragility of information cascades raised by Bikhchandani *et al.* (1992). At a bifurcation point, the dynamic process shifts and the mapping $F(\cdot)$ becomes expansive for one movie and contractive for the other. The equilibrium converges toward an all-or-nothing point where one movie captures all the market and the other one fails. We find in the following section that the box-office revenue dynamic is fragile and that the mapping bifurcates into a hit branch and a non-hit branch.

3.3 The model

We construct the conditional expectation analogue of the demand mapping equation (3.1). Define R_{i+1}^{+} as total revenue earned by a film from week $i + 1$ to the end of its run. Define R_{i}^{-} as the total revenue earned by a film from its opening through week i. We ask the following question: what is the expectation at the beginning of week 2 of total revenue for the duration of the run? This is

$$E[R_2^{+} \mid R_1^{-}].\tag{3.2}$$

We can extend that to ask what is the expectation of revenue at week 3 and beyond conditional on revenue earned through week 2 and through week 1.

$$E[R_3^{+} \mid R_2^{-}, R_1^{-}].\tag{3.3}$$

More generally, we want to know the form of the expectation of future total revenue at week i conditional on total revenue earned through each of the prior weeks of the run:

$$E[R_{i+1}^{+} \mid R_i^{-}, R_{i-1}^{-}, \ldots, R_1^{-}] \quad i = 2, \ldots, n.\tag{3.4}$$

We can put this in a form to be estimated by exploiting the fact that the box-office revenue distribution is a stable distribution with a Paretian upper tail (De Vany and Walls 1997). The Paretian distribution has the form $\text{Prob}[R \geq x] = (k/x)^{\alpha}$, where $x \geq k$ and $k, \alpha > 0$. The constant α is the coefficient of tail weight, which measures the amount of probability mass contained in the upper tail. The Pareto distribution has a mean equal to $k\alpha/(\alpha - 1)$. In the upper tail of the distribution, the conditional probability of an event, $P[R_{i+1}^{+} \geq R_i^{-}]$, is $(R_{i+1}^{+}/R_i)^{\alpha}$. The conditional expectation is

$$E[R_i^{+} \mid R_i^{+} \geq R_i^{-}] = R_i^{-}\alpha/(\alpha - 1).\tag{3.5}$$

This well-known property of the Paretian distribution implies that at each week of a film's run its expected future revenue is proportional to the revenue it has already earned. Thus, a film does not exhaust its revenue potential and success breeds further success.[7]

Using the proportional expectation property, we can write our estimating equation in a linear form relating expected to past revenues for each week of the run.

$$\sum_{t=s+1}^{\infty} \text{Revenue}_t = \beta_0 + \beta_i \sum_{i=1}^{s} \text{Revenue}_i + \mu, \quad s = 1, \ldots, 10. \qquad (3.6)$$

The regression equation (3.6) conditions future revenue on revenue earned from the opening through each of the prior weeks of the run. The β_i are the impacts of revenue earned through each week on revenue earned in subsequent weeks for every week of the run. The values of these coefficients quantify the impact of past revenue on future revenue. If the opening is the best estimator of total revenue, then our estimate $\hat{\beta}_1$ should be large and significant and dominate the $\hat{\beta}_i$ of other weeks.

3.4 Estimates

The β_i coefficients are estimated separately for both hit and non-hit movies for a sample of movies that had runs of at least 10 weeks in duration. The results are shown in Tables 3.1 and 3.2. The diagonal elements of both matrices estimate how well the revenue earned through week i forecasts the revenue earned in all the weeks after week i. The off-diagonal elements estimate the impact of revenue through week i on revenue earned subsequent to each week after week i. Clearly, for both sorts of movies, the diagonal elements dominate the off-diagonal elements. Thus, the immediate past is the best predictor of the revenue the movie will earn beyond that week. In contrast, the revenue earned through week i is not a good forecaster of the future revenue that will be earned at weeks $i + 2, i + 3, \ldots, i + 9$. Note that the goodness of fit, measured by R^2, is lower in the early weeks and higher in the later weeks for hits than non-hits. For hits, the opening revenue accounts for only 12 percent of the variance whereas for non-hits opening revenue accounts for 36 percent. The R^2 values are pretty flat over all the non-hit equations, ranging from 0.32 to 0.47. In contrast, the hit equation R^2 values range from 0.12 to 0.81 and they increase at week 4 and then more dramatically at weeks 6 and 7. By weeks 8 and 9, the regression equation is explaining a very high proportion of the variation in future revenues of hit movies. In the latter weeks, revenue through week 8 also predicts revenue subsequent to weeks 8, 9 and 10 for hit movies.

The regression evidence shows that the opening and early weeks are not as good at predicting subsequent revenues for hits as for non-hits. But, it is also evident that a change in the mapping occurs during the runs of hit movies at about the fourth or fifth week. This is a possible bifurcation point because from then on the β values rise sharply and in a manner consistent with the mapping

Table 3.1 Impact of past revenues on future revenues: hit movies

$$\sum_{t=s+1}^{\infty} \text{Revenue}_t = \alpha + \sum_{i=1}^{s} \beta_i \text{Revenue}_i + \mu, \quad s = 1, \ldots, 10$$

Dependent variable: cumulative revenues after week #

Weekly revenue	1	2	3	4	5	6	7	8	9	10
1	1.047 (0.291)	-1.349 (0.578)	-0.692 (0.367)	-0.523 (0.286)	-0.394 (0.277)	-0.157 (0.145)	0.096 (0.121)	0.228 (0.160)	0.225 (0.192)	0.213 (0.126)
2		3.434 (1.028)	-0.576 (0.991)	-0.612 (0.884)	-1.030 (0.824)	-1.070 (0.549)	-0.562 (0.386)	-0.054 (0.244)	-0.170 (0.216)	-0.130 (0.202)
3			4.348 (1.100)	0.777 (1.200)	1.684 (1.149)	1.093 (0.805)	0.016 (0.556)	-0.323 (0.401)	-0.038 (0.328)	-0.076 (0.279)
4				5.886 (0.976)	-0.071 (1.233)	0.082 (1.129)	-0.084 (0.706)	-0.224 (0.487)	-0.289 (0.402)	-0.104 (0.353)
5					4.804 (1.207)	-0.997 (1.081)	1.374 (0.777)	-0.220 (0.468)	0.087 (0.471)	0.777 (0.447)
6						6.627 (1.243)	-0.420 (1.333)	-0.521 (0.613)	-0.767 (0.548)	-0.427 (0.487)
7							5.095 (1.200)	-0.729 (0.852)	0.238 (0.825)	-0.201 (0.836)
8								9.028 (0.944)	2.697 (2.258)	1.929 (2.641)
9									5.129 (1.720)	2.269 (1.599)
10										3.332 (1.599)
R^2	0.117	0.197	0.257	0.400	0.454	0.578	0.668	0.799	0.810	0.786

Note
All regressions run on common set of movies with box-office gross $\geq$ 50 million and life-length greater than 10 weeks. Estimated standard errors in parentheses.

Table 3.2 Impact of past revenues on future revenues: non-hit movies

$$\sum_{t=s+1}^{\infty} \text{Revenue}_t = \alpha + \sum_{i=1}^{s} \beta_i \text{Revenue}_i + \mu, \quad s = 1, \ldots, 10$$

Weekly revenue   Dependent variable: cumulative revenues after week #

	1	2	3	4	5	6	7	8	9	10
1	1.370 (0.074)	−0.937 (0.169)	−0.357 (0.107)	−0.206 (0.063)	−0.002 (0.051)	0.021 (0.041)	0.031 (0.037)	−0.054 (0.026)	−0.072 (0.034)	−0.069 (0.031)
2		2.763 (0.260)	−0.463 (0.231)	−0.067 (0.165)	−0.052 (0.125)	0.012 (0.113)	−0.058 (0.071)	−0.245 (0.072)	−0.230 (0.071)	−0.179 (0.072)
3			2.923 (0.194)	−0.930 (0.397)	0.656 (0.285)	−0.337 (0.240)	−0.252 (0.172)	−0.274 (0.154)	−0.220 (0.139)	−0.149 (0.133)
4				3.809 (0.486)	−0.435 (0.387)	−0.309 (0.365)	−0.187 (0.278)	−0.228 (0.252)	−0.138 (0.235)	0.007 (0.185)
5					4.142 (0.501)	0.216 (0.576)	0.097 (0.549)	0.050 (0.426)	0.063 (0.327)	−0.141 (0.261)
6						3.423 (0.653)	−0.311 (0.856)	0.045 (0.670)	0.050 (0.511)	−0.034 (0.427)
7							4.367 (0.936)	−0.187 (1.078)	−0.170 (0.664)	0.213 (0.589)
8								4.050 (0.916)	0.391 (0.910)	0.520 (0.730)
9									3.737 (0.873)	−0.982 (1.117)
10										4.411 (1.086)
R^2	0.362	0.379	0.429	0.430	0.447	0.466	0.421	0.355	0.347	0.318

Note
All regressions run on common set of movies with box-office gross <50 million and life-length greater than 10 weeks. Estimated standard errors in parentheses.

becoming expansive (because the βs far exceed unity). The non-hit movies also enter into an expansive range, but the magnitudes and the cumulative pattern are less dramatic. Of course, these are longlasting and among the more successful movies that were chosen to have runs comparable to the hit movies for estimation purposes. We expect the mapping to be nonexpansive, with a diagonal coefficient below unity, for short-running, relatively unsuccessful movies.

Reading across the rows of Tables 3.1 and 3.2 one sees that revenue has a negative but small impact on subsequent total revenue, and the impact is higher for non-hits than hits. For hits, opening revenue has a negative impact on revenues earned after week 2 and then a vanishing effect for the middle weeks of the run, rising to a positive impact at week 8 and beyond. For non-hits, the impact is negative or zero throughout the weeks of the run. For example, $100,000 of opening revenue earned by a hit *increases* the revenue earned after week 10 by $21,000. In the case of a miss, each $100,000 of opening revenue *decreases* revenue after week 10 by $7,000.

The diagonal elements in the hit matrix exceed the diagonal elements in the non-hit matrix. This is evident in Figure 3.1 where the diagonal revenue impact coefficients are plotted for each week of the run. Opening revenue predicts less well total subsequent revenue for hit than for non-hit movies; this is shown by the lower value of the diagonal element for hits than for non-hits. After week 2, the impact is fairly flat for non-hit movies throughout the run.[8] On the other hand, the impact grows for hit movies more or less throughout the course of their run. The cumulative impact over the course

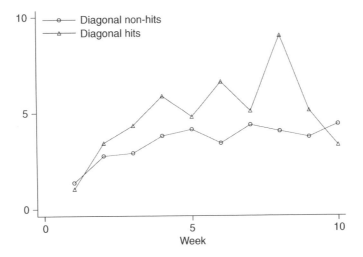

Figure 3.1 Marginal revenue impact of prior revenue on subsequent revenue throughout the run.

of the 10 weeks of past revenue on future revenue is also greater for hit than non-hit movies; their respective cumulative values are 40.26 and 19.62 for hits and non-hits. The crossover at the tenth week is reasonable because very few non-hit movies play longer than this, so revenue through the tenth week is a good predictor of total revenue. This is less true of hit movies, some of which have runs of 40 weeks or longer.

3.5   The stable Paretian model

The β_i values in the regression equation are the estimates of $\alpha_i/(\alpha_i - 1)$ at each week of the run. Solving $\beta_i = \alpha_i/(\alpha_i - 1)$ at each week along the diagonal gives us an estimate of the value of Paretian tail weight coefficient α_i for each type of movie and at each week of the run. All these implied values for α_i, which are shown in Table 3.3, are in the range $1 \leq \alpha \leq 2$ as implied by the stable Paretian hypothesis. Given these α values, the distribution has a finite mean and an infinite variance at each week of the run. Moreover, the tail weight is increasing (α is decreasing) through the run in a manner consistent with models of Bayesian filmgoers who share information (De Vany and Walls 1996; De Vany and Lee 2001). The hit movies have "heavier tails" as is evident in the fact that they tend to have smaller α values.

Here is the remarkable thing. Both these sets of long-running movies in some very real sense do not exhaust their revenue potential the longer they run. This is contrary to what one would expect from a Gaussian point of view, where a movie that has grossed a large amount would have a diminished expectation of future revenues. In the stable Paretian world in which movies live, a movie does not exhaust its revenue potential even though it has already earned an extraordinary revenue. And, this is more true of the

Table 3.3 Implied values of the Paretian tail coefficient

Week	Hit movies	Non-hit movies
2	1.41	1.58
3	1.30	1.52
4	1.20	1.35
5	1.26	1.32
6	1.17	1.41
7	1.24	1.29
8	1.12	1.32
9	1.24	1.36
10	1.43	1.29

Note
Calculated from the coefficient values reported in Tables 3.1 and 3.2.

hit than of the (relatively longlasting) non-hit movies as shown by the β and α estimates. The movies that do not last 10 weeks seldom make it into the Paretian upper tail and they have completely different expectations; they play off in a Gaussian fashion as they exhaust their revenue potential (or turn off their audience).[9]

The importance of "legs" now is apparent and dramatic because it has a strong (and deep) statistical under pinning. To get legs means to run long enough and to gross enough to reach the upper Paretian tail of the distribution where the conditional expectation of forward revenue is linear in past revenue. At this point, past success breeds further success, as is demonstrated by the linearity of the conditional expectation of box-office revenue. The infinite variance of the upper tail and the slow decay of probability in this region of the distribution make the phenomenon of "legs" possible and even probable.

There are some contractual reasons for the appearance of the peaks that appear in Figure 3.1 at weeks 4, 6 and 8 for hit movies. These coincide with typical contracted minimum runs. A movie contracted for a minimum guaranteed run of four weeks might show a decline at week 5 as theaters will have by then fulfilled their obligation and may put another movie on the screen. Similarly for weeks 8 and 10, although few movies are exhibited under licenses that call for minimum runs of that length. More importantly, the hold-over clause, which binds a movie to a theater if it is grossing more than a contracted amount, extends the run, resulting in the peaks and valleys as some theaters hit their holdover and others do not. The hit movies keep many of their screens through much of their run because they earn enough theater revenue to trigger the hold-over clause, extending the run week-by-week. The non-hit movies are losing screens and at a higher rate than the hits and this adjustment makes the forward impact of past revenue more constant.

3.6 Bifurcation at phase transitions

Figure 3.1 already hints at a transition point where hit and non-hit movies diverge in revenues. At this transition point, the movies go onto different paths and the hit movies separate from the others at an increasing rate. A variety of recursive mappings are capable of generating this kind of bifurcation; Arthur (1995) uses a "tent map" to generate a bifurcation. The mappings we have been exploring describe iterative processes which are continuous dynamical systems that can be approximated using discrete systems of difference equations.

3.6.1 *The path to a stable equilibrium*

Consider the logistic map: $f(x, r) = rx(1-x)$. The parameter r is determined by the base birth rate of moviegoers; x is the number of moviegoers in the prior period; and $1 - x$ is the population remaining from which to draw new viewers (the source is normalized to a value of 1 here). The term *rx* is the number of new viewers, conditional on the number who saw the movie last week.

When the population becomes large, the net birth rate decreases nonlinearly because the population of viewers becomes exhausted. When $0 \leq r \leq 4$, the logistic function satisfies these conditions. When the parameter is increased from 0 to 4, there are different regimes of behavior, divided by so called bifurcations. For *r* sufficiently small, the population would die out and the movie would not be a hit. There is a set of values of *r* where the population approaches a stable equilibrium. Beyond these values, the population goes into seemingly random behavior. In Figure 3.2 we use the logistic map to illustrate a phase transition where the movie revenues branch into distinct paths. Beyond this point the distance between hit and non-hit movies widens at an increasing rate and other transitions may occur as well.

This illustration is mirrored almost exactly in the data. First consider the information in Figure 3.3 where the coefficient of variation (mean ÷ standard deviation) of weekly box-office revenues is graphed over the course of the run for hit and not-hit movies. This ratio approximates the degree of consensus of opinion with the mean, and the consensus on hit movies grows throughout the run. For non-hit movies, the consensus rises briefly and then falls precipitously at the fourth week and throughout the remainder of the run. Second, consider the differences between these movies and the levels and changes in the weekly revenues. In Figure 3.4 the weekly differences between mean weekly cumulative revenues of hit and non-hit movies is shown through out the run. Clearly, the difference widens throughout the run in the manner illustrated by the logistic mapping.

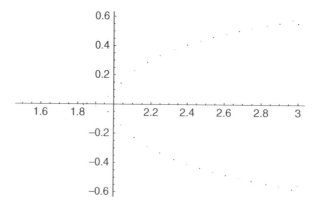

Figure 3.2 Bifurcation of revenue mapping into hit and non-hit paths.

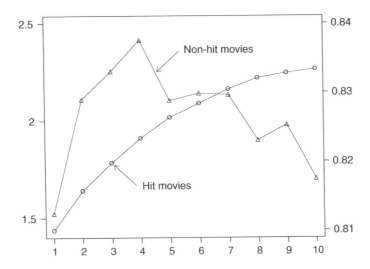

Figure 3.3 Coefficient of variation by week of run.

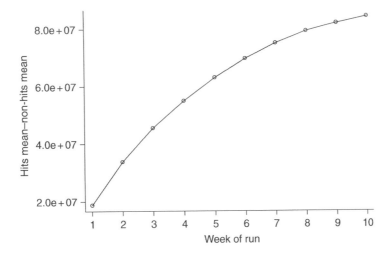

Figure 3.4 Weekly differences between hit and non-hit movies of mean cumulative revenues.

3.6.2   *Evidence from correlations*

In Table 3.4 we report the correlations between all pairs of weekly revenues for hit movies and non-hit movies. The upper triangle of the correlation matrix has the weekly correlations for non-hit movies, while the lower triangle has

Table 3.4 Week-to-week revenue correlations

	Week 1	Week 2	Week 3	Week 4	Week 5	Week 6	Week 7	Week 8	Week 9	Week 10
Week 1		0.936	0.801	0.709	0.605	0.527	0.451	0.399	0.314	0.293
Week 2	0.778		0.906	0.822	0.723	0.641	0.572	0.514	0.404	0.378
Week 3	0.581	0.847		0.943	0.867	0.786	0.714	0.655	0.538	0.498
Week 4	0.346	0.512	0.654		0.943	0.874	0.808	0.756	0.647	0.593
Week 5	0.011	0.152	0.281	0.677		0.942	0.883	0.840	0.745	0.694
Week 6	−0.115	0.071	0.209	0.527	0.862		0.954	0.910	0.830	0.774
Week 7	−0.314	−0.077	0.097	0.364	0.716	0.892		0.950	0.874	0.811
Week 8	−0.388	−0.179	0.014	0.303	0.697	0.838	0.942		0.927	0.858
Week 9	−0.385	−0.231	−0.039	0.252	0.655	0.796	0.911	0.965		0.919
Week 10	−0.400	−0.243	−0.063	0.184	0.573	0.745	0.876	0.927	0.960	

Note

Week-to-week correlations for hit movies (grossing ≥$50 million) in lower triangle. Correlations for non-hits in upper triangle.

the correlations for hit movies. For non-hit movies, the correlations taper off gradually. However, for hit movies the correlations fall to zero and in some cases are even negative for the correlations between weeks 1–3 and weeks 6–10. This is an indication of turbulent dynamics—the equivalent to bifurcation points in chaos—and is consistent with the model of private information sharing. At the onset of chaotic phase transitions there is usually an increase in volatility. To examine if the revenue dynamics in our sample are consistent with this, we computed the variance of weekly revenues by the week of a movie's theatrical run. The results, displayed in Table 3.5, show that for non-hit movies the standard deviation of revenues declines consistently over a movie's lifetime. For hit movies, the standard deviation of revenues first declines and then increases. This is evidence of turbulent dynamics for the hit movies but not for the non-hit movies. Another way to examine the importance of opening revenues on cumulative box-office gross would be to calculate the simple correlations between them. Table 3.6 presents the correlations between week 1 revenue and cumulative revenue for hits and non-hits by life-length of the movie. The correlation for non-hits is quite high, while the correlation for hits is quite low. This is further evidence that the opening event is not very important in making a movie a hit.[10]

3.7 Discussion of results

The evidence rejects the model that a big opening is necessary or sufficient for a movie to gross large box-office revenues.[11] The statistical herding on which this model relies is also clearly rejected. At the point where hit movies distance themselves from others, they are propelled by a dynamic that is completely different from the dynamic that drives other movies. By week 4 or 5, and even sooner for movies that are " turkeys," movies that are not getting the legs necessary to become hits are dropping off theater screens and their revenues are decaying at close to an exponential rate.[12] Some of these movies may have

Table 3.5 Mean and standard deviation of weekly revenues by week of run

Week	Non-hit movies			Hit movies		
	Observation	Mean	Std dev.	Observation	Mean	Std dev.
1	1,839	13.508	2.259	176	16.199	1.304
2	1,713	13.386	2.066	176	16.158	0.869
3	1,597	13.169	1.967	176	15.971	0.690
4	1,491	12.922	1.862	176	15.766	0.540
5	1,399	12.639	1.772	176	15.594	0.470
6	1,298	12.412	1.702	176	15.374	0.510
7	1,202	12.172	1.636	176	15.108	0.599
8	1,096	11.995	1.580	176	14.811	0.702
9	795	11.918	1.453	141	14.492	0.876
10	736	11.736	1.428	140	14.249	0.961

Note
Mean and standard deviation are for the logarithm of weekly revenue for movies that survived to the week indicated.

Table 3.6 Correlation of week 1 and cumulative revenues by life length

Life length	Non-hit movies	Hit movies
2	0.939	—
3	0.955	—
4	0.821	—
5	0.927	—
6	0.944	—
7	0.935	—
8	0.821	0.108
9	0.863	—
≥10	0.798	0.200

Note
Most hits lived ≥ 10 weeks. Only 35 died at week 8.

been early box-office leaders, but when they receive poor word-of-mouth evaluations they slip rapidly down the revenue rankings and are soon gone.

Movies that achieve extraordinary box-office success engage an information dynamic that appears to be statistical herding, but it is not. At week 5 these hit movies move away from long-lived, non-hit movies; at week 6 they hit a peak conditional expectation, whereas other movies experience a fall. By this time, there is plenty of quality information around to guide the filmgoer to good movies and away from bad ones. Hit movies go on to a higher peak conditional expectation at week 8 and from week 2 to the end of the run they have a higher expectation of forward revenue conditional on revenue earned through each week.[13] If statistical herding were driving this process, then audiences should ignore quality information to imitate the choices of others. In this instance, the peak of the conditional expectation should occur at the first or second week of the run. But, the evidence is to the contrary; the peaks occur later in the run after quality evaluations are readily available and widely distributed. And the separation of hits from non-hits indicates that they are driven by a different dynamic, one that is not consistent with the neglect of quality information by filmgoers.

There is a clear warning here about contagion and herding models; many processes can look like herding when they are not. That many people make similar choices need not be evidence of herding; when people make informed choices, they may also make similar ones as they reach some consensus of opinion on the quality of a movie. And, it is an error to look only at successes; one must note the failed products to find the differences in the dynamics of success and failure. In an informed choice dynamic, future choices are conditional on past choices. Early leads at the box office are readily overcome by good movies when the audience discovers and accepts them.[14] Poor movies cannot ride an early lead for long because within two to three weeks the audience discovers and rejects them.

3.8 Conclusions

Big budgets, wide openings, stars, sequels and prequels are all attempts to set the initial conditions to favor an information cascade. Prior empirical analysis has shown that these factors raise the lower portion of the cumulative revenue distribution, but that they are ineffective in moving significantly the upper portion of the revenue distribution—the part that matters. The dynamics of weekly movie revenues show us why this is the case: quality information eventually overcomes an early lead at the box office.

Moviegoers are able to break out of the imitative behavior of an information cascade by sharing information through reviews and word-of-mouth transmission. The only path dependence that can be found is in movies that fail or only succeed moderately. In this case, the revenue they eventually earn is highly correlated with their opening revenue. Indeed, a movie's opening week and total revenue are identical when it is so unsuccessful that it dies before its second week. Successful movies move off the typical path and leap to new ones as they get "legs" when enough people have seen the movie to transmit a large flow of positive quality information about them. From that point on, all correlation with opening revenues vanishes and they assume a life of their own. Week-to-week correlations of revenue then become very large, which is evidence that the audience is growing because of a positive information flow to new viewers.

A wide release is neither necessary nor sufficient for a motion picture to earn high revenue. The dependence of current demand on past demand (which statistical herding models imply) is stronger early in the run for low-grossing movies than for high-grossing movies. And the contrary is true after the fifth week of the run. The mapping of demand on past demand bifurcates into an expansive branch and a contractive branch by about the fifth week. Statistical herding models fail to explain the revenue dynamics of highly successful motion pictures whose revenues dominate total industry revenues.

Part II

"Wild" un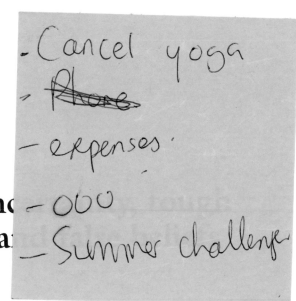
decisions a

The chapters in Part I showed that the dynamics of choice are so complex that they can't be predicted. Movie consumers choose from a vast array of movies (and other entertainment). Every individual is immersed in a private life of many tangled relationships and influences. Consumers rely on a mixture of private and public signals in choosing movies. When you consider how complex is the aggregation of these decisions it would be almost miraculous if it could be predicted or managed.

Yet, the business tries. Executives are always trying to reduce risk (at least the risk they personally bear). The "formula" for managing risk may differ among executives in how they choose to lay off risk to investors or foreign markets, in their preferred financing arrangements, or in the kinds of movies they make. But, a common element in strategies for managing risk is the use of stars. Stars are thought to offer some sort of guarantee for a movie. Is this true?

How could it be true? If it were true, making movies would be easy. Put a star in a movie and stand back and let the profit roll in. To the contrary, Chapters 4, 5 and 10 show that star movies do not increase expected profit and carry high risks (Ravid 1999 shows this result too). If it were this easy, there would be no reason to pay executives very much; the agents, who control access to the stars would have all the power and would take the rewards for themselves and the stars they represent. If it were true that stars guaranteed profits, studio executives would be paid no more than insurance vice presidents.

Why are executives paid so much, far beyond what the studio moguls of the past received? Maybe it is because they are good at deciding which movies to greenlight. But, it isn't true. Nobody has a formula, not even top executives.

Maybe Irving Thalberg had that kind of skill, but, in today's motion picture market, no one does. Put the Thalberg mystique aside for a moment and think about what it would require for an executive to have superior skill in judging movies. Whether it is intuitive or formal, an executive who is good at predicting motion picture success has to have a model, a formula to predict box-office grosses from the attributes of the screenplay, cast and other elements. The formula can't be obvious or well known or the executive's skill would not be unique and could not command premium pay.

There are plenty of models around; consultants will forecast movie grosses (for a fee), studio financial analysts must use some kind of model to evaluate the movies a studio makes, and there are even web sites that claim to be able to forecast box-office revenues of movies based on the "creative" elements (without so much as a script!). One can find forecasts of movie grosses in many publications. For example, every week the *Los Angeles Times* has a column that forecasts the total theatrical world gross of individual movies based on their opening week or the first few weeks of its run. Studios forecast earnings in their financial reports. To do that they have to be able to predict how much their movies will gross. With so many predictions around, why is it so hard to make successful movies? It should be easy.

The answer is simple: there is no formula. No one knows what a movie will gross. Nobody knows anything. As you will see upon reading the chapters in Part II, motion picture revenues are unpredictable. Producers or studio executives who think they can predict movie grosses are prone to errors and suffer decision biases that limit their effectiveness.

Next to deciding which movies to make, everything else is easy; timing the release date, booking screens, making trailers, buying ads, cutting distribution deals and so on, pale in comparison to this most difficult and central question "which movie shall I make?". Paradoxically, all these decisions, and the "greenlighting" decision would be less error-prone if Hollywood understood that outcomes cannot be predicted. The false belief that grosses can be predicted is a source of some of Hollywood's most self-destructive errors. The three chapters that follow in this part of the book challenge Hollywood's belief system.

In Chapter 4 David Walls and I discover that the movies are very risky indeed. It may not come as a surprise given how peculiar the business is if I say that the movie business is not "normal." Remember the bell-shaped curve the teacher always used to assign grades in middle school? That is the normal or Gaussian distribution that many people draw upon when they think about probability distributions and forecasts. If motion pictures and studio executives lived in a Gaussian world, then most movies would be "average" and few would stray too far from the middle ground. But there is little that is normal about this business. Events that are five standard deviations above the mean occur in the movies even though they would never occur in a billion reruns of a Gaussian world.

Why is this so? Well, the chapters in Part II show that box-office revenue dynamics follow a Lévy stable process (which is a Bose–Einstein process with big leaps) and are asymptotically Pareto-distributed with infinite variance. This means that when a studio predicts a movie's revenue before it is released, the error is essentially infinite. A prediction that Movie A will gross X million plus or minus infinity, is no prediction at all.

The average revenue earned by all motion pictures is dominated by a few rare blockbuster movies, movies so improbable in a Gaussian world that they should never occur. These movies are those rare events, four or five or more standard deviations above the mean of a normal distribution, that are located in the far right upper tail of the distribution. There is no typical movie because box-office revenue outcomes do not converge to an average, in fact they diverge over all scales.

In the first chapter in this part, Chapter 4, David Walls and I look to see how stars alter the odds that a movie will be a hit or a bomb. Some readers will be surprised to find that conventional beliefs about stars are not right. Stars do not make a movie open, but they do have some influence in how long it runs. Stars increase the least revenue a movie might earn, but have almost no influence on the most it might earn. The relationship between stars and screens is subtle; stars are more closely associated with screens in the later weeks of a run than with opening screens. Star movies show more variance than other movies. If they were less risky, by conventional measures, they would have less variance. But, variance is good in the movies; it means there is a diversity of creative products out there and a few of them have a shot at being blockbusters.

Given the kind of "wild" uncertainty implied by these infinite variance probability distributions, motion picture portfolios are important ways to mitigate risk. In planning a production slate, a studio should not choose movies independently, one at a time. They should diversify their risks over movies of different types. In the face of uncertainty that stems from each movie's uniqueness and the wild behavior of audiences, a portfolio is one of the few ways a studio can mitigate its risk.

In Chapter 5, David Walls and I look at Hollywood's portfolio from the vantage point of the MPAA ratings. We group movies into ratings categories and see if the industry portfolio is balanced with respect to return and risk. First we find and estimate the probability distributions of budgets, revenues, returns and profits to G-, PG-, PG13- and R-rated movies. Then these distributions are ranked according to a criteria called stochastic dominance. This criteria allows an unambiguous ranking of movie ratings categories by their returns and risks.

The evidence shows that the industry's critics and its shareholders can agree that Hollywood does make too many R-rated movies. The profit distributions have aysmmetric tails which means that Hollywood could trim its "downside" risk while increasing its "upside" possibilities by shifting production dollars out of R-rated movies into G, PG and PG13 movies. Stars who are willing to appear in edgy, counterculture R-rated movies for their prestige value may induce an "illusion of expectation" leading studios to "greenlight" movies that have biased expectations.

It is hard to say why the business makes more R-rated movies than is profitable. Among the reasons I could conjecture are the following: G movies take more talent, they are harder to make, investors have lurid tastes, R-rated movies earn the highest Hollywood insider praise, moviemakers are wimps and feel more macho when they make an R movie, they are ignorant of the odds, third world tastes drive investment and production, executives care more about their jobs than investor returns, stars are more likely to appear in an R-rated movie for the prestige (they overwhelmingly do so), screenwriters submit screenplays that are predominantly R rated and executives must choose among what is submitted. In short, almost everyone in Hollywood contributes in some way to the problem, but they blame it on the audience when they say "we are just giving them what they want." Chapter 5 shows pretty convincingly that they aren't giving the audience what it wants and it is costing them.

It takes a big mind to stand back and look over a portfolio of films to see what common sense and the research says is there for the taking—the lower risk and higher earnings potential of G and PG movies that is being passed over for the higher risk and lower earnings potential of R movies. The studio moguls of the past era could see the big picture as they planned a season's worth of production. They owned the studios and had reasons to consider the true risk. Now, it is an executive protecting a job ("I got you Cruise and Palthrow, it's not my fault it bombed") loading risk onto the studio. Stars increase the least outcomes, so they confer a measure of failure resistance on an executive who "greenlights" a star film. But only "legs" can make a hit.

There are limits to knowledge, and it is difficult to come to grips with uncertainty. Even risk is hard to fathom in this industry. With risk, the odds are known. With uncertainty the odds are not known and are impossible to measure. So, the "wild" uncertainty of the Pareto stable distributions lies somewhere between risk and uncertainty. Whole distributions are not observable, so we are capable only of reaching a tentative degree of belief in a distribution, based upon finite observations. Our knowledge is always subject to revision. This is a real warning that what worked yesterday may not tomorrow. The movies are always exploring new territory.

Aside from choosing which movies among many possible projects to make, a studio has booking and marketing decisions to make. In Chapter 6, David Walls and I test the blockbuster model. The blockbuster marketing model is predicated on a big opening. The idea is to use a big budget, high-profile stars, and massive advertising to create high opening week box-office grosses. The theory behind this expensive and risky marketing strategy seems to be that motion picture audiences choose movies according to how heavily they are advertised, what stars are in them, and how high they are in the revenue rankings reported on the evening news.

This "herding" theory was already shown in Chapter 3 to be wrong. Herding lasts only a few weeks and then gives way to a more informed kind of choice after which hit and non-hit movies rapidly diverge from one another. We take that analysis farther in Chapter 6 to examine the blockbuster strategy in more detail. Contrary to the belief in the efficacy of the blockbuster strategy, we find that the opening is less critical for successful than for unsuccessful films. For example, the median box-office revenue a film earns is best predicted by the revenues it earns in week 5 or week 8 than by what it earns in week 1. These latter weeks are nearly twice as important as the opening week (this is based on quantile regressions I did but do not report here for the sake of brevity). We conclude that the opening can only take a film so far; you can put a movie on lots of screens (appealing to booking agents and their expectations), but you can't make people like it. If they like it, they will come.

We show in this chapter that there are decreasing returns to budgets, opening screens and stars. These variables have an interesting effect on the probability distribution (always the object of choice in the movie business). Budget, screens and stars lift the least revenue a film might earn, but have little effect on the most revenue it might earn. In other words, they place a prop under the revenue a film might earn, but do not have much influence on whether it will be a blockbuster or not.

.

4 Uncertainty in the movie industry
Can star power reduce the terror of the box office?

Nobody knows anything.

Screenwriter William Goldman (1983)

4.1 Introduction

Everyone knows that motion pictures are uncertain products. In this chapter we show that filmmakers must operate under such vague and uncertain knowledge of the probabilities of outcomes that "nobody knows anything." The essence of the movie business is this: The mean of box-office revenue is dominated by a few "blockbuster" movies and the probability distribution of box-office outcomes has infinite variance! The distribution of box-office revenues is a member of the class of probability distributions known as Lévy stable distributions. These distributions are the limiting distributions of sums of random variables and are appropriate for modeling the box-office revenues that motion pictures earn during their theatrical runs.

Lévy stable distributions have a "heavy" upper tail and may not have a finite variance. Our parameter estimates of the asymptotic upper tail index reveal that the variance of box-office revenue is in fact infinite: Motion pictures are among the most risky of products. Theoretically, the skewed shape of the Lévy distribution means there is no natural scale or average to which movie revenues converge. Movie revenues diverge over all possible values of outcomes. One can forecast the mean of box-office revenue since it exists and is finite, but the confidence interval of the forecast is without bounds. The far-from-normal shape of the Lévy probability distribution of box-office revenue and its infinite variance are the sources of Hollywood's "terror of the box office."

Our results explain heretofore puzzling aspects of the movie business. The average of motion picture box-office revenues depends almost entirely on a few extreme revenue outcomes in the upper tail whose chances are extremely small. Success is tied to the extremal events, not the average; the average is driven by the rare, extremal events. The mean and variance of the distribution drift over time and do not converge or settle to an attractor. Movie projects are, in reality, probability distributions and a proper assessment of their prospects requires one to do a risk analysis of the probabilities of extreme outcomes. The normal distribution is completely unsuited for this kind of analysis because when outcomes are normally distributed, the probability of extreme outcomes is vanishingly small. The movie business is not "normal" because outcomes do not follow a normal probability distribution. The probability distribution is highly skewed with a "heavy" upper tail with a theoretical variance far beyond the sample variance. Our estimates of the theoretical Lévy distribution permit calculation of the probability of box-office revenues that have never before been realized.

There are no formulas for success in Hollywood. We find that much conventional Hollywood wisdom is not valid. By making strategic choices in booking screens, budgeting and hiring producers, directors and actors with marquee value, a studio can position a movie to improve its chances of success. But, after a movie opens, the audience decides its fate. The exchange of information among a large number of individuals interacting personally unleashes a dynamic that is complex and unpredictable.[1] Even a carefully managed and expensive marketing program cannot direct the information cascade; it is a complex stochastic process that can go anywhere.[2]

We conclude that the studio model of risk management lacks a foundation in theory or evidence. Revenue forecasts have zero precision, which is just a formal way of saying that "anything can happen." Movies are complex products and the cascade of information among filmgoers during the course of a film's theatrical exhibition can evolve along so many paths that it is impossible to attribute the success of a movie to individual causal factors. In other words, as Goldman said, "nobody knows anything." The audience makes a movie a hit and no amount of "star power" or marketing hype can alter that.[3] The real star is the movie.

4.2   Related literature

Three strands of literature are relevant to our topic: one dealing with motion pictures and uncertainty, one dealing with stars and another dealing with power law probability distributions.

4.2.1   *Motion picture uncertainty*

De Vany and Walls (1996) modeled the motion picture information cascade as a Bose–Einstein statistical process and they argued that it converged on a Pareto distribution; Walls (1997) and Lee (1998) replicated these findings for another market and time period, respectively. In a rank tournament model of the motion picture market, De Vany and Walls (1997) modeled a film's theatrical run as a stochastic survival process with a rising hazard rate; Walls (1998) replicated these results for another market. De Vany and Eckert (1991) portray motion pictures as a market organized to deal with the problem that filmmakers "don't know anything" and showed that the studio system and block booking were adaptations to uncertainty.[4] In a related context, where outcomes are uncertain, Chisholm (1996, 1997) and Weinstein (1998) examine the use of share contracts versus a fixed payment contract for compensating stars.

4.2.2 Stars

Wallace *et al.* (1993) estimate regression models of the relationship of actors and actresses to film rentals and associate stars with positive or negative residuals. Prag and Casavant (1994) also estimate film rentals as a function of production cost, a measure of quality and an index of star power and find that these variables are significant only when advertising costs are omitted. Ravid (1999) examines a signaling model of the role of stars and estimates rental and profit equations, concluding that stars play no role in the financial success of a film.

4.2.3 *Pareto and Lévy distributions*

Pareto (1897) found that income was distributed according to a power law that was subsequently named after him. Atkinson and Harrison (1978) found wealth to be Pareto distributed. Ijiri and Simon (1977) found the size distribution of firms in the United States and in Britain to be Pareto distributed. Lévy showed that there is a class of distribution functions which follow the asymptotic form of the law of Pareto which Mandelbrot defined as

$$1 - F_X(x) \sim \left(\frac{x}{k}\right)^{-\alpha} \quad x \to \infty \tag{4.1}$$

Such distributions are characterized by the fact that $0 < \alpha < 2$ and they have infinite variance. The Lévy is a generalization of the normal distribution when the variance is infinite. Mandelbrot (1963) found that the distribution of cotton price changes is approximated by the Lévy distribution. Fama (1963) described an information process (similar to Bose–Einstein information updating) that could lead to a Lévy stable distribution. Both S&P 500 stock index and NYSE composite index returns are well fitted by a Lévy distribution (Mantegna and Stanley 1995; Lévy and Soloman 1998). This chapter adds motion pictures to the list of processes that follow a Lévy distribution.

4.3 Modeling star power

One has to be humble in approaching this subject—the movie business is complicated and hard to understand. There are many reasons for this difficulty: motion pictures are complex products that are difficult to make well; no one knows they like a movie until they see it; movies are "one-off" unique products; their "shelf life" is only a few weeks; movies enter and exit the market on a continuing basis; movies compete against a changing cast of competitors as they play out their theatrical "runs"; most movies have but a week or two to capture the audience's imagination; a rare handful have "legs" and enjoy long runs; weekly box-office revenues are concentrated on only three or four top ranking films; most movies lose money.

These characteristics of the business led us to model movies as stochastic dynamic processes in our earlier work (De Vany and Walls 1996, 1997). This work has convinced us that a fruitful way to model the movies is to treat them as probability distributions. We model the distribution of probability mass of movie outcomes on the outcome space and strive to uncover how the mass is shifted when certain conditioning variables are changed.[5] Among the variables that we consider as potential probability-altering variables are sequels, genres, ratings, stars, budgets and opening screens. We also consider individual stars and how much power they have to move the movie box-office revenue probability distribution toward more favorable outcomes.

4.3.1   *Modeling probability mass*

Formally our strategy is to characterize the unconditional probability distribution of movies, without qualification as to genre, stars or other variables that may condition the distribution. Then we characterize the distribution conditional on a list of choice variables that potentially alter the location of the distribution's probability mass. We consider the cumulative density function and the probability density functions in continuous and discrete form. Symbolically, we examine the conditional cumulative density function

$$F(x \mid \vec{Z}) \tag{4.2}$$

where x is a random variable, and $\vec{Z}$ is a vector of conditioning variables on which F depends. We seek to find the form of F and the conditional distribution of F with respect to $\vec{Z}$. The random variable x will be box-office revenue, profits or some other variable of interest and the components of the vector $\vec{Z}$ will be budgets, stars, sequels, ratings and other variables that might alter F. We follow a similar modeling strategy for the probability density function. We also rely on a discrete version of this model by estimating the effects of changes in the conditioning variables on the quantiles of the distribution or the probability of a specific event.

In our other work, we found that the dynamics of audiences and box-office revenues follow a Bose–Einstein statistical process. It is known that the Bose–Einstein allocation process has as its limit a power law and this holds for the rank order statistics (Hill and Woodrofe 1975) and for the density function (Ijiri and Simon 1977). Based on these considerations, we expect the probability distribution of motion picture outcomes for extreme outcomes to follow a power law. The Bose–Einstein information process is a stable process (see Fama (1963) on stable information processes and De Vany (1997) on the stability of the Bose–Einstein process), so we expect the distribution generated by a Bose–Einstein process to be in the Lévy stable class. In this class of distributions, only the normal distribution has finite variance. Probability mass in the tail decays as a power of $x, x^{-\alpha}$, in the power law but exponentially,

e^{-x} in the normal distribution. Second and higher moments, therefore, may not converge for a power distribution: $\text{Var}(x) = \int x^2 f(x)\mathrm{d}x = \int x^2 x^{-\alpha}\mathrm{d}x$. If $\alpha < 2$, the product $x^2 x^{-\alpha}$ diverges as $x \to \infty$.

Analysis of the data further sharpens our expectation that a Lévy stable process is at work. As we show later, the mean is dominated by extreme outcomes, which is characteristic of Lévy stable processes. The sample variance is unstable and less than the theoretical variance, also indicators of a Lévy stable process. A further indicator of a stable process is our discovery that the rank and frequency distributions are self-similar.

4.3.2 *Risk and survival*

If, as we expect, the variance of the probability distribution of movie outcomes is infinite, then it is not useful or even well defined to rely on the second moment to make probability-based judgments about movies. One can, however, do risk and survival analyses. In a risk analysis, we consider the probabilities that are attached to certain outcomes in the upper tail. This is a well-defined exercise, even when the variance is infinite. In a survival analysis, we consider the conditional probability that a movie will continue to earn more, given that it already has earned a certain amount. This also is well defined for the Lévy distribution and the conditional probability of continuation of a movie's run can be calculated from the distribution function. These kinds of analyses require that we model quantiles, extremals and probability mass over motion picture outcomes rather than rely on traditional measures like mean and variance.

The question then becomes this: How do stars, genre, release patterns, etc. alter the quantiles, extremals, probability mass and survival functions of motion pictures? These are well-defined questions even for infinite variance distributions, where mean–variance analysis fails. An example will make this clear. Consider the cumulative distribution function and its associated continuation function.[6]

In Figure 4.1, the x axis—the outcome space—corresponds to motion picture revenue outcomes and the y axis corresponds to the cumulative probability of all outcomes up to each point in the outcome space. Once we identify the functional form and parameters of the cumulative probability function, we can calculate the probability of an outcome or set of outcomes. If, in addition, we are successful in identifying how this function is shifted when movies have stars or different release patterns and so on, then we can calculate how these decision variables alter the probabilities of specific outcomes.

In fact, we do identify the conditional probability density functions of movies with stars and without stars. These distributions are shown as the solid (with stars) and dashed lines (without stars) in Figure 4.1. It is a simple matter to calculate the probability that a movie will earn box-office revenues equal to or greater than \$50 million and then to further calculate how stars alter those probabilities. Using the continuation function, we can also calculate

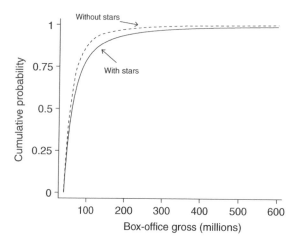

Figure 4.1 Cumulative probability for movies with and without stars.

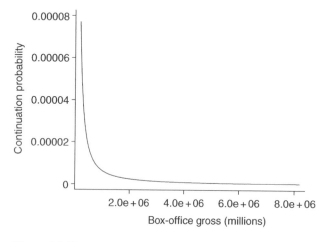

Figure 4.2 The continuation function for movies with stars.

the probability that a movie will continue its run, given that it already has earned $50 million or $300 million, or any amount. The continuation function shown in Figure 4.2 is what we find for movies with stars. The curve shows for every revenue outcome the probability that the movie will continue to go forward to higher revenues, conditional on it having earned some amount. Note that the continuation probability declines very slowly, an indication of the infinite variance and the power law decay in the upper tail. Using risk and continuation analyses we can predict the probability of events never before experienced. The slow decay of the continuation probability predicts unheard of successes like *The Full Monty* or *Titanic* and shows that the probability of even more striking successes does not vanish.

4.4   The movie data

4.4.1   *Data sources and definitions*

The data include 2,015 movies that were released in the closed interval 1984–96. Information on each movie's box-office revenue, production cost, screen counts by week, genre, rating and artists were obtained from Entertainment Data International's historical database. The box-office revenue data include weekly and weekend box-office revenues for the United States and Canada compiled from distributor-reported figures. These data are the standard industry source for published information on the motion pictures and are used by such publications as *Daily Variety* and *Weekly Variety, The Los Angeles Times, The Hollywood Reporter, Screen International* and numerous other newspapers, magazines and electronic media.

Each actor, producer or director appearing on *Premier*'s annual listing of the hundred most powerful people in Hollywood or on James Ulmer's list of A and A+ people is considered to be a "star" in our analysis.[7] In our sample, 1,689 movies do not have a star, meaning they do not feature an actor, director, writer or producer whose name appears on the lists of stars. The number of movies featuring a star is 326, about 20–40 movies a year. Sources list 15 genres—action, adventure, animated, black comedy, comedy, documentary, drama, fantasy, horror, musical, romantic comedy, science fiction, sequel, suspense and western. There are four ratings: G, PG, PG13 and R. The most common genre is drama, followed by comedy. R is by far the most common rating—accounting for more than half—followed by PG13. The least frequent rating is G.

4.4.2   *A film's theatrical run*

Dynamics are an essential feature of motion pictures. Movies open, play out their run over the course of a few weeks and then are gone. Demand and supply are dynamic and adaptive processes. Initially, a movie is booked on theater screens for its opening. The contract will usually call for a minimum run of 4–8 weeks. During the run demand is revealed and the supply of theatrical engagements is adjusted. On a widely released movie, the number of screens on which it is shown will typically decline during the run. But, that is far from certain; some widely released movies become so popular that the number of screens may not decline and might even increase during the run.

Motion picture runs are highly variable. Figure 4.3 plots the temporal pattern of screen counts for several films that were widely released. The upper panel shows the run profile of *Waterworld*, a highly promoted film (starring Kevin Costner) with a production budget of $175 million: it opened on over 2,000 screens and had fallen to 500 screens by the tenth week of its run. In contrast to *Waterworld* is *Home Alone*, a film with a much smaller budget and which featured no stars: it opened on just over 1,000 screens and grew to peak at over 2,000 screens in the eighth week before starting a slow decline. The box-office gross for *Home Alone* was nearly four times as large

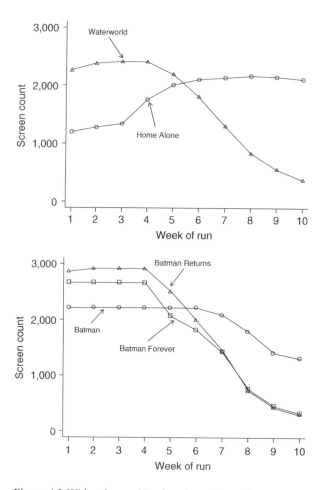

Figure 4.3 Wide releases: hits, bombs and sequels.

as the box-office gross for *Waterworld*. The lower panel of Figure 4.3 shows the run profiles for the series of *Batman* films: *Batman*, *Batman Forever* and *Batman Returns*. Successive *Batman* films cost more to make, opened more widely, played out more rapidly and earned less at the box office.

Figure 4.4 shows the run profiles for smaller budget films. The top panel of the figure shows the run profiles for four films that opened on about 500 screens. *Excessive Force* fell rapidly from the first week, while *Nixon* rose to peak at 977 screens in the third week and then fell rapidly. *Serial Mom*'s run profile was flat to the fifth week and then it fell, while *House Party* had "legs" and was still playing on 430 screens at the tenth week of its run. *Excessive Force, Serial Mom, Nixon* and *House Party* grossed about 0.8, 5, 9 and 20 million dollars, respectively, at the box office. The lower panel of Figure 4.4 shows

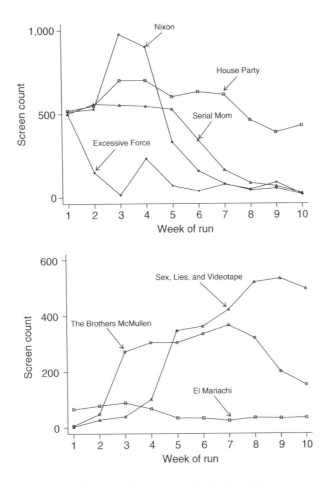

Figure 4.4 Narrow releases: growth, death and legs.

the run profiles of three highly successful micro-budget films. *Sex, Lies, and Videotape* and *The Brothers McMullen* went through tremendous growth after their initial releases. *El Mariachi* got "legs" even though it opened on only 66 screens; it was still showing on 35 screens ten weeks after its release.

About 65–70 percent of all motion pictures earn their maximum box-office revenue in the first week of release; the exceptions are those that gain positive word-of-mouth and enjoy long runs. The point of widest release for most movies is the second week, but the maximum revenue is in the first week. However, if a movie had good revenues in the first week, other exhibitors may choose to play it in the following week, or exhibitors currently showing it could add screens. This is an attempt to accommodate growth in demand after the film's initial exhibition.[8]

Table 4.1 Screen statistics by number of weeks survived

Week	Obs.	Min.	25%	50%	75%	Max.	Mean	Std dev.
1	1,500	1	99	994	1,472	3,012	927	746
2	1,461	1	145	1,032	1,536	3,012	977	752
3	1,400	1	141	881	1,453	3,012	906	744
4	1,326	1	158	706	1,314	2,977	818	708
5	1,246	1	135	520	1,164	2,901	717	658
6	1,165	1	119	449	1,020	2,808	639	610
7	1,081	1	107	377	898	2,532	569	558
8	997	1	96	332	817	2,384	510	510
9	915	1	82	292	710	2,316	454	463
10	853	1	78	264	596	2,331	409	427

Table 4.2 Weekly box-office gross relative to total gross

Year	Week 1/Total		Week 2/Total		Week 3/Total	
	Mean	Std dev.	Mean	Std dev.	Mean	Std dev.
1984	0.289	0.230	0.199	0.099	0.123	0.058
1985	0.303	0.262	0.188	0.104	0.125	0.078
1986	0.312	0.290	0.188	0.108	0.118	0.075
1987	0.306	0.277	0.177	0.107	0.115	0.066
1988	0.325	0.295	0.163	0.117	0.108	0.076
1989	0.364	0.304	0.187	0.126	0.115	0.071
1990	0.323	0.278	0.181	0.111	0.122	0.073
1991	0.316	0.279	0.174	0.104	0.121	0.070
1992	0.320	0.273	0.178	0.109	0.121	0.068
1993	0.346	0.272	0.193	0.106	0.120	0.071
1994	0.339	0.265	0.193	0.103	0.123	0.073
1995	0.364	0.271	0.206	0.102	0.126	0.063
1996	0.350	0.265	0.194	0.100	0.129	0.075

Table 4.1 shows the distribution of screen counts for films that survived to a given week of run.[9] Average (mean and median) screen count is at a maximum in the second week and it falls quickly as films play out. The median screen count fell from 994 in week 1 to 264 in week 10 for the 57 films that lived that long.[10] The distribution of screen counts across films becomes more skewed along the run profile. In week 1 the mean and median screen counts are 927 and 994, respectively. By week 10, the median screen count is 264 while the mean screen count is 409.

Table 4.2 shows the average proportion of cumulative revenues earned in each of the first three weeks of release. About 35, 19 and 12 percent of all box-office revenues are earned in the first, second and third weeks of a film's release, respectively. About 70–72 percent of each week's revenues are earned during the weekends, which account for 3/7 (43 percent) of the week. Nearly 85 percent of all films open on the first day of the weekend, Friday.

About 13 percent open on Wednesday, 1 percent on Thursday, and less than 1 percent opening on the remaining days combined.

4.4.3 *Revenues, budgets and profits*

Revenues Table 4.3 shows box-office revenues in constant 1982–84 dollars for our sample motion pictures. The table also shows the composition of the sample by rating, genre and the presence of a star. The mean revenue in the sample was $17 million and this was much larger than the median of $6.9 million. In fact, the mean was the 71st percentile of the revenue distribution, an indication of its rightward skew. Median revenues varied from $1.14 million for black comedies to $16.1 million for sequels. Movies without stars had a median gross revenue of about $20.9 million, while movies

Table 4.3 Box-office revenue quantiles by rating, genre and stars

	25%ile	*50%ile*	*75%ile*	*Mean*	*Std dev.*
Genre					
Action	1,761,336	8,204,898	2.06e+07	1.79e+07	2.81e+07
Adventure	1,208,704	9,946,237	2.04e+07	1.62e+07	2.16e+07
Animated	2,828,827	1.53e+07	4.40e+07	3.55e+07	4.78e+07
Black Comedy	253,088.4	1,136,575	4,480,660	6,105,246	1.48e+07
Comedy	1,477,800	7,621,048	2.37e+07	1.82e+07	2.74e+07
Documentary	401,100.4	605,250.8	4,051,058	6,786,864	1.51e+07
Drama	653,955.9	3,591,933	1.48e+07	1.15e+07	1.96e+07
Fantasy	4,389,282	1.07e+07	3.23e+07	1.95e+07	1.96e+07
Horror	2,336,693	6,693,621	1.28e+07	1.12e+07	1.45e+07
Musical	1,415,524	5,688,928	9,952,050	9,537,546	1.36e+07
Romantic Comedy	1,412,779	7,589,359	2.26e+07	1.72e+07	2.46e+07
Sci-Fi	4,327,004	1.20e+07	2.86e+07	2.87e+07	4.79e+07
Sequel	7,092,087	1.61e+07	4.10e+07	2.98e+07	3.36e+07
Suspense	450,647	5,043,216	1.56e+07	1.54e+07	2.67e+07
Western	2,994,923	1.45e+07	3.80e+07	2.87e+07	3.70e+07
Total	1,169,457	6,943,376	2.06e+07	1.70e+07	2.68e+07
Rating					
G	2,574,743	1.01e+07	2.28e+07	2.58e+07	3.99e+07
PG	1,566,304	1.12e+07	2.88e+07	2.15e+07	3.00e+07
PG-13	1,922,510	8,439,733	2.39e+07	1.96e+07	3.15e+07
R	814,088.6	5,171,729	1.58e+07	1.35e+07	2.11e+07
Total	1,169,457	6,943,376	2.06e+07	1.70e+07	2.68e+07
Star					
No	852,226.1	4,867,325	1.50e+07	1.22e+07	2.09e+07
Yes	1.33e+07	3.25e+07	5.87e+07	4.15e+07	3.82e+07
Total	1,169,457	6,943,376	2.06e+07	1.70e+07	2.68e+07

Note
All monetary magnitudes are reported in constant 1982–84 dollars.

with stars had a median gross revenue of about $38.2 million. For movies without stars, the mean revenue was equal to the 70th percentile, while for movies with stars the mean revenue was equal to the 62nd percentile. A Kolmorogov–Smirnov test allows us to reject the null hypothesis of equality of the revenue distributions for movies with and without stars.[11] In fact, movies with stars stochastically dominate movies without stars in terms of box-office gross.

Production budgets   The distribution of budgets is highly skewed, but not as skewed as the revenue distribution: The mean is $11.8 million and this is the 62nd percentile. Median budgets varied widely from about $1.9 million for documentaries to $17.4 million for movies in the science-fiction genre. Movies without stars had a median budget of about $9.7 million and movies with stars had a median budget of about $22.8 million. The mean budget for movies without stars was the 61st percentile and the mean budget for movies with stars was the 57th percentile. The results of a Kolmogorov–Smirnov test indicated that we could reject the null hypothesis of equality of distributions of budgets for films with and without stars.[12] Budgets of films with stars also stochastically dominate films without stars.

Profit and returns   Most movies are unprofitable. Large budgets and movie stars do not guarantee success. Even a sequel to a successful movie may be a flop. Figure 4.5 shows a plot of gross profits versus budgets.[13] The figure makes clear that large budgets and star presence can create the biggest of flops, like the film *Waterworld*. Much smaller budget and lack of star presence do not prevent a film from becoming a box-office hit, like *Home Alone*. And while sequels often do well, the series of *Batman* films were successively more costly and less profitable. The median movie lost about 3.8 million (1982–84) dollars and a film had to reach all the way up into the 78th percentile of the gross profit distribution before it broke even in its theatrical run.[14]

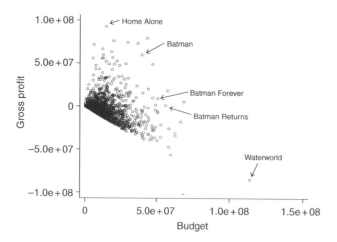

Figure 4.5 Gross profit versus budget.

Most micro-budget films die after a few weeks, but the ones that survive earn very high rates of return, the highest earned by any motion pictures. *The Brothers McMullen, El Mariachi* and *Sex, Lies, and Videotape* grossed 417, 292 and 25 times their respective production costs. Even though their rates of return are often high, small films earn small absolute profits. *The Brothers McMullen* earned only $3 million in gross profits. Only the big budget films have the potential to earn large absolute profits or losses.[15]

Table 4.4 provides a simple measure of the gross return to budget, revenue/budget. Since film rentals are approximately one half of box-office revenues, a gross return of 2 would be equivalent to a film breaking even.[16] The mean return for a film was 1.86, and this is the 76th percentile of the return distribution. Median gross returns varied substantially from 0.23 for

Table 4.4 Gross rate of return quantiles by rating, genre and stars

	25%ile	50%ile	75%ile	Mean	Std dev.
Genre					
Action	0.3010083	0.7120492	1.55091	2.509167	19.15657
Adventure	0.1493989	0.7709483	1.653367	1.497567	2.494713
Animated	0.4113783	1.430555	2.907783	2.493915	4.037561
Black Comedy	0.0938839	0.2253235	0.8240139	0.561118	0.7452614
Comedy	0.2322339	0.8109995	2.153059	2.011794	6.131053
Documentary	0.2499935	0.3140427	1.614872	1.358023	2.09954
Drama	0.1207789	0.470878	1.402122	1.817234	17.19662
Fantasy	0.332314	0.8350364	1.610353	1.06931	0.9841618
Horror	0.4426316	1.057498	2.129804	1.567043	1.603036
Musical	0.1553027	0.5582432	1.242419	1.712149	3.944996
Romantic Comedy	0.230322	0.9349518	1.789731	1.875962	4.298891
Sci-Fi	0.3854831	0.5862929	1.432238	1.193644	1.319072
Sequel	0.718936	1.59978	2.418147	1.926481	1.746636
Suspense	0.0995366	0.5138524	1.436905	1.26147	2.590913
Western	0.2402916	0.6231414	1.788678	1.761385	2.725234
Total	0.2040572	0.7273786	1.789731	1.861434	11.8195
Rating					
G	0.4009847	1.245965	2.381497	2.152229	3.46846
PG	0.25854	0.8924453	2.153059	1.662702	2.23019
PG-13	0.2193686	0.6822007	1.786988	1.390128	2.20324
R	0.1832344	0.6636797	1.669723	2.142445	16.16777
Total	0.2040572	0.7273786	1.789731	1.861434	11.8195
Star					
No	0.152384	0.5983311	1.632457	1.809556	12.86658
Yes	0.6385626	1.365121	2.775173	2.130216	2.406703
Total	0.2040572	0.7273786	1.789731	1.861434	11.8195

Note
Gross return is defined as Revenue/Budget. Since rentals are about half of box-office gross, an approximate return to the studio is 0.5*(gross return) −1. The breakeven gross return is 2.

black comedies to 1.6 for sequels. For movies without stars, the median gross return was about 0.6; assuming that film rentals are half of box-office gross this translates into a net rate of return of about −70 percent. For movies with stars, the median gross return was about 1.37, and this corresponds to a net rate of return of about −32 percent.

Stars increase the median of the returns distribution much more than the mean; they make the distribution less skewed. The mean return with no stars is at the 78th percentile, while the mean return with stars is at the 67th percentile. We performed a Kolmogorov–Smirnov test for equality of distributions and could reject the null hypothesis that the returns distributions were equal for movies with and without stars at a marginal significance level of practically zero. However, movies with stars do not stochastically dominate movies without stars in terms of gross return to budget. The largest gross return to a movie with a star was 16.7 times production cost for (*Beverly Hills Cop*); this movie also had a large box-office revenue. However, most movies with very large gross returns did not have stars, had low revenues and tiny budgets.[17] The successful micro-budget non-star movies have tremendous returns on budget, but they earn a less absolute profit than a big-budget production with a gross return of three times production cost.

4.5   Estimation results

4.5.1   *The size distribution of box-office revenue*

One of the ways star power might work is in moving a movie up in the money rankings by getting it booked on many screens at the opening. Once there, more viewers might be drawn to it if the ranking is taken by moviegoers to be an indicator of entertainment value. Figure 4.6 plots the box-office revenue and rank for each year in our sample. It is clear that the size distribution of revenue is uneven and highly convex in rank. This is consistent with the distribution of box-office revenues following the Pareto rank law: $SR^{\beta_2} = \beta_1$, where S is the size of box-office revenues, R is the rank (1 = highest), and β_1 and β_2 are parameters. The exponent β_2 is an indication of the degree of concentration of revenues on movies because it indicates the relative frequency of large grossing movies to small grossing movies.[18]

The Pareto rank law can be written as

$$\log \text{Revenue} = \log \beta_1 + \beta_2 \log \text{Rank} + \beta_3 \text{Star} + \beta_4 [\log \text{Rank} \times \text{Star}] + \mu.$$

$$(4.3)$$

This is the form we estimate. Table 4.5 shows our estimates of the Pareto rank law regressions. Column 1 shows the results restricting the Pareto parameters to be equal for all movies, with and without stars. In this case, we get a value of $\hat{\beta}_2 = -1.825$ indicating a very high degree of inequality. In column 2 the estimates allow the Pareto rank parameters to differ for movies with and

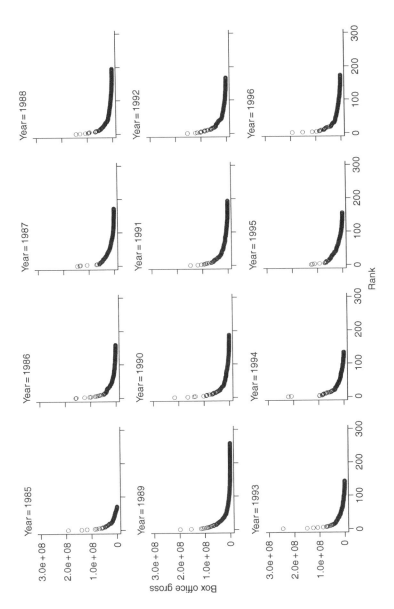

Figure 4.6 Pareto rank distribution by year.

Table 4.5 The Pareto rank distribution for movies

$$\log \text{Revenue} = \log \beta_1 + \beta_2 \log \text{Rank} + \beta_3 \text{Star} + \beta_4 [\log \text{Rank} \times \text{Star}] + \mu$$

Estimator	*(1)*	*(2)*	*(3)*	*(4)*
	LS	LS	MAD	Robust
log Rank	−1.825	−2.149	−2.161	−2.147
	(0.029)	(0.037)	(0.024)	(0.024)
	[0.047]	[0.072]	[0.059]	
log Rank x Star		1.153	1.327	1.284
		(0.069)	(0.046)	(0.046)
		[0.090]	[0.066]	
Star		−4.086	−4.918	−4.798
		(0.258)	(0.171)	(0.172)
		[0.354]	[0.281]	
Constant	22.865	24.276	24.797	24.667
	(0.128)	(0.162)	(0.108)	(0.108)
	[0.196]	[0.316]	[0.258]	
R^2	0.650	0.692	0.515	—

Notes
Dependent variable is log Revenue.
Estimated standard errors in parentheses.
LS is least squares. White's robust standard errors in brackets.
MAD is median regression. Bootstrapped standard errors with 100 replications in
   brackets. Pseudo R^2 reported.
Robust is the version of robust regression implemented in STATA and described in
   detail by Hamilton (1991).
All regressions run on common set of 2,015 observations.

without stars. The estimates indicate that the intercept term is a little smaller and that the slope is much flatter among movies with a star: $\hat{\beta}_2 = -0.996$ for movies with stars versus −2.149 for movies without stars. As we have seen, star movies have larger budgets, wider releases and quite likely better scripts, so these differences in distributions cannot be solely attributed to stars.

Table 4.6 shows estimates of the Pareto rank law regressions for six two-year intervals. With the exception of 1985–86, the Pareto rank parameters show little change. The Pareto rank law has remained quite stable over the years in spite of escalating production and advertising budgets. Independence of form on the time scale of the data is a feature of power law distributions that describe processes that are self-similar on all scales; this is revealed in the similarity of the rank-revenue curves plotted in Figure 4.6. The Pareto rank distribution is a remarkably good fit for all movies, with or without stars. Hence, the distinguishing factor that causes movies to be strongly ranked in terms of revenue cannot be traced to stars. It is a natural order, durable over time and place.[19] The steep decline in box-office revenue share with declining rank has remained stable during a decade of change in advertising and production budgets, the use of stars and changes in opening release patterns.[20]

Table 4.6 The Pareto rank distribution of movies by year

$$\log \text{Revenue} = \log \beta_1 + \beta_2 \log \text{Rank} + \beta_3 \text{Star} + \beta_4 [\log \text{Rank} \times \text{Star}] + \mu$$

Year	(1)	(2)	(3)	(4)	(5)	(6)
	1985–86	1987–88	1989–90	1991–92	1993–94	1995–96
log Rank	−1.649	−2.189	−2.487	−2.058	−2.247	−2.130
	(0.075)	(0.085)	(0.086)	(0.078)	(0.109)	(0.107)
	[0.113]	[0.180]	[0.199]	[0.153]	[0.198]	[0.227]
log Rank x Star	0.810	1.181	1.204	1.115	1.333	1.145
	(0.165)	(0.168)	(0.188)	(0.136)	(0.170)	(0.175)
	[0.138]	[0.286]	[0.242]	[0.185]	[0.211]	[0.253]
Star	−2.441	−4.392	−4.492	−4.043	−4.729	−4.095
	(0.570)	(0.611)	(0.719)	(0.521)	(0.623)	(0.680)
	[0.494]	[0.984]	[1.028]	[0.762]	[0.847]	[1.057]
Constant	22.024	24.519	25.647	24.135	24.634	24.333
	(0.305)	(0.380)	(0.378)	(0.348)	(0.460)	(0.474)
	[0.442]	[0.804]	[0.930]	[0.685]	[0.824]	[0.996]
R^2	0.724	0.698	0.708	0.730	0.686	0.635
Observations	231	369	447	361	278	329

Notes
Dependent variable is log Revenue.
Estimated standard errors in parentheses.
White's robust standard errors are in brackets.

4.5.2 *Opening and staying power*

Do stars give a movie opening power or staying power? Stars might increase a movie's prospects by getting it booked on more theater screens at its opening. Conventional wisdom in Hollywood is that star power is opening power. Another way that stars might affect a movie is by bringing a level of performance to it that lifts the movie above the ordinary.

We estimated screen counts of movies at week 1, week 5 and week 10. These were chosen because week 1 corresponds to the opening, though not always if the film is given an initial "pre-release" before it opens. Week 5 is chosen because it would be the week after a contract requiring a four week minimum run would no longer bind the movie to a theater. If a movie is still grossing high numbers at the end of its minimum contracted run, the hold-over clause will keep it in the theater until revenue drops below the hold-over amount. Week 10 was chosen for similar reasons for movies that might have an eight week minimum run contract.

Table 4.7 contains the results of the estimation. Holding budget and other factors constant, the estimates indicate that a star increases the number of opening screens by around 126 or about 18 percent. By week 5, a star raises screen count by even more—359 screens. And by week 10 a star still increases screen count by 160. The estimates retain high statistical significance throughout. Even though the coefficients decline, they become larger relative

Table 4.7 Regressions of screens at weeks 1, 5 and 10

Dependent variable	Week 1		Week 5		Week 10	
Budget (millions)	34.175 (1.639)	67.630 (2.949)	28.160 (1.671)	39.692 (3.164)	7.267 (1.070)	12.881 (2.017)
Budget^2		−0.746 (0.050)		−0.244 (0.052)		−0.116 (0.031)
Star	133.455 (42.231)	122.958 (40.430)	359.456 (42.405)	348.966 (42.206)	160.265 (27.298)	155.373 (27.332)
Genre	yes	yes	yes	yes	yes	yes
Sequel	781.557 (200.707)	638.136 (191.931)	455.897 (206.715)	408.134 (205.417)	235.792 (138.200)	202.162 (137.520)
Rating	yes	yes	yes	yes	yes	yes
Year	yes	yes	yes	yes	yes	yes
Constant	251.844 (235.194)	123.508 (224.561)	332.926 (243.948)	291.491 (241.977)	199.476 (166.832)	184.680 (165.484)
Observations	1,500		1,246		853	

Notes
Parameters estimated by robust regression.
Estimated standard errors in parentheses.

to the median number of screens. Consequently, stars give more kick to screen counts later than at the opening of a movie's run. In its first week, a movie with a star will have about 20 percent more screens than a movie without a star. By its fifth week a movie with a star will have nearly twice as many theaters as a movie without a star. And by the tenth week nearly three times as many. The effect becomes more pronounced later in the run.

The estimates also show that bigger budgets produce more opening screens: an increase in the budget of one million dollars corresponds to an increase of 36 screens in the opening week. Given that the median production budget for a film was less than $10 million and the mean was $32 million, the effect of a big budget on opening screens does not rise to economic significance. In terms of opening screen counts, a star is worth as much as an extra six million dollars in the production budget. By week 10 the size of the budget has a small effect on the number of screens; a million dollars of production cost only buys seven more screens in the tenth week.

Sequels open on nearly twice as many screens as the average movie. However, during the remainder of the run the sequel advantage declines. By week 10, the sequel advantage is statistically not different from zero.

In modeling screens in week 1 we are primarily modeling the behavior of theater bookers who select films to exhibit. By week 5 we are closer to seeing what the audience likes and not what the booking agents think. And, by week 10 we have a pretty clear vision of what counts with the audience. By then, sequels and budgets become unimportant which suggests that booking agents do not always share the tastes or perceptions of the audience.[21] Star movies have more staying power than opening power.

4.5.3 *The asymptotic Pareto distribution*

The Pareto rank distribution estimated here is an excellent model of the inequality of motion picture revenues, but it tells us little about the probability distribution of revenues. In order to fix probabilities so that we are able to assess the box-office expectations of a movie before it is released, we must estimate probability distributions. As we discussed in Section 2, the Lévy stable process converges asymptotically to a Pareto distribution as $x \to \infty$. To estimate the asymptotic Pareto law of equation (2), we set the minimum revenue at $k = \$40$ million. We fit the Pareto distribution for all movies whose box-office revenues equaled or exceeded $40 million and obtained an estimate of the tail coefficient α of 1.91. Since $1 < \alpha < 2$ the mean is finite and the variance is infinite. A Kolmogorov–Smirnov test of equality of the empirical distribution and the theoretical Pareto law $F(x) = 1 - (x/40)^{-1.91}$ does not reject the Pareto distribution at the 5 percent significance level. Figure 4.7 is a plot of the empirical cumulative distribution against the fitted Pareto distribution. The fit is extraordinary over a wide range of values running from $40 to $250 million in box-office revenues.

We proceeded to estimated α separately for movies with and without stars with k fixed at $40 million. For movies with stars $\alpha = 1.72$ implying a finite expected value and infinite variance. For movies without stars $\alpha = 2.26$ implying that both the expected value and variance are finite. The small value of α and infinite variance of star movies indicates they have more probability mass in the upper tail than movies without stars.

Note how different the Pareto distribution looks relative to the normal distribution that most people unconsciously draw on when they think about

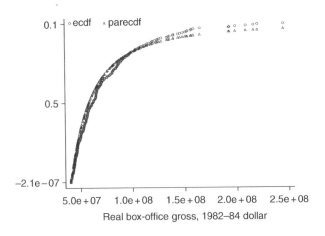

Figure 4.7 Asymptotic Pareto law for box-office gross.

movies. The probability density of the Pareto is "piled up" on the small box-office revenues because most movies earn small revenues. Unlike the normal distribution, where there is a piling up of density in the center around the mean, there is no central tendency in the Pareto distribution. The probability slopes away to the right, where the rare and big grossing films are. The Pareto distribution for values of $\alpha < 2$ (the star movies) has more upper tail mass than the normal distribution.

The Lévy distribution implies that forecasts of box-office revenues for movies lack a foundation. The expectation of the distribution for movies without stars is 7.17×10^7 but the variance is 8.76×10^{15}. The variance is 122 million times as large as the expectation. The expectation of the star movie distribution is 9.55×10^7 but its variance is infinite. In practical terms, forecasting revenue is futile because the magnitude of the forecast variance completely overwhelms the value of the forecast.

4.5.4   The probability of a hit

Because forecasting expected revenue is imprecise and lacking in foundation, we examine another approach. How are stars, budgets, genre, rating and opening screens associated with the probability that a movie will be a hit? These are all variables that can be chosen; if their impact on the probabilities of certain outcomes can be predicted, then better choices might be possible. The problem is that the subtle shifts in probability distributions are difficult to measure and we still face the infinite or nearly infinite variance.

Our attack on this problem is to examine the probabilities of extreme outcomes. We examine the probability that a movie will be a hit, which we define as earning a box-office revenue of fifty million or more. Even with a Pareto distribution of unbounded variance, this exercise is meaningful because we are discretizing the distribution and can easily calculate the probability that revenue will equal or exceed $50 million. We carry this exercise out by modeling the conditional hit probability as a function of the film's budget, star presence, genre, rating, year of release, survival time and number of opening screens.

Column (1) of Table 4.8 contains the parameter estimates and the associated marginal probabilities—the change in the probability that a movie becomes a hit for a unit change in the corresponding independent variable. The individual parameters are all statistically significant. The estimates indicate that a higher budget is associated with a higher hit probability. The star variable has a higher marginal probability than the sequel variable. The same pattern is observed in the results shown in column (2) where we have estimated on the subset of 1,500 observations for which we have screen count and life-length data. These data show that our estimates are not sensitive to the sample selection.

In column (3) two additional variables appear that indicate whether or not the movie survived for at least ten weeks and whether or not the movie was released on not less than 2,000 screens. Now the highest marginal probability

Table 4.8 Estimating the probability of a hit

Variable	(1)		(2)		(3)	
	Coeff.	Marg. Prob.	Coeff.	Marg. Prob.	Coeff.	Marg. Prob.
Budget	$4.8e - 08$	$4.05e - 09$	$4.45e - 08$	$4.09e - 09$	$3.19e - 08$	$8.10e - 10$
	$(5.06e - 09)$	$(4.98e - 10)$	$(5.50e - 09)$	$(5.84e - 10)$	$(6.58e - 09)$	$(4.03e - 10)$
Star	0.92979	0.13440	0.98622	0.15379	0.84640	0.04306
	(0.11414)	(0.02380)	(0.12896)	(0.02809)	(0.14556)	(0.01995)
Sequel	0.62257	0.08146	0.64563	0.09126	0.47506	0.01920
	(0.20406)	(0.03753)	(0.22433)	(0.04404)	(0.26592)	(0.01757)
Genre	yes		yes		yes	
Rating	yes		yes		yes	
Year	yes		yes		yes	
Life $\geq$ 10 weeks					2.16755	0.07339
					(0.47415)	(0.01118)
Wide Release ($\geq 2,000$ screens)					0.90980	0.05519
					(0.19715)	(0.02974)
Constant	-2.24740		-2.20408		-3.59036	
	(0.44974)		(0.51816)		(0.73964)	
Pseudo R^2	0.306		0.303		0.428	
log Likelihood	-414.221		-326.052		-267.306	
Observations	$2,015$		$1,500$		$1,500$	

Notes
Dependent variable = 1 if revenue $\geq$ 50 million, 0 otherwise.
Marginal probability is for discrete 0 to 1 change for dummy variables.
Estimated standard errors in parentheses.

is on a run of at least ten weeks, followed by the number of opening screens, then by star and sequel in that order.[22] That a long run is the most important factor associated with a movie becoming a hit is clear evidence that the audience decides a movie's fate at the box office and no amount of star power, screen counts or promotional hype is as important as the public's acceptance of the film. Controlling for screens and life length, a star has the same effect on the average movie's chances of grossing at least $50 million in theaters as an additional $40 million on production cost. Heavy spending on special effects or "production value" is the most risky strategy for making a movie a hit. Making a movie that audience loves is the surest way to making a hit, but that takes talents that are more rare than the ability to spend money. Next in importance to making a good movie in achieving a box-office hit is to have the movie booked on a large number of opening screens. But this is no simple task either as booking managers are no doubt influenced by their highly profitable concession sales.

A big opening is a double-edged sword (De Vany and David Walls, 1997). Opening on many screens preempts screens from other movies and gives a film a shot at a high rank. High rank movies are more likely to engage the information cascade and draw positive or negative attention. But, if the critical judgments of the viewers are predominately negative, the flow of negative information can kill a film and more swiftly if it is on many screens. On the other hand, a broad opening may bring large screen revenues in the early weeks of a run. Later, the number of screens is adjusted to fit demand and the initial number becomes less important.

4.5.5   *Stars and hits*

To more closely identify the association of individual stars with hit movies, we re-estimated column (1) of Table 4.8 using binary variables for individual stars in place of the single variable indicating the presence of any star in the movie. The coefficients on most of the individual star dummy variables were insignificantly different from zero at the 5 percent marginal significance level: most stars do not have a statistically significant association with the probability that a movie will be a hit. Only a few stars have a nonnegligible correlation with hit movies.

Who are the stars with real impact on a movie's chances of becoming a hit? Table 4.9 lists the individual stars whose coefficients are statistically significant and the associated marginal probabilities. Only 19 stars had a statistically significant impact on the hit probability. The names on the list are familiar ones. But some stars thought to have box-office power do not make the list; for example, neither Sylvester Stallone nor Robert De Niro were statistically significant. All the male stars that are thought to be "bankable" are there, along with behind-the-camera talents Steven Spielberg and Oliver Stone.[23]

The real surprise, given conventional Hollywood wisdom about star power, is the power of the female stars.[24] Of the top 19 stars, 4 are female. The top 2 stars are females and 3 of the top 5 stars are females. No star is a "sure

Table 4.9 Stars with statistically significant impact on the hit probability

Star name	Coeff.	Std. err.	Marg. prob.
Cher	1.283	0.751	0.264
Bullock, Sandra	2.076	0.646	0.569
Carrey, Jim	1.882	0.590	0.493
Costner, Kevin	1.380	0.417	0.297
Cruise, Tom	2.011	0.470	0.542
Douglas, Michael	1.173	0.412	0.226
Eastwood, Clint	1.091	0.520	0.200
Ford, Harrison	1.268	0.433	0.258
Foster, Jodie	1.820	0.952	0.469
Gibson, Mel	1.091	0.421	0.200
Hanks, Tom	1.378	0.357	0.296
Murphy, Eddie	0.997	0.429	0.172
Pfeiffer, Michelle	2.385	0.909	0.682
Pitt, Brad	1.637	0.792	0.396
Schwarzenegger, Arnold	0.813	0.405	0.124
Spielberg, Steven	1.625	0.534	0.391
Stone, Oliver	1.585	0.484	0.375
Travolta, John	1.380	0.385	0.297
Williams, Robin	1.143	0.379	0.216

Notes
Stars with significant coefficients (10 percent level, two-sided) in probit regression of the form of column (1) of Table 4.8. Marginal probabilities are the change in the probability of a movie being a hit with the presence of the given stars.

thing" however. They all face the infinite variance of the Lévy distribution, so they each bring a measure of risk with them. They also have sizable standard errors of their estimated hit coefficients. Jodie Foster, Michelle Pfeiffer and Sandra Bullock have high standard errors, implying that their positive impact is more variable. Tom Cruise has a small standard error; not only does he have a big impact but his impact is more certain than the impact of all the stars but Tom Hanks. The smallest standard error goes to Tom Hanks, though he has a smaller hit impact than Cruise, Pfeiffer, Foster, Carrey, and Bullock. Steven Spielberg is the top behind-the-camera star, with a marginal impact that is slightly higher but more variable than for Oliver Stone.

Of course, none of these estimates guarantee that a particular star will make a movie successful. In fact, they assure that no star can guarantee any outcome because there is infinite variance in the distribution—every star has a sizable probability of making a bomb. Moreover, it would be an error to attribute causality to what is only an association between stars and movie outcome probabilities. Causality could even go in the other direction—a star might just be someone who is lucky enough to give a fine performance in a terrific movie. Once someone is blessed with the mantle of stardom, it is clear that better projects and bigger budgets come his or her way. Hence, their chances of appearing in high grossing movies go up and their chances of being regarded as stars remain higher than average.

4.5.6 *Stars and profit*

To investigate profits we estimate a simple equation of the form

$$\text{Profit} = f(\text{star, sequel, genre, rating, year}) \qquad (4.4)$$

where Profit = $(0.5 \times$ revenue $-$ budget) is measured in millions of dollars.[25] We have reported least squares and robust regression estimates in Table 4.10.[26] We estimated the equation in levels (and not logs) because Profit is negative for a large proportion of the sample. The equation is a very poor fit, with an R-squared value of just 0.118. That is as it should be, for were profits predictable, everyone would make them. The lack of structure of the profit equation is a confirmation of the rule that "nobody knows anything" when it comes to predicting profits.

To investigate how stars may add structure to this featureless pattern, we re-estimated the equation with binary variables representing the individual stars. The 25 stars with statistically significant coefficients are reported in Table 4.11. Jodie Foster tops the list, followed by Tom Cruise, and

Table 4.10 Profit regressions

$$\text{Profit} = \beta_1 + \beta_2\text{star} + \Gamma[\text{sequel, genre, rating, year}] + \mu$$

Estimator	(1)	(2)
	LS	*Robust*
Star	17.144	2.391
	(1.349)	(0.567)
	[2.054]	
Sequel	10.633	4.818
	(2.270)	(0.953)
	[2.510]	
Genre	yes	yes
Rating	yes	yes
Year	yes	yes
Constant	10.394	1.786
	(4.759)	(1.999)
	[5.058]	
R^2	0.118	—

Notes
Dependent variable is profit = $(0.5 *$ revenue $-$ budget) in millions.
All regressions run on common set of 2,015 observations.
LS is least squares regression.
Robust is the robust regression implemented in STATA.
Estimated standard errors in parentheses.
Robust standard errors are in brackets (White's estimator for LS).

Table 4.11 Significant individual stars in the profit
regression

Star name	Coeff.	Std err.
Beatty, Warren	13.220	4.454
Bullock, Sandra	37.829	4.474
Carrey, Jim	37.298	4.021
Coppola, Francis Ford	6.372	3.637
Costner, Kevin	35.927	2.607
Cruise, Tom	70.827	2.828
De Niro, Robert	−5.000	2.391
Eastwood, Clint	19.741	3.167
Ford, Harrison	33.080	2.827
Foster, Jodie	83.069	6.320
Gibson, Mel	20.798	2.586
Hanks, Tom	17.057	2.406
Martin, Steve	10.258	2.590
Murphy, Eddie	10.754	2.718
Nicholson, Jack	−12.275	2.985
Pfeiffer, Michelle	15.519	6.298
Pitt, Brad	36.159	5.162
Redford, Robert	9.657	3.680
Schwarzenegger, Arnold	7.045	2.407
Seagal, Steven	21.584	3.407
Snipes, Wesley	7.417	2.995
Spielberg, Steven	48.530	3.383
Stone, Oliver	17.621	3.163
Washington, Denzel	6.287	2.980
Willis, Bruce	−7.903	2.708

Notes
Stars with significant coefficients (10 percent level, two-
sided) in profit regression of the form of column 2 of
Table 4.10.
Coefficients represent the star's impact on profit in millions
of dollars.

now Steven Spielberg moves into the third position on the profit list.
Sandra Bullock and Jim Carrey are about tied for fourth, with Brad Pitt and
Kevin Costner just behind. A few new names appear that did not show up
before such as Warren Beatty, Steve Martin, Francis Ford Coppola and Robert
Redford. De Niro, Nicholson and Willis appear with statistically significant
negative coefficients.

Given all that stated about the nature of the probability distribution, it is
difficult to place an interpretation on these estimates. They primarily reflect
the success that movies with these stars had in the past and do not imply that
these successes will be repeated in the future. Those successes may reflect their
performances or their judgment in choosing movies. It may just be luck in
the matching of actor and movie. A deeper problem is that if the box-office
revenue distribution has infinite or near infinite variance, then no formula

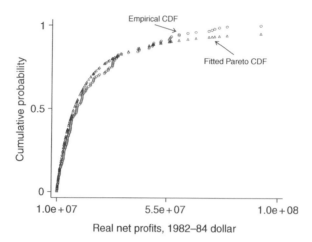

Figure 4.8 Asymptotic Pareto law for profits.

will be able to forecast revenue or profit. Since profit equals some fraction of revenue minus cost, the variance of profit will be infinite if the variance of revenue is infinite. Thus, theory indicates that profits should be asymptotically Pareto-distributed. We find that this prediction is confirmed and that profits in excess of $10 million are Pareto-distributed with an infinite variance.

The estimated Pareto exponent for all movies is $\alpha = 1.357$. For movies without stars, $\alpha = 1.505$. For movies with stars, $\alpha = 1.261$. All the estimates of α are greater than 1 and less than 2, implying that the mean of each distribution exists but the variance is infinite. Kolmogorov–Smirnov tests indicated that we could not reject the null hypothesis of equality of distributions between the fitted Pareto and the empirical cumulative distribution functions. Figure 4.8 plots the fitted Pareto distribution function against the empirical distribution function for all movies. The fit is excellent and this is compelling evidence that profits are Lévy distributed as are revenues.

Stars shift probability mass to higher outcomes. The theoretical mean profits are $38 million for all movies, $48.3 million for star movies, and $29.8 million for non-star movies. The variance of profits for movies that earn high profits ($\geq$ $10 million) is infinite for all movies as a group, for movies with stars, and for movies without stars. A few non-star movies achieve extraordinary profits (*Home Alone*) and some star movies lose extraordinary amounts of money (*Waterworld*). Both these effects contribute to the heavy tails in the profit distribution. Profits are more risky and less predictable than box-office revenues.

4.6   Choosing among movie projects

The movie industry is a small sample business. Studios only get so many chances. If only the very best survive and the competition is intense, then studios need to draw a movie project out of the many that are around that will have an extreme positive result. That is, with just a few draws from the hat the studio has to pull out an unlikely movie to succeed against its competitors. Finishing first in a large field requires doing something far from the average and being lucky enough to have it pay off (Lerinthal and March, 1993).

In such highly competitive situations, experience and learning, which are predictors of success on average, are not closely related to outcomes because success depends on doing something different—getting an extreme draw in a small sample. Experience may be a poor teacher in the movies. Effectiveness or success in the short run and in the neighborhood of recent experience (sequels) interferes with learning and experimentation in the long run. Since success comes from an unlikely event in a small sample, it is not reliable to extrapolate success into the future. This is why it is hard to learn in the movie business.

The movie business encourages selective learning based on extreme events. Ignoring failures and focusing on successes is built into the process. This is so because the statistics of the movies are dominated by a few extreme outcomes. There are a lot of failures and a few rare and unpredictable successes. Individuals tend to attribute causality improperly. They tend to attribute their successes to ability and their failures to bad luck. This error affects how they approach risk in the movie business. If executives attribute poor outcomes to bad luck, then they will overestimate risk. They will be inclined to demand a "bankable" star in a movie before they will make it. If they attribute good outcomes to their ability, then they will be inclined to take too much risk. Hence, one or two successes can lead to too much risk taking and a few failures to too little. Probably, most studio executives overestimate their ability to beat the odds. Of course, not all of them can.[27] In the long run, none of them can beat the odds. The odds are that only about 8 percent of all movies made gross more than $40 million at the domestic box office. Many of these are not profitable in spite of their high revenues.

Experience may not be helpful because it cannot produce a rare event. Most rare events, like the movie *Titanic*, lie outside the sample and are beyond experience. If aspirations are based on highly successful movies, then performance is bound not to match. Even if successful, tying aspirations to successes in the past deters exploration and innovation that are essential to success in the future; too many sequels and copies are made and too few genuinely new movies are produced.

Past successes give executives an illusion of control. They become confident in their ability to manage risk and handle future events. They have difficulty recognizing the role of luck in their achievements. Studio executives and producers have little control in this business. It is a high-skill business because good movies are hard to make. But that very fact fosters an illusion of control. It is such an uncertain business that the distinction between causal factors, luck or the sheer sweep of events is blurred. In a complex system, where there are many interacting parts and complicated stochastic dynamics, there is no simple form of causality. Everything depends, in some way, on almost everything else and it will usually be impossible to attribute an outcome to a cause or complex of causes.

   Managerial errors in judgment are fostered by the very nature of uncertainty in the motion picture industry that is documented in this chapter.[28] The uniqueness of individual movies comes from the underlying probability distribution: because it is a power law, there is no characteristic scale, no central tendency and events on all scales happen.[29] Thus, there is no typical movie. The hold that last year's blockbuster has on the imagination comes also from the power law, a distribution so highly skewed that blockbusters dominate the mean. Only risk and hazard analyses are well defined for this business. The probability that a movie will reach an extreme outcome in the upper tail, which is required for it to be profitable, is small. But, the outcomes associated with extremums dominate total and average revenues and profits. So, risk not only is unavoidable, it is desirable. One wants to choose movies that have a large upside variance. We have only hinted at how it might be done here by investigating a few strategies.[30] Star movies have that kind of variance, but by virtue of that fact they also have unpredictable outcomes. No star is "bankable" if bankers or studio executives want sure things. Stars only increase the odds of favorable events that are highly improbable.

4.7   Conclusions

The movie industry is a profoundly uncertain business. The probability distributions of movie box-office revenues and profits are characterized by heavy tails and infinite variance! It is hard to imagine making choices in more difficult circumstances. Past success does not predict future success because a movie's box-office possibilities are Lévy-distributed. Forecasts of expected revenues are meaningless because the possibilities do not converge on a mean; they diverge over the entire outcome space with an infinite variance. This explains precisely why "nobody knows anything" in the movie business.

   A proper assessment of a movie's prospects requires a risk analysis of extreme outcomes. We have demonstrated that estimates of the Lévy distribution parameters permit calculation of the probability of box-office revenues that have not before been realized. Filmmakers can position a movie to improve its chances of success, but after a movie opens the audience decides its fate: There are no formulas for success in Hollywood. The complex dynamics of personal interaction between viewers and potential viewers overwhelm the initial conditions. The difficulties of predicting outcomes for individual movies are so severe that a strategy of choosing portfolios of movies is more sensible than the current practice of "greenlighting" individual movie projects.

   The most uplifting of our results is that the most important element in a movie's financial success is a long run. Weeks after the opening, after the effects of marketing hype, wide-release openings, special effects and big budgets are gone, what remains is the most important determinant of a movie's success of all—that the audience loves it.

5 Does Hollywood make too many R-rated movies?

Risk, stochastic dominance and the illusion of expectation

When the typical "PG" film generates nearly three times the revenue of the typical "R" bloodbath or shocker, then the industry's insistence on cranking out more than four times as many "R" titles must be seen as an irrational and irresponsible habit.

Michael Medved (1992, p. 290)

5.1 Introduction

Film writer Michael Medved (1992) has criticized Hollywood's fascination with R-rated movies on both cultural and economic grounds. From the White House to Main Street, many individuals share his view that Hollywood makes too many R-rated movies. His argument goes beyond a cultural and critical judgment; he makes a sophisticated economic argument to the effect that Hollywood is missing a profit opportunity by making too many R-rated movies and too few G-rated movies. He accounts for this neglect of profitability as a search for prestige bestowed by Hollywood insiders on edgy, counterculture movies. Could the movie portfolio that brings such criticism of Hollywood on the grounds of taste and morals also be costing it money? Could a search for prestige among peers lead Hollywood to neglect its bottom line?

This chapter shows that Medved is right: there are too many R-rated movies in Hollywood's portfolio. An executive seeking to trim the "downside" risk and increase the "upside" possibilities in a studio's film portfolio could do so by shifting production dollars out of R-rated movies into G, PG and PG13 movies.

Putting tastes and morals aside, even a casual look at the evidence does suggest that there are too many R movies compared to G, PG and PG13 movies. More than half of all the movies released in the past decade are R rated and only 3 percent are G rated. About 20 percent are PG rated, and 25 percent are PG13 rated. Based on the production numbers alone, the R category does seem to be crowded while the G category is almost empty. Medved's evidence suggests that R-rated movies are less successful than G-rated movies: he showed that R films were less than half as likely as PG releases to reach

$25 million in domestic box-office revenue, and the median gross of PG movies was nearly triple the median gross of R pictures while G pictures earned the highest median gross of all.[1]

But, Medved's comparisons do not take budgets or profits into account. More importantly, they fail to account for risk. Motion pictures are among the riskiest of products; each movie is a "one-off" innovation with highly unpredictable revenues and profits.[2] Unless one accounts for the differing risks among the rating categories, it is impossible to reach a conclusion that R-rated movies are overproduced. For example, the low median or average return on R-rated movies could just reflect lower risk (if they had lower risk). In addition, median revenue will generally differ from expected revenue. Moreover, expected values may be so heavily influenced by a few blockbuster movies as to misstate prospects. Even the use of mean/variance analysis can be misleading (as we shall show) in motion pictures, so one must look to more basic considerations to assess the prospects of motion pictures.

Fundamentally, until it is released, a motion picture is just an unknown, uncertain prospect in the eye of the producer and in the studio's statistical model. Even the conviction that a movie is a "sure thing" is no more than a belief that its probabilities are skewed toward successful outcomes. So, in order to compare motion pictures as prospects, one must compare their probability distributions.

The objectives of this chapter are to characterize the uncertainty that movies face by identifying their probability distributions and to bring the right decision tools to the task of evaluating them as economic prospects. This means that the primary task is to get the statistical model right and when we do that we see that there is little scientific justification for a studio to interfere with the producer's "vision." On the other hand, there is plenty of justification for a studio to be wary of a "pitch" for an R-rated, big-budget movie that a major creative talent is eager to make. There is also more room than the "numbers" might suggest for hunches and intuition that center on the non-quantifiable intangibles of story and character. These points follow from, and will be demonstrated in what follows, from the nature of the statistics.

The statistics of the movie business are a bit exotic and much of our work is in characterizing the statistical distributions of the relevant variables. The distribution of motion picture revenues is highly skewed and its mean is related to its variance. Moreover, the distribution may not have finite moments and so may have neither convergent expected values nor variances (De Vany and Walls 1999). The sample mean is unstable and does not converge to the population mean. The distribution of profits is not symmetric—it has a heavier positive than negative tail—so that conventional portfolio choices that reduce variance for a given expected return are dominated by choices that exploit the asymmetry of the tails.[3]

Once we establish the correct statistical model by estimating the probability distributions for revenues, budgets, returns and profits in the four MPA ratings categories—G, PG, PG13 and R—we then rely on the unambiguous ranking concept of stochastic dominance to investigate Medved's hypothesis. One movie (first-degree) stochastically dominates another if it has a higher probability of exceeding the other's revenues for all possible revenue outcomes. This applies also to budgets, with a reversal of the ranking. With regard to profits, stochastic dominance is more subtle because the distribution has two parts that correspond to profits and losses and the distribution is not symmetrical. Accordingly, we rank the positive and negative parts (tails) of the asymmetrical profit distribution. We show that R-rated movies are dominated by G, PG and PG13 movies in all dimensions of revenues, costs, return on production cost and profits. We also show that there are dramatic differences in risk, budget variability and profits among the ratings.

In Section 5.2, we summarize the production levels, revenue and returns to movies in each of the ratings categories. In Section 5.3, we discuss the stable Paretian and normal distribution statistical models and show that the stable Paretian model is the right model of the movie business. In Section 5.4, we discuss the success rates among movie ratings classes. In Section 5.5, we determine the probability distributions of box-office revenues, success rates, returns and profits, and show how R, G, PG and PG13 movies are stochastically ranked with respect to one another. We conclude in Section 5.6.

5.2 Motion picture production by ratings

Our sample of data includes 2,015 movies that were released in North America from 1985 to 1996, inclusive. The data were obtained from ACNielson EDI, Inc.'s historical database. The data were compiled from distributor-reported box-office figures and estimated production budgets. These data are the standard industry source for published information on the motion pictures and are used by many major industry publications including *Daily Variety* and *Weekly Variety*. The EDI data were augmented by creating a variable indicating whether or not a "star" was associated with each film: Each actor, producer or director appearing on *Premier*'s annual listing of the hundred most powerful people in Hollywood or on James Ulmer's list of A and A+ people was considered to be a "star" in our statistical analysis. This definition of star is broader than most since it includes artists on both sides of the camera and includes producers as well. For example, Oliver Stone and Steven Spielberg are included in the list as well as stars in the more conventional sense of Tom Hanks and Sandra Bullock.

A cross-tabulation of the sample of movies is given in Table 5.1. The table shows the sample of movies disaggregated by year, rating classification and the presence of a star. From 1985 to 1996, inclusive, Hollywood made 1,057 R-rated movies; just 60 G-rated movies were made during that same period.[4] R-rated movies dominate G-, PG- and PG13-rated movies and comprise just over half of all the movies made. Of the 2,015 movies in the sample, 326 featured stars either in front of the camera or behind it. R-rated movies

Table 5.1 Tabulation of all movies by rating, star presence and year

Year	No Star				Star				Total
	G	PG	PG13	R	G	PG	PG13	R	
1985	2	20	17	21	0	3	4	4	71
1986	6	28	34	71	0	8	5	8	160
1987	2	42	32	74	0	4	1	18	173
1988	5	45	32	91	0	4	1	18	196
1989	6	36	56	134	0	7	6	14	259
1990	6	25	40	86	0	5	9	17	188
1991	5	27	44	89	0	5	6	18	194
1992	3	25	38	66	1	5	11	18	167
1993	5	34	27	56	0	1	8	13	144
1994	5	21	29	50	0	3	11	15	134
1995	7	17	30	62	1	5	10	24	156
1996	6	24	37	71	0	5	11	19	173
Total	58	344	416	871	2	55	83	186	2,015

Table 5.2 Percentage of movies featuring stars by rating and budget

Rating	Low budget		Medium budget		High budget	
	Star		Star		Star	
	No	Yes	No	Yes	No	Yes
G	100.00	0.00	100.00	0.00	89.47	10.53
PG	99.12	0.88	89.26	10.74	76.97	23.03
PG13	98.18	1.82	90.76	9.24	70.24	29.76
R	97.17	2.83	89.94	10.06	55.25	44.75

Note
This table reports, in each row, the percentage of movies featuring a "star" for that particular rating and budget category. Low-, medium- and high-budget categories correspond to the lower, middle and upper 1/3 of the distribution of movies by budget.

accounted for 52 percent of the 1,689 movies that did not feature a star and they accounted for 57 percent of the movies that did feature a star. The 100 stars of the "A-list" appear in, produce or direct more often in R-rated movies than in any other rating.

The preponderance of stars in R-rated movies is even greater in high-budget movies. Table 5.2 documents the distribution of stars over ratings categories by budget. As is evident, stars are present in 45 percent of R-rated movies with large budgets, a much higher percentage than for any other rating/budget classification. By contrast, they appear in only 30 percent of high-budget PG13 movies, 23 percent of PG movies, and only 10.5 percent of G movies. In medium- and low-budget films, the distribution of stars is similar

Table 5.3 Box-office revenues, budgets and returns

Rating	25%ile	50%ile	75%ile	Mean	Std dev.
The revenue distribution					
G	2,574,743	1.01e + 07	2.28e + 07	2.58e + 07	3.99e + 07
PG	1,566,304	1.12e + 07	2.88e + 07	2.15e + 07	3.00e + 07
PG13	1,922,510	8,439,733	2.39e + 07	1.96e + 07	3.15e + 07
R	814,088.6	5,171,729	1.58e + 07	1.35e + 07	2.11e + 07
Total	1,169,457	6,943,376	2.06e + 07	1.70e + 07	2.68e + 07
The budget distribution					
G	4,917,709	9,510,306	1.47e + 07	1.19e + 07	9,617,613
PG	5,287,010	1.10e + 07	1.71e + 07	1.28e + 07	9,625,494
PG13	6,324,666	1.13e + 07	1.70e + 07	1.39e + 07	1.14e + 07
R	3,776,435	7,285,343	1.41e + 07	1.05e + 07	9,797,142
Total	4,399,749	9,096,816	1.56e + 07	1.18e + 07	1.03e + 07
The returns distribution					
G	0.4009847	1.245965	2.381497	2.152229	3.46846
PG	0.25854	0.8924453	2.153059	1.662702	2.23019
PG13	0.2193686	0.6822007	1.786988	1.390128	2.20324
R	0.1832344	0.6636797	1.669723	2.142445	16.16777
Total	0.2040572	0.7273786	1.789731	1.861434	11.8195

Notes
All monetary magnitudes reported in constant 1982–1984 dollars. Gross returns are defined as Revenue/Budget. Since rentals are about half of box-office gross, an approximate rate of return to the studio is 0.5∗(gross return)-1. The break-even gross return is 2. Revenue is domestic theatrical revenue only and does not include foreign theatrical revenues or other revenue sources.

between PG and PG13 movies. No stars appeared in G-rated medium- or low-budget movies and they appeared predominately in R-rated movies among low-budget movies.[5]

It is notable that R-rated films—portrayed by critics as attacks on conventional social values and morals—attract a disproportionately large share of Hollywood's on-screen and behind-the-camera stars and this is even more true of high-budget R-rated movies.

Table 5.3 shows the mean, standard deviation, median, and upper and lower quantiles for revenues, budgets and returns in our sample of movies disaggregated by rating category. All monetary magnitudes reported in the table, and in the remainder of the chapter, are in constant 1982–84 US dollars. These data are consistent with Medved's calculations as we shall show in Section 5.5. But, it is essential to evaluate the probability distributions to consistently rank the ratings. We estimate and rank the distributions of revenues, budgets, returns and profits in Section 5.5.

5.3   Choosing among uncertain movie projects

Movies are uncertain prospects. Before a movie is released only probabilities of outcomes may be known. Even these probabilities may be difficult to ascertain since each movie is unique. When a studio considers a motion picture for its production slate, it is, in effect, evaluating it as a probability distribution over various profit or return outcomes and comparing it to other projects it might choose to produce. How should an executive decide among the many projects which ones to produce? Here we abstract from considerations of story value, character or originality to focus on the quantifiable economic aspects of the decision. How then should information about the odds of different outcomes be factored into the decision to "greenlight" a movie or to choose the movies that will go into a studio's production portfolio?

Since a motion picture will not incur costs until production begins or earn revenues or profits until it is released, the studio executive must form expectations about these highly uncertain elements of a motion picture as an investment. A first cut at formulating this decision might be making an estimate of the expected budget, box-office revenues and profits given what is known about the movie's "creative elements" such as the screenwriter, director, stars, story line, genre and rating. But, this first cut would be very incomplete for the error in forecasting the expectation must be taken into account. Two movies having the same forecasted expected return on investment may have very different risks associated with them.

A more sophisticated approach would be to consider the expectation and risk associated with a motion picture. To do this, one must know the probabilities of the different outcomes the movie might realize when it is released. This means one must know how probability is distributed over different outcomes. A budget, box office, return or profit outcome for a movie is a random variable X that has a probability distribution $Pr[X \leq z] = F[X \leq z]$. A movie's prospects are given by the probability distributions of its budget, revenues and return on investment, all of which are random variables.

To make informed decisions, the studio must get the statistical model of these random variables right. This is where the first critical error is often made. A studio might assume that the probability distribution is a normal distribution and use estimates of the mean and variance (or standard deviation) from past movies to form an estimate of the expectation and risk (variance or standard deviation) of a movie's prospects. Using the mean and standard deviation estimated from a sample of movies that have been produced and released, a financial analyst might forecast expected revenues, say, as the mean and use the standard deviation to describe the error of the forecast.

As we show here, the normal distribution model is the wrong model for the movie business. In the movies the average is unstable and does not converge to a stationary value over the course of time. The variance of outcomes increases with the size of the sample. These points are dramatically indicated by Figures 5.1 and 5.2 which show the average and variance of box-office revenues as the number of movies released increases over time. Both series are volatile and self-similar (segments of the graphs look like the whole graph).[6] Since the average rises and falls randomly and often quite dramatically as the number of movies in the sample increases, an average calculated from

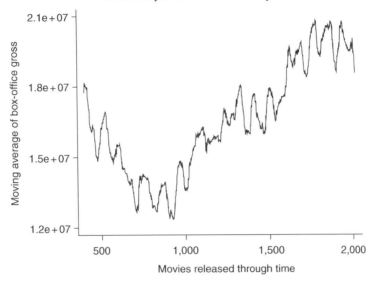

Figure 5.1 Moving average gross over number of films.

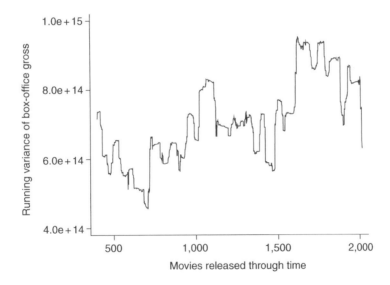

Figure 5.2 Moving average variance about the mean over number of films.

a given sample of movies will be a poor forecast of the average in the future. Because the average is unstable, it is a poor estimator of the expected next event. The average is also a poor forecast of the most probable future outcome. This is because the average is not the most probable outcome. The average is dominated by rare, extreme outcomes and is quite far above the most probable outcome. Because extreme outcomes dominate the average, the expectation will differ from the average and it need not even exist mathematically. The variance is unstable as well; it grows with the number of movies included in the sample. The variance may be unbounded and need not have a finite value (and usually does not).

All these points follow from the nature of the probability distributions that describe the movie business. We shall briefly describe the features of the statistical models that capture the movie business and show that the stable Paretian model is the right model. The stable Paretian model was proposed by Mandelbrot for economic data (Mandelbrot 1963a,b, 1997) and by Fama (1963, 1965) for returns to financial assets.[7] The estimates of the parameters of the stable Paretian model are presented later. Here we focus on the contrast between the stable Paretian model and the normal distribution model.[8]

Consider the probability distributions shown in Figure 5.3. In this figure the probability density function of box-office revenues of all movies in the sample is estimated and overlaid with a normal distribution. The probability distribution is highly skewed and is not symmetrical. Most of the probability mass is placed on the most frequent, low revenue outcomes. These low outcomes far exceed the frequency predicted by the normal distribution. The correct model has a long, thick tail to the right, showing that the probability of extremely high revenues exceeds what is predicted by the normal distribution model. The mean revenue is much higher than the most probable revenue, which is where the density function reaches its highest level. This is

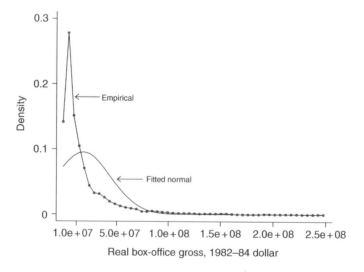

Figure 5.3 Normal and stable Paretian distributions of revenue.

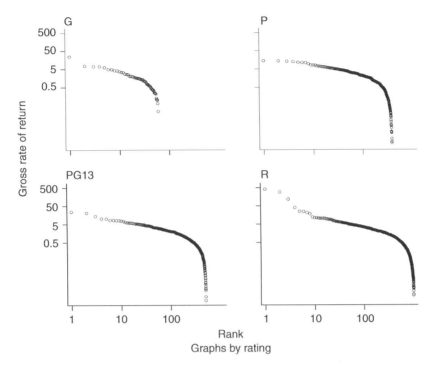

Figure 5.4 Rank Pareto distribution of revenue.

different from a normal distribution where the most probable outcome (the peak of the probability density curve) equals the mean.

The stable Paretian model implies that, in the upper tail, the probability density function converges to a Pareto distribution. Figure 5.4 shows the Paretian property of the upper tail. In this figure we are looking at the highest grossing movies and ranking from high (a value of 1) to low. Hence, we are looking at the upper tail of the probability distribution, but through the lens of its extreme outcomes. The figure shows that the dominant share of the box-office revenues is taken by the high ranking movies so the tails of the distributions are where the important events are located. The central part of the distribution is relatively unimportant relative to the tails. The approximate linearity of the graph at the higher ranks (upper tail) is a signature of a stable Paretian distribution. The significance of the Paretian upper tail property is that a few extreme outcomes dominate total and average revenue and that these extreme outcomes are more probable than the normal distribution model would predict. The heavy tails and the dense probability mass on low outcomes of the stable Paretian model is a most important feature of the movie business.

5.4   Ranking by success rates

Based on the foregoing discussion, it is clear that one must compare whole probability distributions rather than averages or expectations and their variance. A simple first test of which movies are better prospects, then, may rely on simple calculations of the probabilities of outcomes. If movie A has a higher probability of "success" than movie B, then A is weakly preferred to B. By using different measures of success we can rank movie prospects in each rating class.

5.4.1   *Box-office revenue success rates*

In Table 5.4 we tabulate the number of movies that earned cumulative box-office revenues in excess of $50 million; we term these movies "revenue hits." The number of R-rated movies that are box-office revenue hits is 68 while there are far fewer revenue hits in the other rating classifications: only 8 G-rated movies are hits. A Chi-square test indicates that we cannot reject the null hypothesis that the revenue hits in each rating category are independent of the year of theatrical exhibition. In other words, this relationship appears to be independent of the year of release.

However, the number of hits for each rating class should be scaled by the number of movies made in each rating classification each year. Success rates are a more representative measure of revenue earning power than is the number of high grossing films. Table 5.5 displays the proportion of films in each rating classification, by year, that were revenue hits. The success rate for R-rated movies is just 6 percent, whereas 13 percent of G- and PG-rated movies are hits and 10 percent of PG13 movies are hits. The box-office success rates for all non-R-rated movies (G, PG and PG13) are twice the rate for R-rated movies.

Table 5.4 Tabulation of revenue hits. Tabulation of movies with box-office gross > $50 million

Rating	1985	1986	1987	1988	1989	1990	1991	1992	1993	1994	1995	1996	Total
G	0	0	0	0	1	0	1	1	0	1	2	2	8
PG	6	3	2	5	8	6	4	3	2	4	5	3	51
PG13	2	4	3	2	6	5	6	2	4	5	5	5	49
R	1	4	8	6	5	6	7	11	5	5	4	6	68
Total	9	11	13	13	20	17	18	17	11	15	16	16	176

Note
To make expected cell frequencies not less than 5, we grouped years into six two-year categories and deleted the G rating. The resulting $\chi^2(10) = 10.622$ with a marginal significance level of 0.388.

Table 5.5 Success rate for revenue hits. Movies with box-office gross >$50 million

Rating	1985	1986	1987	1988	1989	1990	1991	1992	1993	1994	1995	1996	Total
G	0	0	0	0	17	0	20	25	0	20	25	33	13
PG	26	8	4	10	19	20	13	10	6	17	23	10	13
PG13	10	10	9	6	10	10	12	4	11	13	13	10	10
R	4	5	9	6	3	6	7	13	7	8	5	7	6
Total	13	7	8	7	8	9	9	10	8	11	10	9	9

Note
Percentage success rate calculated as 100 times the ratio of Table 6.3 to Table 6.1.

Table 5.6 Tabulation of returns hits. Movies with box-office gross/budget >3

Rating	1985	1986	1987	1988	1989	1990	1991	1992	1993	1994	1995	1996	Total
G	0	1	1	2	1	0	2	1	1	1	1	1	12
PG	8	5	4	7	8	6	4	8	7	2	3	3	65
PG13	4	8	13	2	7	2	7	1	3	4	5	4	60
R	7	16	14	16	10	5	8	9	8	8	9	11	121
Total	19	30	32	27	26	13	21	19	19	15	18	19	258

Note
To make expected cell frequencies not less than 5, we grouped years into six two-year categories and deleted the G rating. The resulting $\chi^2(10) = 6.094$ with a marginal significance level of 0.807.

5.4.2 *Return on production cost success rates*

Table 5.6 displays a tabulation of movies by year and rating classification that had box-office revenues in excess of three times the production budget; we refer to these films as "returns hits." Here again, we are using a point on the distribution of outcomes to calculate probability. The table shows that there are about twice as many R-rated returns hits as PG- and PG13-rated returns hits, and about ten times as many R-rated hits as G-rated hits. The composition of hits across rating classes is independent of the year of theatrical exhibition according to the standard Chi-squared statistical test for independence of rows and columns.

To calculate the success rate of returns hits we must control for the number of films released each year in each rating class. In Table 5.7 we show the success rates for the ratings. Overall, 20 percent of G-rated films are returns hits in terms of their rate of return on the production budget.[9] The proportion of returns hits is 16 percent in PG, 12 percent in PG13, and 11 percent in R-rated films.

These tabulations of rates of box office and returns hits make clear that the G-, PG- and PG13-rated films had a higher proportion of hits than R-rated movies at the box office and in returns on production dollars. The *number* of R-rated successes is high because more of these movies are made (and that

Table 5.7 Success rate for returns hits. Movies with box-office gross/budget >3

Rating	1985	1986	1987	1988	1989	1990	1991	1992	1993	1994	1995	1996	Total
G	0	17	50	40	17	0	40	25	20	20	13	17	20
PG	35	14	9	14	19	20	13	27	20	8	14	10	16
PG13	19	21	39	6	11	4	14	2	9	10	13	8	12
R	28	20	15	15	7	5	7	11	12	12	10	12	11
Total	27	19	18	14	10	7	11	11	13	11	12	11	13

Note
Percentage success rate calculated as 100 times the ratio of Table 5.6 to Table 6.1.

may blind decision makers who do not pay attention to the odds), but the *success rate* of R-rated movies is much less than the success rates of G, PG and PG13 movies.

5.5   Ranking by stochastic dominance

The foregoing comparison is informative, but, by focusing on the probabilities of a fixed outcome rather than the whole probability distribution, it does not use all the available information. A more sophisticated comparison is to use the probability distributions (which include probabilities of all outcomes) to rank movie prospects. The relevant criterion in this case is stochastic dominance.[10] Intuitively, the idea behind the stochastic dominance ranking is that if probability distribution A has less mass on low value outcomes and more mass on high value outcomes than probability distribution B, then A is preferred to B. This means that for any outcome z the probability that movie A will, say, gross more than z is higher than the probability that movie B will gross more than z. When this holds for all values of z then movie A first-order stochastically dominates movie B. More formally, random variable X first-order stochastically dominates random variable Y if

$$Pr[X > z] \geq Pr[Y > z] \text{ for all } z. \tag{5.1}$$

Since $Pr[X > z] = 1 - F[z]$ where $F[z]$ is the cumulative probability density function, first-order stochastic dominance of distribution $F[z]$ over distribution $G[z]$ be written in the form

$$F[z] \leq G[z] \text{ for all } z. \tag{5.2}$$

So, $F[z]$ first-order stochastically dominates $G[z]$ if there is less probability of lower outcomes than z under $F[z]$ than under $G[z]$ and this holds for all outcomes z. It can be shown that all expected utility maximizing decision makers would prefer $F[z]$ to $G[z]$ when $F[z]$ first-order stochastically dominates $G[z]$.[11]

In what follows we focus on characterizing the statistical distributions and stochastically ranking them. We are not analyzing why these distributions might be different or what factors might account for whatever differences we may find. In this respect, the analysis is comparable to determining the return distribution on common stock in various categories, for example, small cap or large cap funds, or real estate versus energy equities. This is a necessary first step in developing an analysis of Hollywood's film portfolio. Individual differences among movies may be great and can be influenced by budgets, release strategies, stars, story and many other factors. It is this difference within each rating that we capture in the probability distribution and the question is do ratings matter, do the distributions differ for revenues, costs, returns and profits? We have previously shown (De Vany and Walls 1999) that ratings, as a group, are statistically significant in revenue outcomes and that each rating is also individually significant. This is not true of genre, which are significant together but not individually (Simonoff and Sparrow 2000 also find genre as a group to be statistically significant). So, rating does condition the distribution of outcomes and, as we shall show, there are important differences among rating classes.

5.5.1 Revenues

The upper panel of Table 5.3 shows the median, mean, standard deviation, and lower and upper quartiles of box-office revenue in constant 1982–84 US dollars for the movies in our sample. Mean revenues of G-, PG- and PG13-rated movies dominate mean revenues of R-rated movies. The respective average revenues are $25.8, $21.5, $19.6 million versus $13.5 million for R-rated movies. The standard deviations of non-R-rated movie revenues are greater than the standard deviation of R-rated movie revenues, an indication of their long upper tails.

The high degree of rightward skew in the distributions (as implied by the stable Paretian model) can be verified by looking at the revenue percentiles. The percentiles in Table 5.3 indicate that the probability mass of G-rated movies is skewed far to the right. The average revenue exceeds revenue at the 75th percentile. The other ratings classes do not show this extreme skew, but they are highly skewed to the right nonetheless—their means lie above the median and near the 75th percentile.

As Mandelbrot (Mandelbrot 1963a) showed, the stable Paretian model implies that the asymptotic distribution function of extreme values is a Pareto distribution. The Pareto cumulative distribution function is

$$\text{Prob}[X > x] = F(x; k, \alpha) = 1 - \left(\frac{k}{x}\right)^{\alpha} \tag{5.3}$$

where $x \geq k$ and $k, \alpha > 0$.

We have estimated the exponent α of the Pareto distribution by maximum likelihood for the upper tail of the box-office revenue distribution.[12] The

Table 5.8 Estimated Pareto exponents for revenues by rating

Rating	Pareto exponent	K–S test p-value
G	1.591	0.712
PG	1.814	0.252
PG13	1.661	0.617
R	2.274	0.234

Notes
The Pareto exponent was estimated by maximum likelihood conditional on a minimum revenue of $40 million. The Kolmogorov–Smirnov *p*-values reported are for the null hypothesis that the data are consistent with the theoretical Pareto distribution.

data are well-fitted by the Pareto distribution—Kolmogorov–Smirnov tests do not reject the hypothesis that box-office revenues are Pareto-distributed in the upper tail, above $40 million revenue.[13]

The estimated values of α for revenues of $k \geq 40$ million are shown by rating class in Table 5.8. The values of α in all ratings but R lie in the interval $1 \leq \alpha \leq 2$. These estimates imply that for every rating class but R the expected value of box-office revenue is finite but the variance is infinite! Only the R-rated movies have an α value greater than 2, implying that the distribution of R-rated movie revenues has less probability mass at high outcomes and a finite variance (this was shown also by the lower standard deviation over the sample of R-rated movies discussed earlier). The Pareto revenue distribution of movies in the non-R categories places so much probability on extremely high revenue outcomes that their revenues have infinite variance.

It follows from the estimated values of α that box-office revenues of G-rated movies are more skewed to the right than are revenues in the other ratings categories. The G distribution, therefore, has a higher probability of extremely high revenue outcomes. This follows from the small value of α for the G-rated revenue distribution relative to other movies (probability in the upper tail declines as $x^{-\alpha}$ so the smaller value of α for the G distribution implies that probability declines less rapidly as revenues move into the upper tail of the G distribution). This property is important for ranking the revenue distributions.

From the formula for the Pareto distribution in equation (5.3), it follows that the random variable having the smallest value of α puts more probability above every revenue outcome exceeding $40 million. Set $z = 40 million. Denote the cumulative probability distributions of revenues of each rating class as $G[z], PG[z], PG13[z], R[z]$. Then from the α values in Table 5.8 we can succinctly state the relationship as $G[z] \geq PG13[z] \geq PG[z] \geq R[z]$ where $\geq$ denotes first-order stochastic dominance. In other words, G movies dominate PG13 movies which dominate PG movies which dominate R movies.

Using the estimated values of the tail weights, α, we can use the relationship of the expected value to the most probable value for the Pareto distribution to highlight the difference among ratings of the most probable and the expected value. For the Pareto distribution the expectation of revenue conditional on revenue $z \geq \$40$ million may be written as a function of the tail weight α as

$$E[x \mid z] = z \left[\frac{\alpha}{(\alpha - 1)} \right] \quad \alpha > 1. \tag{5.4}$$

The ratio $\alpha/(\alpha - 1)$ is the ratio of the expected value to the most probable value of box-office revenue. The most probable value is the peak of the probability density function for each distribution. Since we used a common value of \$40 million for the estimates, this is the most probable outcome for box-office revenue in each rating. Then the multiple by which the expected value exceeds the most probable value is a direct way to rank the revenue distributions among the ratings. Using the estimated values of α, the ratios of the expected to the most probable value of box-office revenues for G, PG13, PG and R movies respectively are 2.69, 2.51, 2.22, 1.78. R-rated movies rank last.

5.5.2 *Production budgets*

Production budgets also face a kind of uncertainty. Each movie has a planned budget, but anecdotal evidence indicates that many, if not most, movies go over budget. There is always a risk that a movie will exceed its planned budget, and there are examples where they go far over budget: *Heaven's Gate*, *Titanic*, *Cleopatra*. There is no way in our data to judge how the actual production budget compares with the intended budget, but we can still learn something of the variability of budgets by examining the distribution of production budgets. Some of this variability is intended, but some is unintended and is the result of a movie going over the intended budget. Consequently, the variability that we can measure consists of an intended part and an unintended part. By holding rating constant, we control to some extent for the intended part. And by comparing the variability relative to the average we also control for some (unknown) portion of the intended variation.

The middle panel of Table 5.3 presents the statistics on production budgets; this does not include print or advertising cost. The statistics reveal that R-rated movies are generally cheaper to make than G-rated movies. The average budget of an R-rated movie is \$10.5 million whereas the average budget of a G-, PG- and PG13-rated movie is \$11.9, \$12.8 and \$13.9 million, respectively. PG13 movies were the most expensive movies during our time frame and the highest budget movie (*Waterworld*) also was PG13-rated.

The budgets corresponding to the 25th and 50th percentiles of the R-rated production budgets are less than the budgets associated with these percentiles in the other rating categories. By the 75th percentile, however, R-rated movies budgets are nearly equal to G-rated movie budgets. One reason for this is accounted for by the fact that more stars appear in R-rated than in G-rated movies. High budget R-rated movies are also more likely to feature expensive special effects. On the other hand, PG-rated movies have the smallest standard deviation of budget and, by far, the lowest ratio of standard deviation to average budget followed by G, PG-13 and R. This suggests that G-rated movie budgets are easier to control than high-budget star movies.

A picture of the stochastic dominance rankings of movie ratings by budget can be derived from Figure 5.5. Note that we are now ranking in terms of "unfavorable" events, so it is desirable to have a low probability that a given budget level will be exceeded. Consequently, the inequality in equation (5.1) is reversed. In the figure, the so-called survival probability is shown; this is the probability that a movie will exceed each budget level. The survival probability is plotted in the figure for each rating class. Except for the very low-budget R-rated movies, G-rated movies are the least expensive to make. This is illustrated in the figure by the fact that the probability that a G-rated movie budget will exceed the amount on the horizontal axis lies below the probabilities that the budgets of other movies will exceed these amounts. The exception to this rule is the very low-budget R movies.

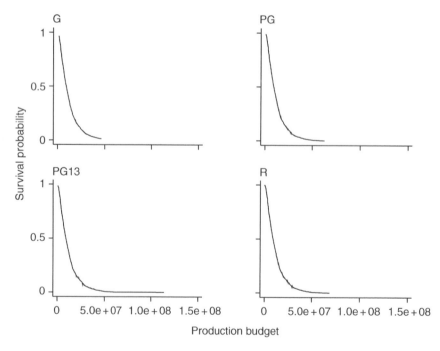

Figure 5.5 Survival probabilities of budgets by rating.

The PG13 movies are more expensive and have more variable budgets than other movies, with the exception of high-budget R-rated movies. The survival probability of PG13 movies is above the others until the R-rated movies cross over the other curves at a budget just below $60 million. With the exception of one PG13-rated movie, R-rated movies become the most expensive movies to make in the above $50 million budget category. The low average budget for R movies primarily is due to the large number of low-budget movies in this rating. As R-rated movies move into the upper budget tier they become more expensive than other movies and their budgets become more variable. This seems to reflect a fact we have already mentioned: high-budget, R-rated films feature a disproportionate number of stars and special effects. In the upper budget categories R-rated movies are first-order stochastically dominated by G, PG and PG13 movies.

5.5.3 *Rates of return*

We now examine rates of return because it is the rate of return that drives investment. We take as a measure of the rate of return the ratio of box-office revenue to production budget in each class. This measure is shown in the bottom panel of Table 5.3. G- and R-rated movies have nearly equal average rates of return: G-rated movies on average earned box-office revenues 2.15 times their production cost; R-rated movies earned 2.14 times their production cost. The PG- and PG13-rated movies have average ratios of revenues to costs of 1.66 and 1.39.

Note how the quantile values compare among movies. At the 25th, 50th and 75th percentiles the return is greater in G, PG and PG13 than in R movies. This means that for every level of return, but the very highest returns, all other ratings categories dominate the returns to R-rated movies. A few extreme returns in the R-rated category pulls the average return in this category to near-equality with the return in G-rated movies and above the average returns in the other categories. The large differences in the standard deviations reflect this dominance in the R category of a few extremely high returns. R-rated movies have a standard deviation of returns that is about 5 times that of G-rated movies and about 8 times that of PG and PG13 movies.

R-rated movies are stochastically dominated by non-R-rated movies in the gross rate of return up to the 75th percentile of the high-grossing movies. Well beyond this point, the R-rated probability distribution crosses the PG and PG13 distributions but is almost everywhere dominated by the G distribution. This situation is depicted graphically in Figure 5.6 in which we plot the empirical cumulative distribution functions of the gross rate of return in each rating category. Only a tiny fraction of R-rated movies make rates of return higher than non-R-rated movies, but this handful of movies earns such high rates of return that they pull the mean R-movie return above PG and PG13 movies but not above the G movies.

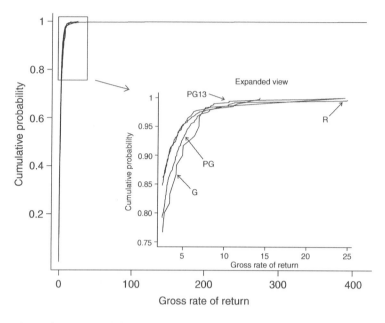

Figure 5.6 Empirical distribution of return by rating.

R-rated movies account for virtually all of the gross rates of return in excess of 25. But, these high returns are concentrated in low-budget movies and the total profit associated with these movies is small. The distribution functions for G- and PG-rated films lie substantially below the distribution functions for PG13 and R-rated movies up to a gross rate of return of about 7. There is, therefore, a higher probability that G and PG movies will earn a gross rate of return greater than 7 compared to R and PG13 movies.

We estimated the probability distributions of returns in each ratings class. In each case, the return distribution is a Pareto distribution. The estimated parameter values are given in Table 5.9 for all movies. The extreme values of return are overwhelmingly earned by low-budget movies.

But, these low-budget movies earn small aggregate profits and make a negligible contribution to Hollywood's bottom line. To look at returns for films that do contribute large absolute profits, we estimated the α parameters in the rating classes for movies that grossed more than \$40 million. These are reported in Table 5.10. Since all the estimates are in $1 \leq \alpha \leq 2$ the mean is finite and the variance is infinite for all ratings.

Applying first-order stochastic dominance to these values is equivalent to ranking the distributions according to their α values, where a distribution with a lower α dominates one with a higher value. Using the values from Table 5.10 we obtain the returns ranking $G[z] \geq PG[z] \geq PG13[z] \geq R[z]$.

Table 5.9 Estimated Pareto exponents for returns
by rating

Rating	Pareto exponent	K–S test p-value
G	1.357	0.495
PG	1.559	0.165
PG13	1.787	0.620
R	1.480	0.838

Notes
The Pareto exponent was estimated by maximum likelihood conditional on a minimum value of 2. The
Kolmogorov–Smirnov *p*-values reported are for the null
hypothesis that the data are consistent with the theoretical Pareto distribution.

Table 5.10 Estimated Pareto exponents for returns by
rating for high-grossing films

Rating	Pareto exponent	K–S test p-value
G	1.241	0.128
PG	1.331	0.194
PG13	1.322	0.306
R	1.406	0.214

Notes
The Pareto exponent was estimated by maximum likelihood
conditional on a minimum value of 2 for return and $40
million for box-office gross. The Kolmogorov–Smirnov *p*-
values reported are for the the null hypothesis that the data
are consistent with the theoretical Pareto distribution.

The fitted Pareto distributions are plotted in Figure 5.7 for each rating category of movie. This figure makes clear that G-, PG- and PG13-rated movies
stochastically dominate the R-rated movies over the range of outcomes that
makes the most important contribution to profits in the industry.

5.5.4 *Profits and losses*

We measure profits as one half of theatrical revenues minus production
cost. This is an approximation to "real" profits, which are known only to
the producer. One half of theatrical revenues approximates rentals that are
paid to the distributor. Certain costs are excluded, such as print, advertising and distribution costs. North American theatrical revenues do not
include world revenues (which are reported in a less complete fashion but are
known to also follow the Pareto distribution Walls (1998), Ghosh (2000) and
Sornette and Zajdenweber (1999), and to be correlated with North American

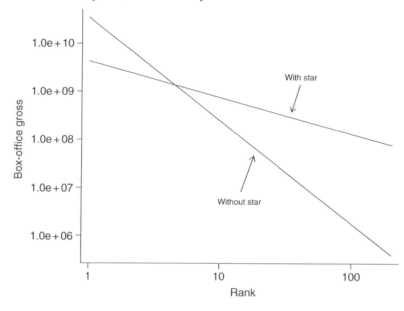

Figure 5.7 Fitted Pareto distribution of return by rating.

revenues); domestic revenues averaged (but what does that mean when the variation among movies is so great?) about 40 percent of total world revenue from all sources during the time period of our sample.[14] Nor do our revenue figures include video, TV or ancillary revenues. So, the "profits" we are able to measure with accuracy do not conform to the total profit of a movie. Nonetheless, the profit measure is a good approximation to the North American profitability of a movie. The importance is threefold: (1) the North American theatrical market is the "launching point" for the markets that follow (2) it is of interest to know the relative profitability of the North American theatrical market to the other markets and (3) the data are more reliable and more movies are reported which permits us to more accurately identify the correct statistical model (which can then be tested on the other "windows" as in Walls (1997) and Ghosh (2000) and Sornette and Zajdenweber (1999)).

In Tables 5.11 and 5.12 we estimate the Pareto exponent α for the upper and lower tails of the profit distribution.[15] In the positive tail of the profits distribution, low α is good because it means there is more probability mass (tail weight) on extreme profits. Consequently, one can stochastically rank the profit tails from low to high α. PG13 movies have the lowest positive tail α, followed by G, PG and then R movies. It is important to note that positive profits have a finite mean and an infinite variance because $1 \leq \alpha \leq 2$. The infinite variance of profits implies there is no natural upper bound on how

Table 5.11 Estimated Pareto exponents for profits

Rating	Pareto exponent	K–S test p-value
G	1.354	0.728
PG	1.431	0.391
PG13	1.118	0.087
R	1.606	0.301

Notes
The Pareto exponent was estimated by maximum likelihood conditional on a minimum value of $10 million for profits. The Kolmogorov–Smirnov p-values reported are for the the null hypothesis that the data are consistent with the theoretical Pareto distribution.

Table 5.12 Estimated Pareto exponents for losses

Rating	Pareto exponent	K–S test p-value
G	2.598	0.516
PG	2.288	0.289
PG13	2.105	0.158
R	2.167	0.123

Notes
The Pareto exponent was estimated by maximum likelihood conditional on a minimum value of $10 million for losses. The Kolmogorov–Smirnov p-values reported are for the the null hypothesis that the data are consistent with the theoretical Pareto distribution.

much profit a movie might earn.[16] In the positive profits tail, R-rated movies are stochastically dominated by other movies.

In the negative tail of profits (the positive tail of losses) large α is good because then there is less probability mass on the extreme losses. So, in this tail, stochastic dominance is ordered from high to low values of α. We see in Table 5.12 that the Pareto exponent for losses is largest for G movies, followed by PG, R and PG13 ratings. Thus, in the loss tail, R-rated movies are dominated by G and PG movies and dominate only PG13 movies. One PG13 movie, *Waterworld*, is responsible for the dominance of the R over the PG13 category. Note that all of the α values in the loss tail are greater than 2, implying that both the mean and variance of losses are finite. Losses are bounded below, but positive profits are not. The lower bound on losses is determined by budget spending since you can't lose more than you spend. There is no upper bound on profits since there is no natural limit to what a movie can earn in box office revenues (De Vany and Walls 1996).

R-rated movies have the lowest tail mass (high α) on positive profits and the second-largest tail weight (low α) on losses. Because losses are finite, this indicates that R-rated movies are stochastically dominated in profits by all the other ratings categories.

For those R-rated movies with high-budgets (there are 295 in the sample), the estimated Pareto exponent for profits is 1.479 and for losses is 1.914. The α values for high-budget R-rated movies imply the distributions are asymmetric in the tails. Profits have infinite variance and losses are near the borderline of finite/infinite variance. But aggregating high-budget R movies with and without stars hides an important difference in the distribution of profits. The positive tail α for high-budget R-rated star movies is 1.589 while the negative tail α value is 1.666; the tails are nearly symmetric. The positive tail α for high-budget R-rated movies without a star is 1.398 while the negative tail α value is 2.232; the tails are not symmetric. An R-rated high-budget movie without a star has less probability mass in the loss tail (with a finite variance there) and more mass in the profit tail (with infinite variance there) than an R-rated high-budget movie with a star. The conclusion is that putting a star in a high-budget R-rated project gives a thinner tail for positive profits (with a finite variance) and it gives a thicker tail (with infinite variance) on losses.[17]

5.6   Conclusions

The estimates of the stable Paretian model of revenue, returns and profit distributions show that they are highly skewed and may have infinite variance. Outliers in the upper tail influence the mean and the distributions are not symmetrical. There is no representative or "typical" movie, nor is there a natural revenue or profit as outcomes of all scales occur; that is to say, the distributions are scale-free[18] and motion picture outcomes have no natural scale. The most probable outcome for a motion picture is located at the peak of the distribution near the lowest values and is well below the average and the expected value. The average and expected values needn't converge and are unstable. Outcomes diverge all over the space of possibilities and the probabilities of extreme outcomes are not negligible. All of these features are implied by the stable Paretian model.

These features of the stable Paretian model mean that a belief that one can make accurate predictions of revenues or profits, even if a star is in a movie, is an illusion.[19] In the decision to produce, release, book or appear in a movie no one can afford to ignore those elements that cannot be quantified—like story and character—because those that can be quantified predict little and often mislead.

A studio seeking to trim the "downside" risk and increase the "upside" possibilities could do so by shifting production dollars out of R-rated movies into G, PG and PG13 movies. Such a reallocation of production dollars over the film portfolio trims the loss tail and expands the profit tail. If Hollywood shifts production from R to other movies, the differences in the returns distributions will narrow as the supply of R-rated movies diminishes relative to G, PG and PG13 movies; in equilibrium, the higher risk R-rated movies might earn a higher return. On the other hand, the skill or talent required to produce G, PG and PG13-rated movies may be rarer than the skill required of R-rated movies, so the differences we see in revenues, returns and profits between non R- and R-rated movies may reflect a return to scarce skills and talent.

Why would Hollywood make too many R-rated movies? Medved argued that there is a strong need for approval among Hollywood producers, executives and stars. He argues returns have taken a back seat to art as artists, producers, and even executives attempt to earn insider praise and esteem by making movies that are "audacious, artistic, and unusual and [they show] a disposition to dislike any piece of work that too obviously panders to the public" (Medved 1992, p. 306). The alternative hypothesis, out of Hollywood, is that they are producing what the audience wants.

Our results reject the "we are giving the audience what it wants" hypothesis and come close to confirming Medved's hypothesis. A portfolio overweighted in R-rated movies raises the probabilities of extreme losses and lowers the probabilities of extreme profits. A studio that accepts this inferior prospect is clearly trading profit for something or does not understand the odds.

6 Big budgets, big openings and legs
Analysis of the blockbuster strategy

6.1 Introduction

Since the nationwide release of *Jaws* in 1975, motion picture distributors have relied on the "blockbuster" strategy to capture movie demand and high revenues. *Jaws* was the first motion picture to use national television advertising in conjunction with a nationwide release. The movie was preceded in the marketplace by a reissue of the best-selling book in paperback that used the artwork from the motion picture on the cover (Wyatt 1998, pp. 78–79).[1] The blockbuster strategy uses heavy advertising, a star or stars who have "opening power," and a large number of opening theater screens to launch a movie's run. The idea behind the strategy is that the opening is the most critical event in a film's commercial life. As Producer Robert Evans said in comparing a movie's opening to a parachute jump, "If it doesn't open, you are dead" (Litwak 1986, p. 84).

The blockbuster strategy is based on the theory that motion picture audiences choose movies according to how heavily they are advertised, what stars are in them, and their revenues at the box-office tournament. The blockbuster strategy is primarily a marketing strategy that suggests the moviegoing audience can be "herded" to the cinema. Were this theory true, then the choices of just a few moviegoers early in a film's run would determine the choices of those to follow. This suggests that the early choosers are leaders or people on whom later choosers base their choices. They choose to follow these "leaders" because they believe they are more informed than they are or because they neglect their own preferences in order to mimic the leaders. Audiences who behave this way are said to be engaged in a non-informative information cascade. It is non-informative because their choices are not based on the opinions of the leaders, only their revealed actions, and the followers do not reveal their true preferences when they choose only what the leaders chose.

If moviegoers follow a non-informed information cascade, then a movie's early lead at the box office is the critical event and a large opening will be leveraged into a dominant position in the film market. This position generally will be independent of the film's quality if the cascade is a follow-the-leader, non-informative cascade. Opposed to this "lemming" theory of choice is the view that quality matters and that through the communication of personal quality information the audience will turn the non-informative cascade of the opening into an informed cascade in which quality signals dominate quantity signals.

In this chapter, we examine empirically the conditional distribution of box-office revenue outcomes in relation to budgets, release patterns and the presence of movie stars. In our analysis we test for increasing returns that would result if the choices of moviegoers feed off the information about early box-office revenues. We know that word-of-mouth information can kill a non-informative cascade, but that takes some time to develop. This means that as the cascade becomes more informative there will be a weakening the influence of budget, stars, opening screens and this is true whether the movie is good or bad. This implies that stars, budgets and opening screens may place a floor under revenues without increasing the upper tail of the revenue distribution very much because those high-grossing movies that rise to the upper tail of the distribution are determined by informed choices.

Big budgets and promotional hype will fail if viewers dislike the movie and spread the word. Private information and learning can give the opening tremendous leverage if the information cascade is positive or they can kill it if the information cascade is negative. In the end, we conclude that the audience decides and nearly all information cascades become informed cascades.

6.2 The blockbuster and the information cascade

In the past decade, competition for cinema viewers has intensified with studios producing and distributing ever more costly films filled with expensive special effects and star actors, and promoted through nationwide advertising campaigns, often with all sorts of tie-in sales.[2] The blockbuster strategy—that big-budget widely-released star-filled films are the key to high profits—is based largely on the creation of an information cascade that causes an avalanche of attendance as herds of potential viewers flock to the cinema. In this chapter we analyze empirically the success of the blockbuster strategy in creating high-grossing films by quantifying the distribution of box-office revenues conditional on big budgets, wide releases and movie stars.

Movies that capture a high share of theater screens crowd out other movies and begin their runs with high box-office revenues. Among some filmgoers high opening box-office revenues may serve as a signal that leads them to choose the leading movies instead of others that are playing. If subsequent moviegoers use the same decision rule, then an early lead at the box-office would ensure high revenues in the following weeks. An early advantage at the opening would then be leveraged into even greater advantage as the run progresses as moviegoers flock to see those movies that previous moviegoers chose. The potential for success of the blockbuster strategy then depends on how audiences use the information gained from prior viewers and other sources to choose films. If a major star can make a film "open" and gain large opening grosses, then those moviegoers who use box-office grosses to choose movies would simply act like "lemmings" following one another to a movie for no other reason than because that is where the first few moviegoers headed.

The lemming-like behavior where some public signal, like box-office revenues, leads people to copy the choices of a few early moviegoers is what economists call an uninformative information cascade. It is a cascade because it is a sequential choice process in which early choices condition the subsequent choices in such a way that a few decisive early choices direct the dynamic to flow onto a few movies whose early box-office leads, decisively, direct the cascade. It is called uninformative because the individuals do not exchange their private information about the quality of the film; the only signals observed are the choices of the previous moviegoers, not their opinions of the movie's worth. The blockbuster strategy is an attempt to create an uninformative information cascade. If audiences really were that simple, then the movie whose box-office gross is the highest in its opening week would expand on that advantage during the following weeks of the run and would capture virtually the whole audience.

Of course, audiences are not like that and this is evident from the facts that movies do not remain number 1 in box-office rank throughout their run and that movies that open "big" can fail. What limits the uninformative information cascade in which only actions are observed is learning about what people think of the movie. An information cascade becomes informative when moviegoers share information regarding their opinion or evaluation of the film. When the purely quantity signals about revenues of the uninformed cascade give way to quality information, the information cascade carries more information and takes on a new dynamic. Conclusive word-of-mouth information can break an uninformative information cascade and even turn it against a widely released movie.

The battle for opening weekend high box-office grosses is a competition for early rank in revenue and represents a belief that the audience can be led into an information cascade where they flock blindly to those films chosen in the first week. Stars and big budgets are the instruments to gain screen counts and early attendance leads. If this strategy succeeds in securing large screen counts and large opening box-office revenues, then there will be increasing returns to budget working through screen counts, an early box-office lead and subsequent herding. But, if the information cascade is broken because of word-of-mouth information, then the early leads will not be decisive. At some point in the run, private information will take over and break the uninformative cascade and then audiences will choose on the basis of quality rather than quantity signals.

One can think of a studio executive's job as making and distributing movies that will land in the upper tail of the revenue distribution. Studio executives can be thought of as choosing the budget, the talent[3] and the release pattern of the film.[4] For a movie to become a hit it must be exhibited widely and survive at the box-office.[5] If the box-office revenue sequence were an information cascade in which only actions and not private information were revealed, then a large opening week's revenue would be all that is required for box-office success. After the opening, movies would play off mechanically and the largest opening would generate the largest revenues in following weeks. The problem is that the actions, but not the experience or private information, of each moviegoer is revealed. In such a case, acting on the basis of the actions of previous choosers becomes an information cascade. The early choices and the opening lead all subsequent agents to select movies with probabilities that correspond to box-office shares.

Movie attendance does not seem to be the result of an uninformed avalanche. Moviegoers share their experiences with others and in this way reveal more than just their attendance. They reveal their private information about how well they liked the movie and this information can be conclusive if it is from a local or trusted source. Thus, word-of-mouth information, to the extent that it is conclusive and dominates summary statistics about box-office revenues, can break a non-informative cascade. If shared information dominates, then audiences make movies either hits or flops, not studio executives; the intrinsic quality of the movie is whether or not viewers like it.[6] The question is to what extent filmgoers depend on signals of prior actions versus signals about private information from someone who has seen the film.

6.3 Movie industry data and summary statistics

Our data set contains information on 2,015 movies that were released in North America in the 10-year interval 1985–96. Information on movie revenue, production cost, genre, rating, actors and behind-the-camera personnel were obtained from Entertainment Data International's historical database. The box-office data cover weekly and weekend totals from the United States and Canada using distributor-reported box-office figures. These data are the primary source for most published information on the motion picture industry and they are also followed closely by industry decision makers. To examine the effect of stars, a binary variable was created to indicate the presence of a star in a motion picture. An actor or director appearing on *Premier*'s annual listing of the hundred most powerful people in Hollywood or on James Ulmer's list of A and A+ actors was considered to be a star in our empirical analysis.[7]

Table 6.1 shows the composition of all movies in our sample by year, genre, rating and whether or not the movie included a star. In the tabulation we report the Chi-squared test statistics for the hypotheses of independence of years and genre, rating and star presence. In each case we can reject the null hypothesis of independence at marginal significance levels of less than 1 percent and this indicates clearly that there is a systematic pattern of movies through time. This is consistent with the common perception

Table 6.1 Tabulation of all movies by genre, rating and stars

	1985	1986	1987	1988	1989	1990	1991	1992	1993	1994	1995	1996	Total
Genre[a]													
Action	16	39	23	19	24	18	27	9	13	17	9	18	232
Adventure	4	10	12	0	4	3	4	2	6	3	5	4	57
Animated	2	4	1	4	3	3	2	6	3	4	4	5	41
Black Comedy	0	0	0	0	8	2	2	3	2	2	5	2	26
Comedy	11	25	37	42	66	43	39	39	21	37	22	37	419
Documentary	0	0	2	1	0	1	3	1	0	0	0	0	8
Drama	13	32	35	61	68	49	60	60	48	39	62	65	592
Fantasy	3	2	2	3	6	2	2	0	2	2	7	3	34
Horror	3	10	8	7	14	9	4	4	3	2	4	3	71
Musical	2	5	2	2	4	2	5	2	4	1	0	3	32
Romantic Comedy	2	12	16	12	15	11	14	13	11	9	5	13	133
Sci-Fi	4	5	3	3	4	5	0	3	2	2	6	5	42
Sequel	9	9	15	23	23	22	20	11	10	8	12	8	170
Suspense	1	6	16	18	20	16	12	13	16	5	12	7	142
Western	1	1	1	1	0	2	0	1	3	3	3	0	16
Total	71	160	173	196	159	188	194	167	144	134	156	173	2,015
Rating[b]													
G	2	6	2	5	6	6	5	4	5	5	8	6	60
PG	23	36	46	49	43	30	32	30	35	24	22	29	399
PG13	21	39	33	33	62	49	50	49	35	40	40	48	499
R	25	79	92	109	148	103	107	84	69	65	86	90	1,057
Total	71	160	173	196	259	188	194	167	144	134	156	173	2,015
Star[c]													
No	60	139	150	173	232	157	165	132	122	105	116	138	1,689
Yes	11	21	23	23	27	31	29	35	22	29	40	35	326
Total	71	160	173	196	259	188	194	167	144	134	156	173	2,015

Notes

a To make expected cell frequencies not less than 5, we grouped years into 6 two-year categories and included only action, adventure, comedy, drama, horror, romantic comedy, sequel and suspense genres. The resulting Chi-square(35) = 110.433 with a marginal significance level of 0.00.

b To make expected cell frequencies not less than 5, we deleted the G rating category. The resulting Chi-square(22) = 41.142 with a marginal significance level of 0.008.

c Chi-square(11) = 29.836 with marginal significance level of 0.002.

that moviemakers systematically copy whatever types of films that have been recently successful at the box-office.

Table 6.2 shows descriptive statistics on box-office revenues and production budgets in constant 1982–84 dollars for our sample motion pictures. The mean revenue in the sample was 17 million and this was much larger than the median of 6.9 million. In fact, the mean was the 71st percentile of the revenue distribution, an indication of its skewness. Median revenues varied from 1.14 million for black comedies to 16.1 million for sequels. Movies without stars had a median gross revenue of about 20.9 million, while movies with stars had a median gross revenue of about 38.2 million. For movies without stars, the mean revenue was equal to the 70th percentile, while for movies with stars the mean revenue was equal to the 62nd percentile. Movies with stars may have high budgets or other movie-specific features that help to explain

Table 6.2 Box-office revenues and budgets by rating, genre and stars

	Box-office revenue			Budget		
	Median	*Mean*	*Std dev.*	*Median*	*Mean*	*Std dev.*
Genre						
Action	8,204,898	1.79e+07	2.81e+07	1.01e+07	1.42e+07	1.31e+07
Adventure	9,946,237	1.62e+07	2.16e+07	8,673,027	1.28e+07	1.02e+07
Animated	1.53e+07	3.55e+07	4.78e+07	1.16e+07	1.55e+07	1.30e+07
Black Comedy	1,136,575	6,105,246	1.48e+07	3,952,569	6,663,847	6,526,101
Comedy	7,621,048	1.82e+07	2.74e+07	9,486,166	1.07e+07	7,512,908
Documentary	605,250.8	6,786,864	1.51e+07	1,951,118	2,570,598	1,968,908
Drama	3,591,933	1.15e+07	1.96e+07	7,905,139	1.04e+07	8,655,036
Fantasy	1.07e+07	1.95e+07	1.96e+07	1.36e+07	1.77e+07	1.29e+07
Horror	6,693,621	1.12e+07	1.45e+07	6,720,430	7,778,079	7,052,512
Musical	5,688,928	9,537,546	1.36e+07	9,403,682	1.07e+07	8,743,469
Romantic Comedy	7,589,359	1.72e+07	2.46e+07	9,063,445	1.08e+07	8,108,445
Sci-Fi	1.20e+07	2.87e+07	4.79e+07	1.74e+07	2.16e+07	2.03e+07
Sequel	1.61e+07	2.98e+07	3.36e+07	1.26e+07	1.58e+07	1.30e+07
Suspense	5,043,216	1.54e+07	2.67e+07	9,484,180	1.08e+07	8,047,906
Western	1.45e+07	2.87e+07	3.70e+07	1.66e+07	1.82e+07	1.22e+07
Total	6,943,376	1.70e+07	2.68e+07	9,096,816	1.18e+07	1.03e+07
Rating						
G	1.01e+07	2.58e+07	3.99e+07	9,510,306	1.19e+07	9,617,613
PG	1.12e+07	2.15e+07	3.00e+07	1.10e+07	1.28e+07	9,625,494
PG-13	8,439,733	1.96e+07	3.15e+07	1.13e+07	1.39e+07	1.14e+07
R	5,171,729	1.35e+07	2.11e+07	7,285,343	1.05e+07	9,797,142
Total	6,943,376	1.70e+07	2.68e+07	9,096,816	1.18e+07	1.03e+07
Star						
No	4,867,325	1.22e+07	2.09e+07	7,797,272	9,702,034	7,855,408
Yes	3.25e+07	4.15e+07	3.82e+07	2.02e+07	2.28e+07	1.37e+07
Total	6,943,376	1.70e+07	2.68e+07	9,096,816	1.18e+07	1.03e+07

Note
All monetary magnitudes are reported in constant 1982–84 dollars.

their high revenues. The distribution of budgets is also highly skewed, but not as skewed as the revenue distribution: The mean is 11.8 million and this is the 62nd percentile. Median budgets varied widely from about 1.9 million for documentaries to 17.4 million for movies in the science-fiction genre. Movies without stars had a median budget of about 9.7 million and movies with stars had a median budget of about 22.8 million. The mean budget for movies without stars was the 61st percentile and the mean budget for movies with stars was the 57th percentile.

Table 6.3 shows tabulations by year and genre, rating and star of movies that earned in excess of $50 million in box-office revenue. Examining the upper panel of the table it is clear that some genres, such as comedy, drama and sequel, account for a disproportionate share of the high-revenue movies; these three genres alone account for nearly 60 percent of all hit movies. However

Table 6.3 Tabulation of revenue hits. Tabulation of movies with box-office gross >\$50 million

	1985	1986	1987	1988	1989	1990	1991	1992	1993	1994	1995	1996	Total
Genre[a]													
Action	0	2	2	1	2	2	2	1	1	3	1	4	21
Adventure	1	1	0	0	0	1	1	0	0	0	1	0	5
Animated	0	0	0	1	1	0	1	1	0	1	2	2	9
Black Comedy	0	0	0	0	1	0	0	0	0	0	0	0	1
Comedy	2	4	4	7	6	2	4	4	1	3	0	4	41
Drama	2	0	1	2	3	2	3	2	5	4	3	2	29
Fantasy	0	0	1	0	1	0	0	0	0	0	2	0	4
Horror	0	0	0	0	0	0	0	1	0	1	0	1	3
Musical	0	0	1	0	0	0	0	0	0	0	0	0	1
Romantic Comedy	0	1	2	1	1	2	0	0	1	0	1	1	10
Sci-Fi	1	0	0	0	0	1	0	0	1	0	1	1	5
Sequel	3	3	1	1	5	5	4	4	0	2	4	1	33
Suspense	0	0	1	0	0	1	3	3	2	0	1	0	11
Western	0	0	0	0	0	1	0	1	0	1	0	0	3
Total	9	11	13	13	20	17	18	17	11	15	16	16	176
Rating[b]													
G	0	0	0	0	1	0	1	1	0	1	2	2	8
PG	6	3	2	5	8	6	4	3	2	4	5	3	51
PG13	2	4	3	2	6	5	6	2	4	5	5	5	49
R	1	4	8	6	5	6	7	11	5	5	4	6	68
Total	9	11	13	13	20	17	18	17	11	15	16	16	176
Star[c]													
No	5	8	5	6	10	6	10	5	1	5	5	7	73
Yes	4	3	8	7	10	11	8	12	10	10	11	9	103
Total	9	11	13	13	20	17	18	17	11	15	16	16	176

Notes

a To make expected cell frequencies not less than 5, we grouped years into three four-year categories and included only action, comedy, drama and sequel genres. The resulting Chi-square(6) = 9.511 with a marginal significance level of 0.147.

b To make expected cell frequencies not less than 5, we grouped years into 6 two-year categories and deleted the G rating. The resulting Chi-square(10) = 10.622 with a marginal significance level of 0.388.

c To make expected cell frequencies not less than 5, we grouped years into 6 two-year categories. The resulting Chi-square(5) = 8.477 with a marginal significance level of 0.132.

over time there has been substantial variation in the composition of hit movies by genre. For example, comedy films accounted for 7/13 of all hits in 1988 but only 1/11 of all hits in 1993. Using a Chi-squared goodness-of-fit test we cannot reject the null hypothesis that the composition of hit movies is independent of time when categorized by genre, rating or the presence of a star.

Table 6.4 shows the success rates of hit movies calculated as the ratio of hit movies to all movies times 100. Success rates vary widely by genre and year: Sequels had a success rate of 19 percent; however this varied from 36 percent in 1992 to 0 percent in 1993 and we cannot reject the null hypothesis of serial independence of the success rate.[8] We also cannot reject the hypothesis of random order in the success rates of movies with and without stars through time. However the success rate for movies with stars is consistently much larger than the success rate for movies without stars.[9]

Table 6.4 Success rate for revenue hits. Success rate for movies with box-office gross > $50 million

	1985	1986	1987	1988	1989	1990	1991	1992	1993	1994	1995	1996	Total
Genre													
Action	0	5	9	5	8	11	7	11	8	18	11	22	9
Adventure	25	10	0	—	0	33	25	0	0	0	20	0	9
Animated	0	0	0	25	33	0	50	17	0	25	50	40	22
Black Comedy	—	—	—	—	13	0	0	0	0	0	0	0	4
Comedy	18	16	11	17	9	5	10	10	5	8	0	11	5
Drama	15	0	3	3	4	4	5	3	10	10	5	3	5
Fantasy	0	0	50	0	17	0	0	—	0	0	29	0	12
Horror	0	0	0	0	0	0	0	25	0	50	0	33	4
Musical	0	0	50	0	0	0	0	0	0	0	—	0	3
Romantic Comedy	0	8	13	8	7	18	0	0	9	0	20	8	8
Sci-Fi	25	0	0	0	0	20	—	0	50	0	17	20	12
Sequel	33	33	7	4	22	23	20	36	0	25	33	13	19
Suspense	0	0	6	0	0	6	25	23	13	0	8	0	8
Western	0	0	0	0	—	50	—	100	0	33	0	—	19
Total	13	7	8	7	8	9	9	10	8	11	10	9	9
Rating													
G	0	0	0	0	17	0	20	25	0	20	25	33	13
PG	26	8	4	10	19	20	13	10	6	17	23	10	13
PG13	10	10	9	6	10	10	12	4	11	13	13	10	10
R	4	5	9	6	3	6	7	13	7	8	5	7	6
Total	13	7	8	7	8	9	9	10	8	11	10	9	9
Star													
No	8	6	3	3	4	4	6	4	1	5	4	5	4
Yes	36	14	35	30	37	35	28	34	45	34	28	26	32
Total	13	7	8	7	8	9	9	10	8	11	10	9	9

Note

Percentage success rate calculated as 100 times the ratio of Table 5.4 to Table 6.1.

The descriptive statistics and cross-tabulations indicate that there is some sort of a systematic relationship between movie revenues, genres, ratings, budgets and star presence. In the following section we develop an econometric model to quantify the relationship between box-office performance and the various attributes of a movie and its theatrical release. Our aim is to separate out the effects of budgets, stars, openings, genres and ratings.

6.4   A structural model of box-office outcomes

Figure 6.1 plots cumulative box-office gross against budget on a log–log scale.[10] With the exception of a few successful low-budget movies, there appears to be a linear relationship between log revenue and log budget, even though this figure lumps together all genres and ratings. Larger budgets appear to be associated with larger box-office revenues, although with much variation. Disaggregating the scatterplot by genre and rating, as is done in Figures 6.2 and 6.3 suggests even more strongly a log–linear functional form.

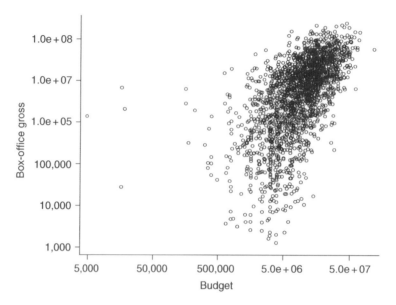

Figure 6.1 Scatter plot of box-office revenues and budgets.

To examine formally if the logarithmic transformation of cumulative revenues is appropriate, we estimated the exponent λ in the Box–Cox (1964) transform of revenues $\text{Revenue}^{(\lambda)} = (\text{Revenue}^{\lambda} - 1)/\lambda$. The Box–Cox transform is particularly useful because it nests a number of common transformations and permits convenient hypothesis tests on them: the Box–Cox transform is log Revenue when $\lambda = 0$; the transform is $\text{Revenue} - 1$, remaining in original units, when $\lambda = 1$; and the transform is $1 - 1/\text{Revenue}$, in inverted units, when $\lambda = -1$. Our estimate of the Box–Cox exponent is $\hat{\lambda} = 0.0088$; we can reject the hypotheses that $\lambda = \pm 1$ but we cannot reject the hypothesis that $\lambda = 0$. The computed test statistics and marginal significance levels are reported in Table 6.5. The evidence from the Box–Cox transform indicates that the logarithmic transformation is appropriate for our data on cumulative box-office revenue.

The structural model of cumulative box-office revenue that we examine empirically has the following form:

$$\log \text{Revenue}_i = \beta_1 + \beta_2 \log \text{Budget}_i + \beta_3 \text{Star}_i + \beta_4 \log \text{Screens}_i$$
$$+ \beta_5 \text{Sequel}_i + \Gamma[\text{Genre, Rating, Year}]'_i + \mu_i \qquad (6.1)$$

where i indexes movies, Star and Sequel are indicator variables equal to unity when a movie contains a star or is a sequel, respectively, and zero otherwise, and Γ is a row vector of parameters corresponding to the coefficients on the sets of explanatory variables indicating particular genres, ratings and years of

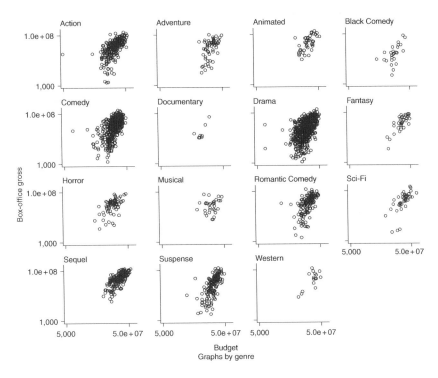

Figure 6.2 Box-office gross versus budget by genre.

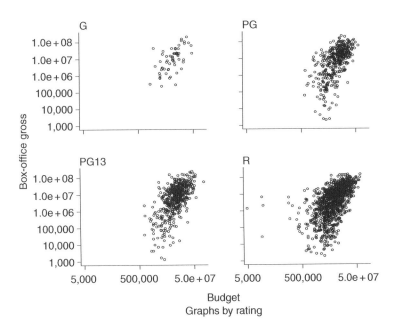

Figure 6.3 Box-office gross versus budget by rating.

Table 6.5 Box–Cox transform of box-office revenue

Transform: $(\text{Revenue}^\lambda - 1)/\lambda$

$\hat{\lambda} = 0.0088$ Log likelihood $= -25419.85$

Hypothesis	χ^2 Statistic	Prob $> \chi^2$
$\lambda = -1$	2,156.66	0.0000
$\lambda = 0$	0.41	0.5213
$\lambda = 1$	2,328.17	0.0000

release.[11] The results of a least-squares regression of equation (6.1) are shown in column 1 of Table 6.6.[12] Given the substantial variation of revenues in our cross-section of movies, the fit of the regression equation seems reasonable.[13] We also report in column 2 estimates obtained by robust regression because the least-squares estimator can be sensitive to outliers such as *El Mariachi* and *Berlin Blues*.[14] The robust regression estimates differ only slightly from the least-squares estimates: the estimates differ by less than 1 standard error for each parameter. Our regression results do not appear to be sensitive to any possible outliers.

Instead of modeling the mean of the log revenue distribution as is done in the least-squares and robust regressions, it may be more illuminating to model various quantiles of the revenue distribution since the effects of budget, star presence and opening screens etc., may differ for the various attributes of the probability distribution. Columns 3–7 of Table 6.6 show results of minimum absolute deviation regressions for quantiles 0.10, 0.25, 0.50, 0.75 and 0.90, respectively. What we are seeking to uncover is how the explanatory variables associated with the blockbuster strategy alter the shape of the revenue distribution; the estimates reported in the next section give us the answers. We will discuss in turn how budgets, sequels, opening screens and stars are related to box-office revenue, using the mean-based estimates and the quantile-based estimates. In each case, we will refer to the robust standard errors reported in brackets beneath the coefficient estimates.[15]

6.4.1 Budgets

The point estimates of the elasticity of mean box-office gross with respect to budget are 0.545 and 0.591 from the least-squares and robust estimations, respectively. Both are significantly larger than zero and significantly smaller than unity. Controlling for stars, sequels, genre, rating, year of release and the opening, bigger budgets generally result in greater production value and in higher box-office revenue.

The quantile regressions are more revealing and they explain the source of the inelasticity of revenue with respect to budget. The quantile regression

Table 6.6 Box-office revenue regressions

$$\log \text{Revenue} = \beta_1 + \beta_2 \log \text{Budget} + \beta_3 \text{Star} + \beta_4 \log \text{Opening Screens}$$
$$+ \beta_5 \text{Sequel} + \Gamma[\text{Genre, Rating, Year}]' + \mu$$

	(1)	(2)	(3)	(4)	(5)	(6)	(7)
Estimator	*LS*	*Robust*	*MAD*	*MAD*	*MAD*	*MAD*	*MAD*
Quantile	—	—	0.10	0.25	0.50	0.75	0.90
log Budget	0.545	0.591	0.636	0.576	0.563	0.527	0.448
	(0.044)	(0.045)	(0.067)	(0.047)	(0.040)	(0.075)	(0.085)
	[0.061]		[0.091]	[0.088]	[0.070]	[0.084]	[0.088]
Star	0.849	0.795	0.775	0.848	0.730	0.752	0.742
	(0.089)	(0.092)	(0.145)	(0.102)	(0.082)	(0.134)	(0.127)
	[0.090]		[0.149]	[0.126]	[0.113]	[0.109]	[0.116]
log Opening Screens	0.205	0.218	0.363	0.303	0.234	0.133	0.070
	(0.016)	(0.017)	(0.021)	(0.016)	(0.015)	(0.025)	(0.024)
	[0.019]		[0.036]	[0.029]	[0.026]	[0.028]	[0.029]
Sequel	0.762	0.709	−0.557	0.452	0.889	0.759	0.605
	(0.447)	(0.429)	(0.309)	(0.400)	(0.390)	(0.652)	(0.257)
	[0.325]		[0.430]	[0.401]	[0.522]	[0.747]	[0.270]
Genre	yes	yes	yes	yes	yes	yes	yes
Rating	yes	yes	yes	yes	yes	yes	yes
Year	yes	yes	yes	yes	yes	yes	yes
Constant	6.005	5.212	3.691	4.705	5.460	6.651	9.239
	(0.797)	(0.818)	(1.037)	(0.830)	(0.719)	(1.257)	(1.215)
	[0.933]		[1.410]	[1.345]	[1.156]	[1.501]	[1.504]
R^2	0.438	—	0.310	0.315	0.279	0.237	0.214

Notes
Dependent variable is log revenue. All regressions run on common set of 1,500 observations.
LS is least-squares regression.
MAD is minimum absolute deviation regression for the quantile indicated.
Robust is the robust regression implemented in STATA and described in Hamilton (1991).

reported in column (3) indicates that the elasticity of the lower decile of revenues with respect to budget is 0.636. This is significantly higher than the mean budget elasticity of revenue in both statistical and practical terms. The budget elasticity of revenue decreases as we model successively higher quantiles: for the lower quartile it is 0.576, 0.563 for the median, 0.527 for the upper quartile and 0.448 for the upper decile. This is the source of decreasing returns to budget. Big budgets push up the lower decile of the revenue distribution and in some sense place a probabilistic floor under box-office revenues controlling for all of the other movie attributes in the estimating equation. Big budgets may proxy for special effects and promotional activities that may help to seed the information cascade. However, big budgets do not push up the upper quantiles as much because viewers have to like a movie and spread the word for cumulative box-office receipts to have high upper quantiles.

The decreasing returns to budget at upper quantiles is also consistent with the nature of competition at the box office. If every movie distributor pursued the blockbuster strategy in order to gain box-office revenue against competitors, the overall probability distribution of revenues may not shift much at the upper quantiles. How much production budget draws a bigger total audience as opposed to merely shifting audiences among movies is not known either. Initially drawing in a larger audience would lead to larger increases in the lower revenue quantiles than in the upper quantiles because the information cascade for any one movie will be truncated by the release of competing box-office attractions. Also, a bigger budget need not show up on the screen as enhanced production value and the chances of this are probably higher the bigger the budget.[16]

6.4.2   Sequels

Sequels are often regarded as the safest movies to make. The positive coefficients on Sequel in columns 1 and 2 do suggest that a sequel has enhanced prospects relative to a non-sequel movie. In the least-squares regression the coefficient on Sequel is statistically different from zero at the 5 percent level; in the robust regression the coefficient on Sequel is not statistically different from zero at the 5 percent level, but it is statistically different from zero at the 10 percent level. Sequels are not a sure thing, but a sequel's prospects are certainly better than a non-sequel when testing at a lower level of significance. The quantile regressions may shed some light on the source of a sequel's impact on mean box-office revenue.

The results from the quantile regressions indicate that being a sequel *decreases* the lower decile of box-office revenue, but the coefficient is not significantly different from zero. Only the median and the upper decile are significantly different from zero at the 10 and 5 percent testing levels, respectively. This indicates that the real impact of a sequel is to push up the top decile of revenue. Presumably this is exactly why sequels are made—they are facsimiles of movies that people liked. That is just what a producer is looking for—a movie that puts more probability mass on the high outcomes.

6.4.3   Opening screens

Opening screens have a significant impact on mean box-office revenue with an estimated elasticity of over 0.2. Since opening screens are the distributor's primary means of directing the information cascade once a movie is complete, it provides a test of how launching the opening affects box-office outcomes. The impact of the opening on mean box-office revenue is not surprising, but we can learn more by examining the quantile estimates. In each quantile regression the coefficient on opening screens is all positive and highly statistically significant. However, the magnitude and thus the economic significance of the coefficient estimates declines continuously in moving from the lower decile to the upper decile: 0.36, 0.30, 0.23, 0.13, 0.07. This indicates that launching the opening through booking a large number of screens helps to place a floor under box-office revenues: the lower decile increases substantially. However, the upper deciles are much less affected because the the opening cascade diminishes as information sharing among filmgoers replaces the inertia of the opening.

6.4.4 Genre and rating

The sets of dummy variables for genres and ratings were each statistically significant, but individual genres and rating variables were not individually significant. Rating and genre, as groups, explain a significant amount of the variation in the mean and various quantiles of the probability distribution of revenue. However, because individual genres are not statistically significant they cannot be used to enhance predictions of box-office revenue for an individual movie, given all of the other information already controlled for in the regression equation. This is where thinking in terms of simple correlations instead of partial correlations can be misleading. We know from the simple tabulations discussed in Section 6.3 that some genres account for a substantial proportion of high-revenue movies. But we also know that this information will not permit us to predict that an individual movie will be a hit because it is in a particular genre.

6.4.5 Stars

Our estimates show that stars shift upward the revenue distribution. The coefficient on the star variable is significantly greater than zero at the 1 percent level in all of the regressions, regardless of estimator or quantile examined. Also, the point estimates are similar across the mean and the various quantiles, the range of estimates being only about 1 standard error in length. This indicates that stars as a group push up the entire revenue distribution, although there is some indication in the point estimates of a U-shaped relation where the median is pushed up less than the extreme upper and lower quantiles.[17] The point estimates are about 0.8 in magnitude and this corresponds to the mean or quantile of revenues being increased by a factor of $e^{0.8}$ or 2.2.

The results reported so far consider all the stars as a group. To examine the impact of individual stars on box-office revenue, we reestimated all of the regressions reported in Table 6.6 with separate indicator variables for *individual* stars instead of using the single indicator variable for the presence of any star. Most of the individual star variables were insignificantly different from zero at the 10 percent marginal significance level, although as a group the set of star variables was significant at the 1 percent level. Even though many actors and directors are on the star list, most do not have measurable individual box-office power when controlling for the attributes of a movie and its release. This is consistent with the idea that movies with stars are successful not because of the star, but because the star chooses projects that people tend to like. A star is born through a successful movie and he or she remains a star by choosing projects that people will like. Signaling quality may be the real power of stars.

Only 12 individual stars were found to have a significant impact on any attribute of the distribution of box-office revenue. The remaining stars on the Top-100 and James Ulmer's lists do not have statistically significant impacts on movie revenues, controlling for budget, sequel, opening screens, genre, year of release and rating. The estimated coefficients and standard errors are reported in Table 6.7 for the individual stars whose coefficients were statistically significant at the 10 percent level in a two-sided test. Exponentiating the coefficient estimates provides the factor by which revenues are expanded: for

Table 6.7 Stars with significant box-office revenue impact

Star name	Mean (OLS)	Mean (Robust)	Quantile 0.10	Quantile 0.25	Quantile 0.50	Quantile 0.75	Quantile 0.90
Bullock, Sandra			2.429 (0.817)				
Carrey, Jim			1.571 (0.880)				
Foster, Jodie			3.571 (0.678)	2.149 (0.915)	1.426 (0.335)		
Hanks, Tom	1.139 (0.503)	1.135 (0.521)			1.070 (0.488)		
Kilmer, Val			−1.458 (0.812)	−2.482 (1.319)			
Pfeiffer, Michelle					1.497 (0.714)		
Ryder, Winona			−2.070 (0.790)				
Spielberg, Steven	1.075 (0.581)	1.054 (0.602)	1.641 (0.749)				
Stone, Oliver	1.062 (0.550)	0.911 (0.529)				1.870 (0.971)	2.017 (0.853)
Stone, Sharon					−0.817 (0.457)		
Streisand, Barbra			1.688 (0.703)				
Williams, Robin		0.973 (0.520)				1.891 (0.959)	

Notes

Only statistically significant coefficients (10 percent, two-sided) are reported.

Estimated standard errors are reported in parentheses.

The coefficients represent the change in the logarithm of the particular attribute (mean, median etc.) of the box-office revenue distribution with respect to the particular star. Exponentiating the coefficient estimates yields the multiplicative effect on box-office revenues.

example, Tom Hanks' performance is associated with mean revenues being expanded by a factor of $e^{1.14}$ or 3.13.

In terms of affecting the mean box-office revenue, only three stars had statistically significant coefficients: Tom Hanks, Steven Spielberg and Oliver Stone. When using the robust regression estimates we can also add Robin Williams to the list. The quantile estimates allow us to see how individual stars move the revenue distribution. Seven stars significantly affect the lower decile of revenue: Sandra Bullock, Jim Carrey, Jodie Foster, Steven Spielberg and Barbra Streisand affect if positively, and Val Kilmer and Winona Ryder affect if negatively, controlling for all of the other variables in the regression equation. Only two stars affect the lower quartile of revenues: Jodie Foster positively and Val Kilmer negatively. Four stars affect the median revenue: Jodie Foster, Tom Hanks and Michelle Pfeiffer positively, and Sharon Stone negatively. Oliver Stone and Robin Williams are associated with higher upper quartile revenues and Oliver Stone alone is associated with higher upper decile revenues.

Overall, the top revenue power actors identified in our analysis are Tom Hanks and Jodie Foster.[18] Michelle Pfeiffer had the largest individual impact on median revenue and Tom Hanks had the largest individual impact on mean revenue. Behind-the-camera talents Steven Spielberg and Oliver Stone also were associated with significantly higher box-office revenues. The estimates for Kilmer, Ryder and Sharon Stone, indicate that these stars are associated with *lower* revenues controlling for the other factors; however, for Kilmer and Sharon Stone, the coefficients are not significantly different from zero when testing at the 5 percent significance level.

Most stars do not have a statistical impact on revenues, but those that do usually are associated with increases in box-office receipts (Table 6.7). No actors are able to move the upper decile of revenues, although several are able to move upward the lower decile of revenues. This means that star power has the same sort of impact as the opening. Stars help to launch the film and in this way they put a floor under revenues by increasing the lower decile of revenues. But audiences decide whether or not they like the film and this explains why stars do not have any statistical impact on the upper decile of the revenue distribution when controlling for genre, rating, budget and a movie's release.

6.5 Conclusions

The theory of observational learning explains a great deal about the movie business. The criticality of the opening weekend box-office revenue and distributors battling for opening theater screens are implied by the use of these signals by filmgoers in making choices of movies. Strict reliance on these signals by moviegoers would lead to non-informative information cascades in which the size of the opening weekend and not the quality of a movie would determine its success. Big budgets and stars are used to gain a large share of opening screens to "seed" the initial conditions of the information cascade in the hope that it will carry a movie to a dominant position in the market, which may even be independent of its quality. Were this an effective strategy, then there would be strictly increasing returns to production budgets and opening screens would be the primary determinant of a movie's success.

We find evidence of decreasing returns to budget, opening screens and stars, in that larger values push up the lower quantiles of the revenue distribution much more than the upper quantiles. This is evidence that moviegoers are able to break the information cascade when they share their private information through word-of-mouth information and reviews.[19] During a movie's theatrical lifetime the sharing of private information eventually can overcome the non-informative cascade from an early lead at the box office.

Part III

Judges, lawyers and the movies

The next three chapters take up some key questions about the influence of the law on the industry. Few know that motion pictures is a judicially supervised industry. Power producers are regulated by utility commissions, but the movies are regulated by the courts. Its long-time supervisor is the Southern District Court of New York acting in response to matters pressed by the US Department of Justice. How has the litigation brought by the Department of Justice under the antitrust law shaped the movie industry? Did the courts and the Department of Justice make the industry more open and competitive, or did they weaken it? What is the magnitude of their influence? Are the antitrust laws relevant to the industry today? Why is it hard to apply antitrust law to this information industry?

Before we began this research it was difficult to understand the extent to which the economics of the motion picture industry was dominated by sterile models inherited from antitrust economics. It seemed that all that the economists who studied the industry could do was calculate concentration ratios and apply antitrust concepts to the deals and contracts and even the structure of the industry. For the Supreme Court to justify its monstrous decision to break up the integrated studios and force Warner Bros., Paramount, Fox and the other integrated studios to sell their theaters, all it had to do was to find that the industry was concentrated and that its contracts were unusual. Of course, the contracts were unusual because the movies was a brand new kind of industry, an information industry. Every aspect of the industry's sophisticated and unusual contracts was seen by the Department of Justice as a sinister attempt to monopolize the market. The Department of Justice persisted so in this attitude that a District Court judge admonished them for it. But, eventually the Department of Justice argument won the day and the courts separated the studios of their theaters and swept aside the seemingly strange contracts the industry had long relied on.

The changes the courts mandated in contracts are hard to fathom; they said that each film was to be licensed individually, theater by theater, play date by play date through one-time, short-term contracts. To draw a parallel to a current situation, it is as though MacDonald's were instructed by the Court that it could no longer own or franchise its restaurants (no circuit deals, no vertical arrangements) and that it was required to license its Big Macs separately from its french fries (no block booking). Applying other aspects of *Paramount* to MacDonald's would require that it sign each of its restaurants to a new deal each week (no track booking). Nor could MacDonald's enter into a long-term relationship with any of their restaurants (no season or multiple season contracts). Imagine the turmoil that would cause. MacDonald's locations would float from restaurant to restaurant according to their bids for one week franchises. One restaurant might have french fries and another Big Macs. Customers would never know where the MacDonald's they ate at last week would be this week, or what they could buy there.

That is the situation the movie business faced in the wake of the Supreme Court's *Paramount* decision. Nearly all of its integrated structures, practices and institutions were swept aside. It was a long time in recovering and some studios went under (RKO) or never reclaimed their luster (MGM). The industry was ill suited to operating under rules meant for dealing in pork bellies or corn futures and it went through difficult times just as it was facing a new competitor: television.

Who would have done this to this marvelous new technology and art form? People who did not understand the movies: the courts, the US Department of Justice, and the antitrust economists who gave them their ammunition.

Ronald Coase called economists, who unthinkingly apply antitrust models to industries they do not fit, "camp followers" to justices and lawyers. I would say that has been true of motion picture economics. Many economists are inspired by a search for "market power," whatever that is, and ingenious only in their ability to find it. As a movie fan and an information economist, I could not bear how sterile and mechanical this antitrust-inspired work was and how little it paid heed to the complex features of the industry. I saw no justification for the conclusions drawn from static calculations of market shares that were typical of this work, and few modern economists do either.

The deeper problem is how little respect these antitrust models gave to the facts or to the sophisticated instruments devised by the industry to deal with its complex problems. Antitrust economics tends to see every departure from a stylized competitive norm as a sinister attempt to manipulate prices or close the market to entry. But, the mainstay of antitrust economics—market structure measured as market shares—is meaningless in this industry where shares are wildly volatile and driven not by the studios but by the unpredictable choices of the audience.

Arguably, the movies was the first information industry. In the 1930s the movies were a new kind of product and a new kind of industry. This is why antitrust economics, modeled after toaster manufacturers, brick factories, warehouses and hard goods stores, got everything mostly wrong about the movies. Every movie is unique. Movies can't be inspected by reading the label; you have to see it to find out if you like it. They are readily copied. Copyright protection had a strong influence on the early organization of the industry and continues to have a powerful influence today.

Chapter 7 examines the landmark *Paramount* antitrust cases. *Paramount* forced the major studios to sell their theaters and restricted the kinds of contracts and methods used to license motion pictures to theaters. It also made the courts and the Department of Justice the industry's supervisors. In Chapter 7, Ross Eckert and I argue that the *Paramount* decision was a mistake. The theory behind it was based on faulty economic analysis and evidence and the restrictions it imposed on the industry made it less efficient.

We show that motion pictures present information and incentive problems because each film is unique, production is difficult to monitor and control, and the public's demand is unpredictable. The hierarchical structure of the major studio and the long-term contracts it used to book films fostered the discovery of information about the public's demand for motion pictures and maintained the flexibility necessary to adapt to new information. The studio system was effective in controlling principal–agent problems. The studios and the theaters they owned were not instruments for the monopolization of the motion picture market. The organization of the industry and the nature of its controversial licensing arrangements, including franchising, block-booking and booking films in advance of production, we show in the chapter, are competitive solutions to the information and incentive problems that plague the industry.

Following up on this analysis in Chapter 8, Henry McMillan and I look at the whole series of cases and decisions that make up the protracted *Paramount* litigation. *Paramount* was eight separate actions brought by the Department of Justice beginning in 1938 and ending in 1949. In the early cases, the Department of Justice succeeded in changing industry contracts. But, it took a decade of litigation to accomplish what the Department of Justice really wanted, which was to break up the studios and force them to sell their theater chains.

In Chapter 8, Hank MacMillan and I use stock market evidence to evaluate the impact of each of the decisions in the litigation on the market values of the studios. We conclude that, by the stock market's assessment, the Supreme Court decision of 1948 was the major event; when it was announced, the studios lost from 4 to 12 percent of their value (adjusted for market level) in two days. The way this and other decisions affected the value of the integrated and nonintegrated defendants, and a nondefendant studio, fails to support the view that the courts dismantled an inefficient monopoly. This and Chapter 7 with Ross make two strikes against *Paramount*.

A key point on which the *Paramount* litigation turned was the market shares of the major studios in first run films. The court came up with a peculiar definition of the market—the first run films of the defendant studios—and found that, sure enough, the defendants had high market shares in the market for their own films. Courts and lawyers do this all the time and many key decisions in antitrust have turned on silly definitions of the market and the defendant's share in that market. *Paramount* is not so unusual in this respect. But, what is unusual is that anyone would think that market shares tell you anything about competition in the movie business. They mean nothing. The movies is a highly concentrated industry, "winner takes all," but its concentration comes from its diversity, which is the true measure of its competitiveness.

How can an industry be concentrated but diverse and highly competitive? It all follows from the stable Paretian distribution. The wild statistics of the business tell us that there is no typical movie; each one is unique and revenues of all scales occur. Thus, the business is diverse and the tail weight of the Paretian distribution is a measure of its diversity (small α, high diversity). It is so diverse that the variance of revenues is infinite and the mean is unstable. So, there is no average movie. Complex dynamics drive the hits into the far tail of the distribution (winner takes all), but no one knows who the winner will be. And winners turn over at a high and random rate. The turnover in rank is continuous and turbulent. Market shares of films, and the studios that produce and release them, are extremely volatile.

We know from chapters on the complex and uncertain dynamics of the business that movie revenues are close to turbulent. A top ranked movie or studio can fall to the bottom in a week. A new movie can ascend to the top in a week or two. How then do market shares behave and what do they tell us about competition? That is the topic of Chapter 9.

This chapter, written with my former graduate student Cassey Lee, shows that concentration ratios and the Herfindahl–Hirschman index HHI commonly used in antitrust give a misleading picture of the degree of competition in the industry. All these indices exhibit wild fluctuations and their behavior makes the common practice of calculating a static measure of market concentration wrong. Market shares follow a stochastic Pareto–Lévy motion in which the shares and positions of the leading distributors change dramatically. So volatile are shares that the expected value of the HHI, a sort of gold standard of antitrust, does not exist; it is infinite! strike three against paramount.

7 Motion picture antitrust
The *Paramount* cases revisited

In the *Paramount* antitrust cases[1] and consent decree[2] of the late 1940s and early 1950s the courts made an ambitious attempt to reorganize the motion picture industry through the Sherman Act.[3] The major studios were required to sell their theaters and change the way they licensed motion pictures to exhibitor. The theater chain newly formed by the sale of the studio theaters were forced to sell some of their theaters, and were prohibited from acquiring or constructing theaters without permission of the United States District Court. Most of the studios were barred from acquiring, owning, or operating theaters unless the court approved the action.

Paramount barred the integration of motion picture distribution with exhibition and it barred terms in industry contracts that were integrative in nature. The court also barred certain methods for allocating motion pictures including block booking and blind selling, and required features to be licensed individually, one theater at a time, to a wider group of exhibitors. Also barred were long-term relationships, franchises, multiple-film licenses and admission price clauses in licenses. The decrees imposed a far-reaching form of judicial regulation on the industry and made the District Court its Tzar.

The courts concluded that the vertically integrated structure of the industry, with the complexity, scope and diversity of its practices and contractual arrangements, did not fit the reigning theory of competition. The courts thought these arrangements were the product of a conspiracy to monopolize the industry that harmed the public and discriminated against the independent exhibitor. Arguments by the defendants that the nature of the business made it both reasonable and efficient for studios to own and operate theaters and to license motion pictures the way they did and arguments by one United States District judge that the industry was competitive were dismissed. The District Court acknowledged that it was unfamiliar with the motion picture business, but that did not dissuade the court from changing the structure of the major firms and the methods by which all firms allocated motion pictures:

> It is argued that the steps we have proposed would involve an interference with commercial practices that are generally acceptable and a hazardous attempt on the part of judges—unfamiliar with the details of business—to

remodel its delicate adjustments which have hitherto provided the public with what is a new and great art. But we see nothing ruinous in the remedies proposed.[4]

This Article concludes that the courts erred in finding the studios guilty of violating the Sherman Act. It shows that vertical integration and the controversial licensing methods that the Department of Justice attacked and the District Court condemned were as old as the feature motion picture itself. They were developed when the industry was intensely competitive. We argue that the industry's structure and its licensing practices were reasonable ways for dealing with the problems that arose from four characteristics of the industry's product: each motion picture is unique and cannot be duplicated; demand is unpredictable; a motion picture needs time on the screen to build an audience; and most of the costs of production and distribution occur before a film can be shown and are sunk.

These characteristics determined how the industry was organized at the time of *Paramount*. The biggest studios integrated production, distribution and exhibition under one authority. Because they had guaranteed outlets for their productions, they could contract upstream for all the productive resources required for filmmaking. The studio system reduced cost and uncertainty because it controlled production by authority rather than contract. Because films must be produced in stages and have no value until they are finished, they are vulnerable to "hold up" attempts by the cast and crew. In the studio system, talent and craftsmen were studio employees who easily could be replaced by other skilled employees. Hold-up attempts could be punished and good performance could be rewarded because the studio system extended beyond the making of a single motion picture. But, it was essential for the studios to retain their theaters for this system to work, and when they were forced to sell their theaters the studio system fell apart.

As this discussion indicates, *Paramount* had implications for how motion picture production could be organized, but it dealt primarily with the relationship between distribution and exhibition. In this chapter, we focus on that relationship and its mediation through vertical controls and exhibition contracts. We show that studio theaters and the form and content of motion picture licenses promoted the search for demand information and encouraged accurate reporting of this information. The contracts contained provisions that allowed the parties to respond jointly to newly discovered information without violating their individual interests. These points are developed in a model of the industry in Section 7.1. In Section 7.2, we show why tile studios became vertically integrated and why exhibition contracts came to have their controversial features. Various restraints and practices that were found to be illegal by the courts are discussed in Section 7.3, where we show that they have a competitive explanation. Then, in Section 7.4, we show the faults in the analysis of the issues by the courts. We discuss the effects of *Paramount* and recent developments in Section 7.5. Our conclusions are contained in Section 7.6.

We argue that the changes wrought by the *Paramount* decision and decrees weakened the industry's basic mechanisms of coordination and distorted the incentives and information on which good filmmaking depended. The decisions and decrees substituted short-term, individual licenses for long-term, multi-film license, and reduced the flexibility of contracts. They closed entry into exhibition and broke down studio control mechanisms, shifting creative control from the studios to exhibitors and "deal-makers." They plunged the industry into a period of costly and counterproductive litigation. We find that the evidence is inconsistent with the court's theory that vertical integration of motion picture distribution and exhibition facilitated monopolistic practices.

The resurgence of vertical integration in today's relatively uncontrolled and highly competitive environment is further evidence that it is a cost-reducing, competitive instrument. It also makes timely a reexamination of the *Paramount* cases, for the same concerns about vertical integration and restraints that were expressed then are being raised again.[5] Between 1981 and 1986, a half-dozen distributors purchased 14 of the theaters in the United States only about 3 percentage points short of what the 5 majors owned when they were forced to divest 4 decades ago. Some of these acquisitions were made by *Paramount* defendants who were not barred from entering exhibition, and the Department of Justice has not opposed them. If the Department of Justice returns to its earlier views and once again seeks to restructure the industry, the parties may re-litigate the issues raised in this Article.

7.1 The essential economics of motion pictures

The essential characteristic of the motion picture was stated succinctly in 1946 by United States District Court Judge Vaught in *United States* v. *Griffith Amusement Co.*:

> Moving picture films are a fluctuating and uncertain product. Until a film has been exhibited no one knows or can accurately estimate its value as a box-office attraction, either as a first-run exhibition or a subsequent run exhibition.[6]

Economic and legal scholars are in substantial agreement on this fact, but no study has explored its implications for the organization of the industry.[7] The industry's structure and practices are linked to this fundamental property of the motion picture. In this section, we discuss the implications of this for the organization of the industry. We divide the discussion into demand, supply and incentive elements and their interaction through contracts and vertical controls.

7.1.1   *Discovering demand*

Each motion picture is unique and the potential audience must be informed of its existence and qualities. It is also perishable and must play long enough to build an audience. Demand is unknown and must be discovered through search. Managing the mechanisms by which information is acquired and transmitted is essential to its success. Consider a feature about to begin its first run. Some of its qualities—cast, director, writer and story—are discoverable by reading advertising or reviews. The elusive qualities that make it a "hit" or a "bomb," however, can be discovered only through direct experience.[8] Viewers spread word-of-mouth information based on their own experience. The distributor provides information to searchers through advertising and promotions. Industry publications provide demand information through box-office reports that are followed closely by distributors and exhibitors.[9]

The distributor chooses an initial release pattern (number and location of theaters in which the film is booked) based on à priori estimates of demand. Information acquired from box-office reports is then used to adjust the release pattern over time and to match demand geographically. New exhibitors are added consistent with contractual commitments to initial exhibitors and the availability of prints (copies) of the picture. Strong demand may lead the distributor to produce more prints. Decisions at each stage of the release depend on box-office information acquired in the previous stage.[10] The "life cycle" of a successful picture illustrates these issues. A film like *Hannah and Her Sisters* had legs. It opened in a few houses, and was released more widely when initial favorable per-screen box-office grosses became evident. Additional screens and favorable word-of-mouth caused weekly total box-office gross to build. Eventually, unfilled demand slackened and the number of screens tapered off slowly in response to declining per-screen grosses. This fall in the number of screens dampened the decline in box-office gross per screen, flattening it over the later weeks of the run.

The general pattern of rise, fall and flattening of gross revenue holds for most films that play well, but individual differences among films are great. For example, *Tootsie* ran about as long as *Hannah and Her Sisters*, but the box-office revenue in some weeks past the peak shot up to levels that exceeded the best weeks before the peak. There is no typical pattern.

Initial exhibitors were protected from competitive showings of the movie in order to protect their grosses in the first few weeks. Continued and growing grosses per screen attracted other exhibitors and total gross grew strongly as more outlets were booked. These additional bookings then caused the per-screen gross to decline. If the distributor had booked *Hannah* in too many theaters earlier in the run, the per-screen gross would have fallen more quickly and fewer theaters would have played it later in the run. Consequently, the film might "play off" too rapidly, and moviegoers who were slower in discovering or reacting to information would find that it was no longer playing or was playing too far away. This segment of demand will go unnerved in a fast playoff.

Some exhibitors were unable to get *Hannah* during its weeks of highest success since the distributor must restrict outlets to protect film rentals and to keep it playing long enough to capture demand. A competitive supply of exhibitors tends to drive the per-screen gross to the minimum necessary to prevent exhibitors from dropping *Hannah* for something else. The distributor adjusts the number of screens subject to the supply of exhibitors willing to show the film, but also subject to the objective of maximizing film rental revenue less distribution cost. If bookings are accepted by the distributor up to the competitive limit, film rentals approach zero because the per-screen box-office gross approaches the exhibitor's fixed cost at the competitive limit.

Adapting to a failure produces a different pattern. Poor box-office grosses in early weeks fail to attract new exhibitors, and initial exhibitors try to cut their losses. Distributors may permit them either to cut short the run or to add a second feature. Box-office per-screen gross falls throughout the run, rather than rising and then falling as with a successful picture. The failure disappears quickly, leaving wounded exhibitors, losses for the distributor and perhaps broken contacts in its wake. George Lucas's *Howard the Duck* opened on many more screens than *Hannah*, but weekly box-office grosses, number of screens and box-office per-screen gross declined rapidly and throughout its short first run. Produced by the maker of *Star Wars* and costing over $35 million, *Howard the Duck* grossed about $15 million and dropped off "the charts" of the top fifty in three weeks.[11]

Most of the box-office receipts arise from first runs of relatively few pictures,[12] and these features must have long runs to maximize rentals. Given a contractual royalty (rental) percentage and customary admission prices, the market clears through quantity adjustments: length of run, number of theaters and timing of later runs. Most features in the pre-*Paramount* era (before integration and after) went through some kind of sequential release that included a preview (which sometimes led to re-editing or even re-shooting of some scenes if audience reaction was poor), a limited first-run engagement, a broader first-run release, and eventually second- and third-run releases. In the days of *Paramount*, 30–50 percent of a picture's theatrical film rentals came from its first run.[13] The last runs were so unimportant that the rental was a flat weekly charge and no longer depended on the box-office gross. (The contemporary search process may begin with a "sneak preview" and end with cable and broadcast television and videotaped cassettes.)

Another important quantity adjustment for extremely popular features is the production of sequels. In the *Paramount* era these included the musicals of Fred Astaire and Ginger Rogers or Dick Powell and Ruby Keeler, and the series of *Thin Man*, *Sherlock Holmes* and *Charlie Chan* mysteries. (Contemporary counterparts are the series of *James Bond*, *Dirty Harry*, *Star Wars*, *Rocky*, *Indiana Jones*, *Crocodile Dundee*, *Rambo* and various horror pictures.)

7.1.2 *Determining supply*

The "output" of a motion picture is entertainment. To entertain, a feature must reach an audience, which requires that it be distributed to exhibitors, who select it from the other features available. In turn, the ultimate judge of a motion picture, the audience, must be given adequate opportunity to reach a judgment. A motion picture must be "supplied" before demand can be discovered, so subsequent supply decisions are guided by the information gained from the audience's judgment early in the run. It is useful, therefore, to think of supply decisions as involving an initial decision, which is designed to elicit information, followed by a second set of decisions that expand or modify the initial release based on the early information. A great deal of flexibility is required to adapt to new information as the run progresses. Since less information is available at the initial release, the earlier bookings are riskier for an exhibitor. They also provide information to subsequent exhibitors.

Since theater sizes are fixed during the run, the supply of seats must be increased by lengthening the run and by supplying additional prints to more theaters. This is complicated by the fact that an exhibitor must make a fixed commitment of a theater to the motion picture for a period of time. When the run at existing theaters is extended, the prints they have are held over and cannot be supplied to other exhibitors. Since the cost of a copy of a motion picture is low relative to its production cost, additional copies can be supplied cheaply. But the cost of a print is high relative to the price of an admission, so it is not practical to provide a copy for every, exhibitor who wishes to show it and, as noted earlier, too much exposure may cause it to play off too quickly to capture demand. By the time new prints can be produced and distributed to theaters, demand may have changed. These factors restrain the response of supply to high demand and are the reasons why varying the length of the run is the primary supply response to a high demand.

7.1.3 *Structuring incentives*

Producing and distributing motion pictures to exhibitors entails a sequence of transactions. At each stage, previously expended resources are nonrecoverable. The economic returns to the participants depend on whether deals can be structured to create the correct incentives and information. Once a feature is finished, its production costs are sunk and are no longer relevant to the producer's decision to offer it for distribution. Knowing that a feature is completed and will eventually be supplied, each individual exhibitor has an incentive to pay only the distributor's marginal cost, leaving the producer with sunk costs. Exhibitors as a group, however, know that producers must recover their costs or they will not produce motion pictures. Both sides, therefore, have an interest in devising arrangements that permit an adequate supply of future pictures to be produced. Increasing the supply of features was a strong incentive for advance contracting and vertical integration in the industry's early years.

Once a distributor and exhibitor agree to a booking, each makes nonrecoverable commitments that are potential hostages in the exchange. The exhibitor may harm the distributor by reneging, pulling the feature off the screen too quickly, under-reporting box-office revenues, failing to pay rentals when due or at all, damaging the print, or failing to return it. The distributor may harm the exhibitor by failing to deliver the picture on time or at all, by supplying a picture of less-than-promised quality, by booking it in more theaters, or by failing to provide the promised level of advertising support.

The quality of theaters also affects film rentals. The distributor who owns screens has an incentive to provide the correct level of service quality, but the individual theater owner may not. Most, but not all, of the benefits of theaters that are more comfortable, cleaner, and have better visual and sound systems are captured by their owners in higher box-office receipts. Through the film royalty, the distributor and producer also share in these benefits. If audiences attracted to higher quality theaters spread the word about a new feature, subsequent exhibitors also benefit. Independent exhibitors, who do not capture subsequent royalties and who also must share a percentage of their own box-office revenues with the distributor, may choose too low a level of theater quality. Controlling theater quality through exhibition contracts would probably be ineffective. No distributor would place much reliance on a clause by which the exhibitor promised to keep the floor clear of chewing gum, to eject rowdies and to clean the toilets. The admission price clauses in the *Paramount* era exhibition licenses may have strengthened exhibitor incentives to maintain quality as we show in Section 7.3.1.

7.1.4 *Contracts versus integration as coordination mechanisms*

As we have shown, transactions must be structured to reveal demand information and transmit it truthfully. But the arrangements must leave sufficient flexibility to adjust to this information, subject to the limits of technology, while maintaining the compatibility of distributor and exhibitor incentives. Since the deal between a distributor and an exhibitor must be made before either knows the demand for the motion picture, their transactions must be structured to manage *ex post* regret and to deal flexibly with the information revealed as the picture plays. At the time of the *Paramount* cases, a full range of control structures and contracts had been developed to deal with these issues. Most of the defendants employed a mixture of arrangements. They owned some theaters outright. They owned a joint interest in others. They leased theaters from independent owners and operated them as their own, both long term and short term. They franchised theaters to exhibit their pictures over a period of years. They had implicit long-term contracts with exhibitors with whom they dealt repetitively. They made contracts covering a full season of pictures. They made master agreements and blanket deals that covered an integrated circuit of theaters or a pool of independent theaters. The overriding characteristics of these arrangements are that they are longer term, that they cover more than one film and theater and that they offer greater flexibility than the short-term, one-picture, one-theater contracts that the courts prescribed in the decrees.

The terms of the motion picture exhibition license reveal the issues that must be dealt with in contracts and shed light on why this range of integrative arrangements came into use.[14] A first-run exhibition contract in the *Paramount* era called for a royalty equal to a percentage of the excess of box-office revenue over the exhibitor's fixed cost. For a typical theater showing a feature for a fixed run, most costs are fixed. Total cost varies slightly over the run with respect to admissions, but the marginal cost of an admission is small. A royalty based on the excess of revenue over fixed cost closely approximates profit sharing between the distributor and exhibitor. The royalty rate on contracts during the period varied, but was around 50 percent. (Today, an exhibition contract might call for a royalty of 70 or 90 percent of the excess of weekly gross box-office revenues over a negotiated fixed cost figure.) Box-office revenue declines in the later weeks of a run and as it approaches the "house nut" the effective royalty rate approaches zero.[15]

Box-office revenues can be observed by the exhibitor but not the distributor. The theater pays royalties on reported receipts, but retains revenue on its true receipts. Profits net of the royalty decline in the exhibitor's reported revenue, so the incentive is to report less than true box-office gross. This can be done by misrepresenting either the price, admissions or both. The incentive to misrepresent revenues is not present when the theater is integrated with the distributor.[16] Similarly, the distributor cannot observe the theaters' true fixed cost and the exhibitor also has an incentive to misrepresent fixed cost. In general, the profit-sharing arrangement creates incentives for the exhibitor that are incompatible with the goals of the distributor and various industry practices are aimed at reconciling their incompatible incentives.

One example of incentive-compatible contracts barred by *Paramount* were the so-called formula deals (analyzed in Section 7.3.3). In these contracts a circuit's royalty payments were tied to national box-office receipts rather than the circuit's reported receipts. The circuit's royalty was almost independent of its report to the distributor, reducing its incentive to report falsely.

Another example of a practice that reconciled incentives was repeat dealing. Repeat dealing builds a historical base against which to compare box-office reports, which strengthens the exhibitor's incentives to give an "honest box-office count." Repeat dealing also facilitates the collection of royalties. If a distributor discovers that an exhibitor pays too little or pays too slowly, as is more common, future bookings can be withheld. Independent movie producers who do not have a continuous line of product to offer exhibitors, are plagued with collection problems when they distribute for themselves. A major distributor, having a stream of product to deny to a balking exhibitor, can collect royalties for the independent producer more readily than the producer can.[17]

Another area where exhibitor and distributor incentives are potentially incompatible is in adjusting to high demand by lengthening the run. The contractual provision that permitted flexibility in the length of the run was the holdover clause. The contract stipulated the minimum weekly gross that held the print at that theater for another week. The clause obligated the exhibitor to show the print and obligated the distributor to leave the print for the exhibitor to show and not pull it to rent it to another theater. Holdovers sometimes created problems, however, such as when a picture continued to meet its holdover figure and played into the Hate scheduled for another picture.

The vertically integrated distributor exhibitor could decide which picture should play without violating contractual obligations. When the distributor and exhibitor were independent of one another, the holdover clause in the picture playing and the play date of the license to exhibit the new picture clashed. We show later how these problems were resolved through "move over" provisions.

Vertical integration substitutes internal control for contracts. It expands the scope of observation relative to contracts since the theater and distributor are no longer separate entities with their own private information. The incentive-compatibility problems that result from incomplete information and limited observability give rise to economies of scope in motion picture distribution (falling average cost with the number of distinct films distributed) that strengthen the economies of scale (falling average cost with increasing volume for a given number of films) that often characterize distribution. Economies of scale and scope are important determinants of the industry's degree of integration now and prior to the *Paramount* decrees.

7.2 The feature motion picture

In this section we show how integrative control structures and contracts developed early in the industry's history when the feature motion picture became its primary product. During this time, the industry's structure was unquestionably competitive and entry was open at all stages; so the emergence of vertical controls suggests that they were innovations associated with the feature motion picture rather than with monopolization.

In the 1890s and 1900s, producers sold copies or prints of motion pictures outright to theater owners. Exhibitors resold their prints when they no longer wished to show them. By 1903, organized film exchanges supported print trading among exhibitors and eventually the exchanges became rental agencies to reduce the cost of transferring prints. Most of these pictures were low-budget one-reelers: 10-minute comedies, 15-minute dramas, 5-minute scenic films, and 20-minute westerns. About 4–8 of them comprised the typical 2-hour program for which patrons paid a nickel. By 1911, about 150 exchanges supplied 11,500 theaters with enough reels to provide almost daily program changes. First-run showings fetched higher rentals than later runs.[18]

Distributing pictures with little pre-selection-on "plain, pipe racks," as A. D. Murphy called it,[19] left many problems unresolved. Copying and bootlegging were common. The flat price that the producer received for selling a print to an exchange did not correlate with the picture's success in the market place. Revenues were too small to justify big productions. Releases had to be changed frequently to stimulate attendance. Without marketing, however, the public had scant information about which pictures to see. There were no great stars or studios in this period, and memorable pictures were rare. Exhibitors faced an unreliable stream of indifferent motion pictures by unrecognized producers and artists.

The multi-reel feature picture, produced as early as 1907, became popular and widespread by 1913.[20] It fundamentally changed the industry. Producers and exhibitors saw the value in moving from daily program changes to 2 or 3 features per week. Demand would rise through word-of-mouth advertising, theaters could become more lavish and charge higher admission prices and distributors could choose among better theaters. The major studios, great stars and memorable features developed as the industry moved away from one-reelers to feature motion pictures.

Features cost more to make than one-reelers, so a nationwide distribution system was formed to lower the cost of distributing them. Higher cost productions required larger financial commitments from producers. To finance a picture, a producer had to show that a distributor would handle it and exhibitors play it. The ability to contract "forward" for the distribution and exhibition of a feature before it was produced was essential to financing. Forward contracting was also the genesis of vertical integration.

At first producers sold exclusive rights to their productions by territory. These were called "states' rights" deals because they encompassed one or more states, although they were often for only part of a state. Copies of the motion picture were sold to a states' rights distributor for a flat fee that was based on the territory's population. As they gained experience, however, distributors learned that population was an unreliable estimator of demand and they began to charge a rental fee, or royalty, that was based on a percentage of the exhibitor's box-office revenue. This arrangement was also acceptable to the exhibitor since it was less risky than paying a flat rental for the feature no matter how well it did at the box office. However, for the royalty to correspond with demand, the exhibitor's box-office report had to be truthful. Anticipating vertical integration, the typical area or states' rights distributors were exhibitors who would show the picture in their own theaters, and then lease it to other exhibitors in the allowed territory for later runs. These later runs were for designated periods and sometimes were exclusive.[21]

For "blockbusters" like *Birth of a Nation*, distribution was preceded by "road shows." In a road show, producers distributed and exhibited their motion pictures themselves. In this antecedent of vertical integration, the producer selected and operated theaters (usually the larger legitimate stage houses); did the advertising; and sold reserved seats at premium prices. Such arrangements have survived to the present.[22]

By the 1913–14 season, producers began to license blocks of a full season's production (perhaps 30 pictures) to a distributor in a single agreement. Paramount Pictures Corporation, formed in May 1914 by the merger of five exchanges and regional distributors, secured the franchise to distribute all the features of three of the leading production companies. The producers agreed to make a number of features each season for Paramount to distribute. In return, Paramount agreed to pay the producers 65 percent of the gross revenue it received from theater rentals and sub-licensing to other distributors. Paramount retained 35 percent as its distribution fee. Paramount guaranteed minimum returns to the producers and advanced cash to them on each features release date, against which film rental payments received from exhibitors were credited.

By these arrangements, Paramount and other distributors helped producers finance their productions. To secure exhibitors for the features they financed, distributors granted franchises to exhibitors. These franchises gradually replaced states' rights deals.[23] By taking a distributor's franchise, an exhibitor agreed to show the distributor's entire line of motion pictures. As Cassady noted, "this early version of block-booking developed out of the need of distributors for a more efficient method of selling films."[24] The franchising exhibitor also bought features "blind," before they were seen, or before they had even been made. The contractual precursors to vertical integration had by now been formed; the distributor contracted upstream with the producer for films and downstream with franchising theaters for outlets in which to show them.

By the end of 1916, features predominated over short films, and movie "palaces" had been built.[25] But, producers and distributors wanted stable outlets and exhibitors wanted assured supplies. The contractual arrangements on which they relied were a more uncertain instrument to that end than vertical integration. Vertical integration of distribution with exhibition offered stronger assurance of outlets for distributors, who could then guarantee play time to the producers with whom they contracted. The theater-owning distributor could also set admission price and get a more accurate box-office count. With guaranteed distribution and exhibition, producers were better able to finance their features.

The impetus for integration came from all links of the production, distribution and exhibition chain: distributors merged upstream and downstream; theater chains merged upstream through distribution to production; producers merged downstream through distribution to exhibition. Paramount the distributor became a producer and also built theaters, owning 303 houses by 1923. Fox and Loew's entered production from the exhibition side of the industry. The producers Warner and Goldwyn formed theater chains and also developed distribution facilities. By 1931, Paramount owned nearly 1,000 houses and its own studios, and the six largest circuits owned 2,437 theaters (1/8 of all theaters).[26]

The driving force behind these changes was the introduction of the high-cost feature motion picture. Within 15 years of its introduction, the structure and controversial practices that the government would later attack: exclusive territorial licensing, a sequence of runs, block booking, franchising, price stipulations and vertical integration had become the industry norm. While this historical evidence does not dispose off the issue, it is more consistent with the view that the structure and practices were mechanisms for more effectively producing and delivering feature motion pictures, than with the view the courts would later adopt, that they were unnecessary and unlawful restraints which were part of a conspiracy to monopolize the industry.

7.3   Integration and licensing

The five major defendants in *United States* v. *Paramount Pictures, Inc.*[27] were Loew's (renamed MGM), Paramount, RKO, Twentieth Century Fox, and Warner Brothers. Each of the major studios was vertically integrated from production through distribution and exhibition. The three minor studio-defendants were not fully integrated: Columbia and Universal produced and distributed motion pictures; United Artists distributed only. The court found that the defendants conspired to fix minimum theater admission prices; engaged in intertemporal price discrimination by charging higher admission prices for first run than for later showings; conspired to fix patterns of exclusive clearances and runs for neighborhood theaters; and restrained trade by licensing motion pictures in blocks. It found that the five major defendants also operated theaters monopolistically through joint ownership or pools. There is another explanation for these practices, made by the defendants, which states that they were essential, given the nature of the motion picture business. The practices are impossible to justify within standard economic models of industry prevalent at the time, but they do have a justification within the model of the industry that we set out earlier. In this section we discuss the five main charges against the defendants in order to evaluate their reasonableness within our model of the industry. We also consider the evidence on the charges.

7.3.1   *Admission price fixing and price discrimination*

The District Court, upheld on appeal by the Supreme Court, found the defendants guilty of two price-fixing conspiracies: a horizontal conspiracy between all the defendants and a vertical conspiracy between each distributor-defendant and its licensed exhibitors.

Evidence for vertical conspiracy consisted of numerous express (licensing) agreements with penalties for violation: cancellation of a particular license, cancellation of all subsequent licenses of the distributor for that theater, or substantial reductions in that theater's clearance for subsequent runs.[28] The total effect, the District Court said, "is that through the separate contracts between the distributor and its licensees a price structure is erected which regulates the licensees' ability to compete against one another in admission prices."[29]

The horizontal price-fixing theory was based on the fact that many distributors stipulated a minimum admission price in their licenses. The minimum price that distributors stipulated in license agreements was the usual and customary price the theater charged. The distributors defended this practice as a device for maintaining revenue in percentage royalty agreements.[30] The District Court misunderstood the significance of the defense, which was undoubtedly correct for three reasons. First, an individual exhibitor can gain revenue from rivals by cutting the admission price. Such rivalry will fail to maximize the distributor's royalty revenue.[31] The distributor can counteract this rivalry by reducing the number of outlets, but must balance the gain from reduced rivalry against poorer coverage of the market.

An exclusive run protects the exhibitor against rivalry, so that the exhibitor's profit-maximizing price will also maximize the distributor's rental revenue. In lieu of an exclusive run, rivalry in prices between exhibitors showing the same motion picture was limited by a minimum admission price clause in the license. But the exhibitor was more familiar with the local market than the distributor and so had an advantage in setting admission price. The license exploited this informational advantage by stipulating that the usual or customary price was to be the minimum admission price, leaving the exhibitor the option of raising the price for motion pictures the exhibitor thought would be in high demand. Vertical integration guaranteed that the admission price would be set to maximize film rental while covering the theater's costs.

A second reason why the exhibitor failed to charge an admission price that maximized film rentals is that the royalty was based on box-office receipts exclusive of concession revenue. The exhibitor could divert revenue from admissions by tying purchases of candy or popcorn to reduced admission prices. The exhibitors' concession profits could exceed what was earned at the box office if the royalty rate was high.[32] Exhibitors have refused to include concession revenues in license agreements. Distributors have sought to limit diversion of admissions to other areas beyond the distributor's reach within limited success.[33] They have, for example, attempted to limit free passes and they have specified minimum royalties per admission independent of the price charged by the exhibitor.[34]

A third reason why distributors stipulated admission prices in exhibition licenses was so that they would be known and constant during the run. The exhibitor has an incentive to misrepresent box-office revenue. When the admission price is stipulated the exhibitor has less scope to misrepresent box-office revenue because the dimensionality of the signal is reduced. Also, since distributors accommodate demand by adjusting the quantity of prints or the number of theaters booked to box-office reports, fluctuating admission prices and free passes would contaminate the demand signal contained in box-office information because variations in the gross could be caused by variation in price as well as admissions. Here was another divergence between the incentives of distributors and exhibitors that vertical integration could cure.

The horizontal conspiracy was, as Justice Douglas put it, "inferred from the pattern of price-fixing disclosed in the record," for which he found an "adequate foundation."[35] The District Court believed that record contained evidence of prices that had "great similarity and in many cases identity."[36] The exhibits of stipulated admission prices that were included in the District Court's opinion do not appear to support its contention.[37] The price data in the exhibits were taken from answers of defendants to plaintiff's inter-rogatories about the first block of 5 features licensed by each of the 5 major distributor-defendants, and about the first 5 pictures licensed by each of the 3 minor defendants, at 40 theaters for the 1943–44 season. The data do not show a high degree of price uniformity in distributors' licenses or among theaters.

The highest distributor's stipulated admission price was almost twice the lowest price in the theater with the largest difference between distributor prices. The average difference between the lowest and highest price stipulated by a distributor was 10 cents. A cartel would establish a uniform price for each distributor's features at the same theater, but that occurred at only 7 of the 40 theaters. At many theaters there were substantial differences. For example, at the Rex in Beloit, Wisconsin, minimum prices for each distributor in cents were: Paramount, 36; Loew's, 27; Warner, 40; RKO, 35; Fox, 36; Columbia, 35; and Universal, 42. The range in distributor prices at the Rex was 15 cents and the mean, standard deviation, and coefficient of variation were 35.87, 4.74, and 13.22 cents, respectively. These data hardly support the thesis that prices were uniform.

The within-theater variation of distributor minimum prices in 1943–44 exceeds the variation of admission prices across theaters in a sample of 28 Los Angeles theaters in 1986, where the mean, standard deviation and coefficient of variation are respectively $5.607, $0.367 and $6.583. The variation across distributor minimums at the Rex was about twice the variation in admission prices across theaters in the Los Angeles sample when distributors were prohibited from stipulating admission prices under the *Paramount* decrees.

It is difficult to see how the distributor-defendants could have set monopoly admission prices. The District Court found that entry into exhibition was open and the growth in theaters in the period preceding the decision supports this finding.[38] Therefore, to monopolize exhibition the defendants would have had to control production. But entry into either production or distribution or both was open as well. It is difficult to see how the eight defendants could have operated as a cartel to restrict motion picture production or distribution.[39] At this time, their production was 62 percent of domestic film production and 49 percent of domestic production plus imports. Their sum of their shares was 52 percent of feature motion pictures distributed in 1940, and no single studio had a share above 8 percent. Market shares are probably not useful indicators of market power in this industry, since the shares change constantly and by large amounts. Individual studio production and distribution shares at the time showed a typical range of 200 percent. In addition to casting doubt on market shares as a measure of market power, the wide range of shares suggests they were not controlled in the manner required for a cartel to be successful.

In the 1943–44 season the five major *Paramount* defendants accounted for 73 percent of domestic film rentals, and in 1948 seven of the *Paramount* defendants accounted for 76 percent of all domestic film rentals.[40] The volatility of gross income and earnings of the studios exceeds the volatility of their production. That suggests that control of production is not even a sufficient instrument for control of revenue, which is a basic requirement of a successful cartel. Stock prices, earnings and domestic box-office revenue shares are not correlated across studios, which would be true of a revenue-sharing cartel.[41]

Coordinating a cartel of film production under these market circumstances would be so difficult as to be implausible and the large volatility of production shares among the defendants indicates that there was no effective coordination of production among them. "None of the defendants herein," the District Court conceded, "has violated the Sherman Act by monopolizing or attempting to monopolize or conspiring to monopolize the production of motion picture films."[42]

The courts also attacked admission pricing in first and later runs. Opening a feature with a limited first run at higher priced theaters and eventually spreading the release through time to lower priced theaters was construed to be intertemporal price discrimination. The District Court said:

> The differentials in price set by a distributor in licensing a particular picture in theatres exhibiting on different rung in the same competitive area are calculated to encourage as many patrons as possible to see the picture in the prior-run theatres where they will pay higher prices than in the subsequent runs.[43]

Some academic studies accepted the court's intertemporal discrimination hypothesis.[44] Others argued that a maximizing solution to the problem was impossible since demand (and thus admission price) at each theater depended on the admission price, run and clearance at all other theaters within a reasonable travel distance that were showing the feature.[45]

Intertemporal discrimination may have been tempting for distributors to try and there is little to object to about it, since if it increases film rentals it will expand the supply of motion pictures. But it required information about audience preferences that could only be learned as the feature played. Since audience preference is private information and is not even known to the moviegoer until after the film is seen and depends on the other films available to the viewer and their locations and prices, no distributor had the information necessary to practice price discrimination. The best a distributor seeking to extract additional revenue from a film could do given this imperfect information would be to put it in a high-priced theater first and allow movie-goers with high preference to "self-select" for that theater. Exclusive, first-run engagements followed by broader release to neighborhood theaters enabled distributors to create a system of priority among filmgoers that would extract more film rental if the film played well. The limited number of prints of a first-run release had to be rationed sequentially in any event. The most efficient procedure for searching for a motion picture's audience is to begin at the centrally located theaters with high-revenue potential that would provide the best demand information and function as a "showcase" for other exhibitors.

7.3.2 *Fixing clearances and runs*

A motion picture played at one or more theaters in a town, usually in sequence. In a large city like Chicago the run sequence might consist of 11 runs.[46] There would be fewer runs in a smaller city. Distributors protected each theater from contemporaneous and later showings by "clearing" an area and time period in which they agreed not to license the same feature to a rival theater. The clearance was a commitment from the distributor to the exhibitor to protect the play date from rival showings of the picture. The clearance included the time period between theater dates and the theaters against which the licensing exhibitor was given clearance.[47]

The District Court recognized that granting a time or area clearance between rival showings of a film was a legal device to protect the value of the royalty while maintaining exhibitor incentives.[48] The clearance also was a device for maximizing the distributor's royalty. By sequentially ordering limited exhibitions of the motion picture, the largest and highest priced theaters which placed the highest value on it would play the first run and pay the highest price for it.

Clearances were essential since it was too costly to make prints for each theater that would eventually show the picture. A distributor maximizes profits by making the number of prints that equates expected marginal revenue with marginal cost. In 1945, a black-and-white print cost $150–300 and a color print cost $600–800.[49] The average length of run in a theater was 2.25 days, so the average run could not return rentals exceeding the cost of a print. The average number of prints made by a major distributor was 300 and each print had about 100 bookings.[50] In order to book a print 100 times, bookings would have to be scheduled sequentially, which required a system of priority. The priority system assigned each theater its run status. Once established, the run status of a theater seldom changed unless its neighborhood changed or its owner made significant improvements that would attract larger audiences.

The stability over time of the run status of theaters was one of the objections to it by the courts. The District Court held that some clearances were unreasonably long, that in many communities they evolved historically with little change over a decade or two, and that new and independent distributors or exhibitors appeared to be frozen out of the clearance system.[51] The court concluded that the distributor-defendants had agreed to a uniform system of runs and clearances. Justice Douglas said:

> The evidence is ample to show, as the District Court plainly demonstrated, that many clearances had no relation to the competitive factors which alone could justify them. The clearances which were in vogue had, indeed, acquired a fixed and uniform character and were made applicable to situations without regard to the special circumstances which are necessary to sustain them as reasonable restraints of trade.[52]

The District Court and the Supreme Court argued that the clearances of a competitive system would not be stable and might even change from one motion picture to the next. But this is not necessarily so. The exhibitors willing to pay the highest royalty on a motion picture may change little from picture to picture since this depends on the theater's location, quality and size which do not change from picture to picture. The courts also neglected the fact that runs are variable even within a fixed run and clearance system. Pictures that play poorly do not run the expected length of time and those that are more successful run longer than expected. The actual run and clearance period was flexible; only the initial allocation of pictures to theaters was stable.

The stability of runs and clearances were reasonable restraints on trade because they enabled distributors and exhibitors to build trust and reputation. Longer term relations with first-run exhibitors supported demand-revealing experimentation and promoted stable expectations of product and play time on both sides.

A theater's position in the tier of runs determined how much information it had about a film prior to its being played. The exhibitors whose runs followed the riskier first run had more information on which to base their booking decision and could select films more carefully. Their rewards were potentially smaller, but so was their risk and the potential gross of their house. A distributor could plan an optimal release for a film on the basis of expected audience response, but the distributor could not continuously optimize runs and clearances to a picture's characteristics and demand until demand had been discovered. Early box-office grosses provided information to subsequent exhibitors, and some theaters were better located to generate reliable data. In the *Paramount* era these were usually the higher quality, downtown theaters that reviewers and targeted audiences were more likely to attend. The importance of these first-run "showcases" explains why many of them were owned by the studio-distributors.

Stable, long-term dealing permitted *ex post* flexibility that was lacking in arms-length, short-term licensing. When a film meets its holdover box-office gross for several weeks, its engagement may extend into a period when the theater is obligated to show another picture. When this happens, the exhibitor is caught between two incompatible contracts and must violate one of them. Uniform clearances usually meant that an exhibitor showed the pictures of one distributor. This would mean that the held-over picture and the scheduled picture would likely be from the same distributor. Thus, there would be no contractual conflict; the distributor and exhibitor would simply decide which picture to play and reschedule the other one. Their interests would be compatible, almost as though they were vertically integrated. Similar flexibility was extended to theater chains that dealt repetitively with the distributor. They were permitted to "move over" one of the pictures to another theater in the chain.

Clearance policy also affects the variety of films exhibited. When a distributor limits the number of theaters exhibiting one of its films, there are more theaters available to show other motion pictures. If distributors did not restrict their films to one or a few theaters, then most exhibitors would want to play one of the top films and there would be less variety available to the public. The courts missed the economic significance of uniform systems of clearance and run, but not because it had not been suggested to them. The District Court, in one of the findings of fact it issued along with the original 1946 decree, framed the issue correctly:

> The successful exhibition of a feature in its initial runs in any area is widely publicized and closely observed by subsequent run exhibitors in that area and success in exploiting a picture in such exhibitions produces increased revenue both for the distributor and for subsequent run exhibitors.[53]

The significance of the point was understood by District Judge Vaught. The importance that he and the District Court attached to the uncertainty of market demand was ignored by the Supreme Court.[54]

7.3.3 *Formula deals, master agreements and franchises*

The District Court classified practices of distributors as restraints if they favored affiliated circuits or the larger independent circuits over smaller exhibitors. Formula deals determined the rental price of a given feature for the circuit as a specified percentage of the feature's national box-office gross receipts. All the circuit's theaters were included in the deal, and the circuit could allocate playing time and film rentals among them at its discretion. Master agreements (or "blanket deals") licensed a feature in more than one theater (usually a circuit). Franchises covered all of a distributor's releases for more than one season. Various provisions of these agreements gave licensees extra flexibility to deal with information as it was revealed. "Moveover" clauses allowed a licensee to exhibit a given picture in a second theater as a continuation of a run in the first theater. "Overage-and-underage" provisions permitted an exhibitor to apply a deficit in the playing time in one theater to another. Some large circuits got better clearances, lower rental rates for double bills, credit for promotional expenses against box-office receipts subject to film rental, freedom in a license to substitute another theater for one closed down, permission to charge lower admission prices, and privileges in selecting and eliminating pictures.[55]

The District Court argued that these practices were restraints because they substituted for competitive bidding by exhibitors individually by picture and by theater. The Supreme Court agreed that formula deals and master agreements were "devices for stifling competition and diverting the cream of the business to the large operators."[56] But they appear to have been devices to share risk, control cheating and reduce exchange costs. Theater circuits reduce exchange costs and supply information to distributors and subsequent exhibitors. Thus, it was economical for distributors to grant these circuits favorable terms—lower royalty rates and better clearances. These arrangements required greater flexibility because more than one theater and motion picture were involved in the deal and because it was essential to adjust play times and motion pictures as demand information became available.

In a formula deal a circuit paid royalties on the nationwide gross box-office receipts rather than its own. Because the royalty was independent of the pattern of play chosen by the circuit, it could move a film among its theaters so as to maximize its earnings. The circuit could not decrease its rental payments by manipulating a film's play pattern among its theaters since its film rentals were tied to the national box-office revenues. In addition to promoting flexibility, the formula deal weakened the circuit's incentive to report box-office figures falsely. Since the rental it paid was independent of its reported box-office revenue, there was nothing to be gained by misrepresenting it.[57] Vertical integration provided directly the flexibility and incentives for honest reporting of the formula deals. The exhibitor's incentive to give an honest box-office report and to pay royalties owed could also be strengthened if the distributor could withhold future bookings.[58]

7.3.4 *Block booking and blind selling*

Licensing features "in 'blocks' or indivisible groups, before they had been actually produced,"[59] was practiced by all but one of the distributor-defendants.[60] Justice Douglas said:

> Where a high-quality film greatly desired is licensed only if an inferior one is taken, the latter borrows quality from the former and strengthens its monopoly by drawing on the other. The practice tends to equalize rather than differentiate the reward for the individual copyrights. Even where all the films included in the package are of equal quality, the requirement that all be taken if one is desired increases the market for some. Each stands not on its own footing but in whole or in part on the appeal which another film may have. As the District Court said, the result is to add to

the monopoly of the copyright in violation of the principle of the patent cases involving tying clauses.[61]

The District Court ordered that pictures could be licensed in groups at the request of exhibitors, but distributors could not refuse to license any picture in the group separately. The courts appear to have improperly characterized block booking. It was not done to tie weak pictures to strong ones. The information necessary to distinguish between weak and strong pictures was not available, and the royalty structure of the license already captured higher rentals from better pictures. Block booking, franchising and other forms of advanced licensing applied to new productions only; they were not used to license previously produced motion pictures, which were readily available at fixed rental prices. We have noted how contracting for a producer's entire season of films in advance of their production became a common practice in the industry with the introduction of the feature motion picture. It developed because exhibitors wanted a reliable stream of motion pictures to show and were willing to contract in advance for them so that distributors could finance producers and assure them of outlets for their products. This practice had developed into the season contract by 1946 when Judge Vaught wrote his opinion in *United States* v. *Griffith Amusement Co.*[62]

In the development of the trade in the industry, the distributor endeavored to have his exhibitor license the produce for a year in a contract called a season contract. The solicitation began about the latter part of July or the first of August in each year. At that time the distributor had determined the kind and character of product he would be able to supply the exhibitor for that season. In most instances, the films had not yet been produced and no one knew how successful they would prove as box-office attractions, and in the season contract the exhibitor would be accorded the privilege of rejecting or eliminating a certain percent of the product during the season.[63]

The essential features of the season contract were its timing and quantity: it was an advance contract for several pictures. The forward nature of the contract made it "blind" in that as yet unmade pictures were contracted in advance. The fact that it covered several pictures made it a "block" contract.

The typical studio-distributor's line-up was between 40 and 60 features a year, released at the rate of about 1 per week. In a season contract the exhibitor agreed to take all these features, subject to certain refusal rights. At the time of the contract there was little information that would distinguish superior from inferior motion pictures. None of them had been produced yet, and many would not be produced. The distributor was able to furnish the exhibitor only with a tentative list of titles and stars.[64] The season contract was made credible by the studio's stable of stars under contract and its reputation for producing and delivering motion pictures. Unless a season or production could be sold under advance contract, it would not be feasible for the studio to maintain its stable of stars and to finance and produce its pictures reliably. The distributor's downstream season contracts with exhibitors were essential to support the studio/distributor's upstream long-term contracts with its stars.

The season contract did not squeeze out other motion pictures, for no studio produced enough pictures to do so. No studio/distributor produced enough features even to fill all the available play dates of its own theaters, and they could not fill the play dates of the theaters with whom they block-contracted for all their production.[65] The typical studio-owned theater-rented pictures from other distributors in blocks, even if it booked the entire season of production of 1 or 2 distributors, and it might take a season of pictures from different distributors in different years.[66] Each studio-owned theater implicitly block booked its studio's entire production; the reason a studio-owned theaters was to ensure they would book all of its films season after season. Vertical integration was the ultimate form of block booking.

A significant number of season contracts were made for less than the studio's entire block of production. The percentage of Loew's contracts that were for blocks is described as "well over half," implying that block booking did not dominate licensing.[67] That block booking was a device to force unwanted features onto exhibitors seems implausible in light of the scarcity of the product and the dependence of studio-owned theaters on other studios for feature films.

In addition to facilitating production of motion pictures, season contracts also reduced distribution costs and reduced the exhibitor's booking costs. Frequent program changes made booking a complicated and costly process.[68] Booking motion pictures in blocks was a reasonable way to reduce booking costs. In considering the economic unit to be the individual motion picture, the court neglected the evidence that the economic units actually exchanged in the market were groups of pictures. That the economic unit was structured to reduce distribution and licensing costs is consistent with the historical evidence discussed in Section 7.2.[69] Following the end of block booking it was found that "[m]any exhibitors...found negotiating for each picture individually too time consuming and preferred to buy films in groups."[70]

Economists have argued that block booking was a device for price discrimination by tying a "B-grade" feature to an "A-grade" feature.[71] Their argument is based on the premise that the demand differs over markets for the films in a block. If the distributor does not know the relative demands, then selling films in a block can increase total revenue so long as they are priced so that each type of exhibitor "self-selects" the blocks. A less sophisticated variant on this theory is that better motion pictures were underpriced and poorer ones were overpriced.[72] But these theories do not fit the facts. First-run motion pictures were not licensed at fixed prices for theatrical play. They were licensed for a percentage of the gross box-office revenue. Through the percentage royalty the distributors directly received a share of the returns from successful films. The royalty, or price, automatically was higher for a successful film than for an unsuccessful one. There was no need to tie B features to capture the value of an A feature. The number of B features demanded is not correlated with box-office grosses on A features, so the B's could not measure the demand for the A features.

Blind selling refers to booking films before the buyer has seen them. The District Court said that blind selling did "not appear to be as inherently restrictive of competition as block booking, although it is capable of some abuse" when a distributor promises a picture of one quality and delivers something less.[73] The District Court's decree gave exhibitors the right to reject 20 percent of features licensed in a block without screenings within 10 days of being allowed to inspect them.[74] In this ruling, the court did not innovate, it substituted its determination of the number that could be rejected for the number which the distributor and exhibitor negotiated.

The court was correct in its view that blind selling was related to block booking because both were necessary to sell a season's production in advance. Advance commitments from exhibitors were crucial so the distributor could guarantee screens for productions they had financed for producers. Blind selling reduced inventory and selling costs, and accelerated royalty income because it shortened the time interval between a film's completion and its eventual exhibition.[75] Blind selling also reduced the number of prints, distribution facilities and staff that a distributor would require for exhibitor previews. Advance bookings for limited playdates in the key summer and holiday seasons could only be made if the films could be booked before they were finished, meaning they had to be sold blind. The court did not recognize that the distributor was also "blind" when the deal was made. The exhibitor's bid and the distributor's reservation price both reflected the fact that neither knew the product's demand. Contracting before seeing the final product is typical of creative productions such as books, works of art, theatrical productions and construction projects.[76]

An additional advantage of blind selling was that it generated more accurate demand information than screenings for exhibitors. The film was immediately put before the ultimate judge of its value—the audience— whose response provided information to the distributor and exhibitors. This explains the substantial resources devoted to the collection and dissemination of picture-by-picture attendance data. Trade screenings were "poorly attended by exhibitors" during 1940–45 when they were required by the District Court.[77] Block booking, repeat-dealing and other integrative mechanisms made screenings less important because they substituted the audience's judgment for the exhibitor's.

7.3.5 *Vertical integration and integrative contracting arrangements*

The studio-defendants employed four measures to build their circuits. They owned theaters outright, they leased theaters from independent owners, they owned a partial interest in some theaters and they franchised others. They owned an interest in 1,501 theaters that they operated jointly with the other owners who were independent exhibitors or other studio-defendants. Joint interests, the District Court said, "enable[d] the major defendants to operate theatres collectively, rather than competitively."[78] Another explanation for distributors to have a joint interest in theaters is that this would prevent monopolization of the distributors by the local exhibitor when there is only one theater in town. The joint interest form of operation was common only in small, one-theater towns, which supports this alternative explanation. Another form of joint operation was a pooling agreement that joined independent exhibitors. In these agreements, two or more independent theaters were operated as a chain managed by a member or a committee of members.

We know details about the operation of one pool of independent theaters from the Griffith case.[79] The pool consisted of four circuits of 200 theaters in New Mexico, Oklahoma and Texas, that were owned and operated by independents and by the Griffith brothers. The pool made two types of agreements. The first type were partnership agreements between the Griffiths and theater owners to operate and divide profits from a theater or theaters in certain towns. The second type was for administrative services: booking and buying pictures, bookkeeping and auditing, and pickup and delivery of reels. Administrative services were provided to the members in exchange for 4 percent of the gross annual receipts. The senior Griffith brother traveled to New York to negotiate directly with distributors. On these trips he bought a season of pictures for the pool managed by the two younger Griffiths.[80] Judge Vaught held that "these pooling agreements were brought about by efforts to economize and more efficiently operate, with no apparent desire to stifle competition, unreasonably restrain, or monopolize."[81] His decision was reversed by the Supreme Court.[82]

The District Court found in *Paramount* that pooling agreements reduced competition between theaters in admission prices and clearances, and were therefore "in clear conflict with the Sherman Act." It required the dissolution of existing pooling agreements and enjoined any future arrangements of that type.[83] Justice Douglas agreed:

> The practices were bald efforts to substitute monopoly for competition and to strengthen the hold of the exhibitor-defendants on the industry by alignment of competitors on their side. Clearer restraints of trade are difficult to imagine.[84]

The argument that pools were monopolizing devices is at odds with the fact that they received preferential treatment by distributors in the form of formula deals, master agreements and franchises. Pools reduced the cost of distributing motion pictures. Booking more than one theater and picture at a time reduced a series of negotiations to perhaps a single transaction. Pooling contracts let the smaller independent theaters form theater circuits to realize the economies of scale and scope that were available to the integrated circuits and the studio circuits. They extended the flexibility of integrated circuits to independent operators. They also substituted a local owner for a paid employee of a large circuit. Managers of theaters in circuits have some scope to shirk or skim box-office revenues, but a local owner is better able to monitor employees and avoid or reduce this problem.[85]

When the independent operators were prevented from entering into pooling contracts, they became less effective competitors to chains, and the way was paved for either their acquisition by chains or their eventual demise.

7.4   The courts' analyses

The District Court found the defendants guilty of restraints of trade but found "no monopoly on any phase of the cases."[86] The integrated defendants owned only 1/6 of the nation's theaters. Their theaters competed with each other or with an independent theater on every run in all but the smallest towns. Moreover, there [was] no substantial proof that any of the corporate defendants was organized or has been maintained for the purpose of achieving a national monopoly. . . . The five major defendants cannot be treated collectively so as to establish claims of general monopolization in exhibition. They can only be restrained from the unlawful practices in fixing minimum prices, obtaining unreasonable clearances, block booking and other things we have criticized.[87]

The court concluded that if the five distributor-defendants remained integrated "[t]he percentage of pictures on the market which any of [them] could play in its own theaters would be relatively small and in nowise approximates a monopoly of film exhibition."[88] Entry would be open, the District Court believed, if the mixed system of runs and clearances were replaced with competitive bidding by picture and theater according to its decree.[89] The District Court enjoined the defendants from expanding their circuits, but concluded that divestiture was unnecessary.

The Supreme Court held that this conclusion was wrong and remanded the case to the District Court to reconsider divestiture. The Supreme Court found three errors in the District Court's ruling. The first error was to require competitive bidding. The Supreme Court said that competitive bidding would make matters worse by strengthening the larger circuits and integrated distributors at the expense of the independents. The District Court's second error was in defining the product market too broadly. The Supreme Court redefined the relevant market as first-run theaters, where the position of the major defendants was strongest. Third, the Supreme Court said that the District Court had failed to consider the ability of integrated distributors to leverage small-town monopoly positions to large-city competitive markets. We discuss each of the points briefly.

7.4.1   *Licensing by competitive bidding*

The District Court wanted product to be allocated through competitive bidding and licensed one picture and one theater at a time. It failed to understand that the optimal deal encompassed more than one picture and more than one theater (discussed in Sections 7.3.4 and 7.3.5). The Supreme Court reversed the ruling requiring competitive bidding, but partly for the wrong reasons.

Justice Douglas expected that supervising a system of competitive bidding would involve the judiciary "in the administration of intricate and detailed rules" governing the "ingredients of each bid's priority, period of clearance, length (and dates) of run, competitive areas, reasonable return," film rental percentages, theater quality and so forth. He compared the problems of such supervision to "a continuous receivership" to which "[t]he judiciary is unsuited."[90] The experience of judicial regulation of entry under the decrees in the 1950s and early 1960s showed that his fears were well taken.[91]

Justice Douglas also argued that competitive bidding would favor bidders "with the longest purse—the exhibitor-defendants and the large circuits." They would "take the cream of the business, eliminate the smaller independents and thus increase their own strategic hold on the industry.[92] But that position misstates the incentives of bidders. No exhibitor would bid more than the expected value of a feature, and it would have the highest value to the exhibitor who could best promote it and gain the most patronage. No exhibitor knows which picture will turn out to be 'the cream of the business' at the time of the auction, nor is there any evidence that they have good prior expectations as to the winners or losers."[93] The "cream" is separated from the rest only after all the features have run.

The Supreme Court explained that the District Court considered competitive bidding as an alternative to divestiture in the sense that it concluded that further consideration of divestiture should not be had until competitive bidding had been tried and found wanting.[94]

Bidding was "the central arch" of the District Court's decree.[95] Eliminating it, as far as the District Court was concerned, made divestiture unavoidable.[96]

Bidding is not essential for competition, and the lack of bidding does not indicate monopoly. Antiques and other unique products where there is some element of monopoly are sold in auctions daily, and many highly competitive markets operate without bidding. One problem with the District Court's auction remedy is that bidding is a complex and costly mechanism for licensing pictures that increases the risks of an already uncertain business.[97] Bids are difficult to compare because different runs and clearances have different implications for demand and film rentals. Bidding by individual picture and theater also precludes repeat dealing and the realization of its beneficial effects in reducing risk and harmonizing incentives. A more important problem is that the District Court's requirement for features to be individually auctioned would have restricted the feasible licensing contracts in the industry to short term, spot contracts for single pictures. The integrated structure of the industry and its rich array of contracts were created to provide additional mechanisms beyond those available in the simple contracts that the court wanted to be used. It is the restriction in contracts implicit in licensing by bidding that is the major problem in the court's proposed remedy.

7.4.2 *Defendant control of first-run exhibition*

The Supreme Court criticized the District Court for finding that the defendants had not sought to achieve a national monopoly:

> [T]here is no finding as to the presence or absence of monopoly on the part of the five majors in the first-run field for the entire country, in the first-run field in the 92 largest cities of the country, or in the first-run field in separate localities. Yet the first-run field, which constitutes the cream of the exhibition business, is the core of the present cases [emphasis in original].[98]

On remand, the District Court applied this narrower market definition and found that if viewed collectively, the major defendants owned in 1945 at least 70 percent of the first-run theaters in the 92 largest cities, and the Supreme Court has noted that they owned 60 percent of the first-run theaters in cities with populations between 25,000 and 100,000. As distributors, they received approximately 73 percent of the domestic film rental from the films, except westerns, distributed in the 1943–44 season. These figures certainly indicate, when coupled with the strategic advantages of vertical integration, a power to exclude competition from these markets when desired.[99]

The share of theaters owned by the five fully integrated defendants was 17 percent when the product market was defined broadly as all theaters in the United States. Their share was about four times larger when the market was defined more narrowly by the Supreme Court as first-run theaters in large cities. It is dubious that majors had market power that related in any way to the Supreme Court's definition of market share. There were too many producers and exhibitors of motion pictures and there were too many substitutes for any single motion picture. It was erroneous to speak of the combined market shares of the five defendants as a collective entity, which surely they were not. Eight firms—and usually even five (counting only the fully-integrated studios)—is too many to treat as a monopoly absent evidence of conspiracy, which was never found. The defendants represented a small fraction of the 110 producers and 71 distributors of feature films in 1939.[100]

The argument that the integrated defendants "controlled" access to first-run theaters and used their "power" to prevent other producers from exhibiting films is central to the judicial theory underlying divestiture. The argument was trivial because it defined first-run theaters as those theaters that exhibited the defendants' pictures. The argument was flawed because it did not explain why other producers and distributors secured first-run exhibition for their pictures in theaters that were owned by the defendants.

The Supreme Court's market control argument was also inconsistent, probably because the Court misunderstood how competitive distribution worked. A studio-distributor who wished to monopolize first-run exhibition would book a feature into many theaters to squeeze out alternative features available to the public. The risk of this strategy was that the feature would play off so quickly that it would not fully capture demand.[101] The sequential pattern of runs in the *Paramount* era gave each exhibitor a limited engagement and left other screens open to show competing pictures. A feature faced contemporaneous competition from other features in a given run and intertemporal competition from itself between phases of its first and later runs. A release pattern that permitted more theaters to show a feature simultaneously would increase contemporaneous competition among exhibitors of the same feature but would reduce contemporaneous competition among different features.

The market control argument was also incomplete because it failed to explain why distributors and exhibitors had been integrated from the industry's earliest days. Vertical integration was neither necessary nor sufficient for monopolization. A monopoly producer of features could have raised film rentals directly without extending "forward" into exhibition. Similarly, a monopoly exhibitor could have charged monopoly admission prices and paid producer-distributors less-than-competitive royalties without merging "backward." A monopoly distributor could exact a monopoly distribution fee.

7.4.3 Extending leverage from monopoly towns

The courts acknowledged that the exhibitor-defendants did not have "a national monopoly," but emphasized that they had many local monopolies in smaller towns. In almost 92 percent of towns having populations under 100,000, some single one of the major defendants was the sole exhibitor.[102] Ownership in monopoly towns, Justice Douglas said, could be leveraged to other towns:

> A man with a monopoly of theaters in any one town commands the entrance for all films into that area. If he uses that strategic position to acquire exclusive privileges in a city where he has competitors, he is employing his monopoly power as a trade weapon against his competitors.[103]

The Supreme Court's theory of leverage failed to recognize that a theater owner did not know in advance which films to use leverage to acquire. And the exhibitor might have to lever against different distributors at different times depending on which film was being sought. Even successful use of leverage would not deny pictures to other exhibitors in towns where there were other exhibitors. If one distributor's picture was acquired through leverage, there were still seven or more other distributors left to supply pictures to the remaining exhibitors. The possibility of using leverage required the operator to own theaters in several towns and so the argument really was against the circuits.

The theory neglects the fact that one-theater towns had to compete with other one-, two-, or many-theater towns for a limited number of prints. Any reduction in film rentals that a monopoly theater obtained would have been disciplined by bids of theaters in other towns. Each theater had to bid at least as much as the print could have earned in another town. Features known to have strong demand offered higher returns to exhibitors but also had higher opportunity costs because of their higher returns in other towns. The theater-owning distributors relied on one another for films. If each of them used leverage to pay the others lower rentals this would not have been a sound foundation for a successful cartel.

It is more likely that individual or joint ownership of the only theater in town by a distributor or distributors was used to prevent the use of local monopoly against the distributors, since vertical integration into a down-stream monopoly by an upstream producer is a well known method of solving this problem.

The Supreme Court also failed to recognize that circuits and pools could not have tied up the pictures of every distributor. Pictures were available to independent, non-pool exhibitors from other distributors. Judge Vaught noted that no more than two major distributors had simultaneously granted franchises to the Griffith brothers' pool, leaving six other major distributors each year as sources of pictures to theaters outside the pool.[104]

7.5   *Paramount*'s effects

The *Paramount* decisions and decrees separated production and distribution from exhibition and prohibited long standing but controversial industry practices that conserved on information and exchange costs. The courts, however, accepted the contrary view that integration and licensing practices were arrangements for collusion and monopolization. What does the evidence show? A definitive test of these opposing views is not possible within the scope of this chapter. The decisions and decrees were fashioned in a period marked by a war and its aftermath, by substantial demographic changes and by the growth of television.[105] It is also difficult to separate the effects of *Paramount*'s changes in vertical integration from the effects of its changes in contracts. Within a stable environment, economic theory predicts different consequences from *Paramount*'s restructuring and altered contractual arrangements under the monopoly hypothesis than under the hypothesis that the structure and contractual arrangements promoted efficiency. All else being equal, if *Paramount* had dismantled a structure that monopolized the industry, then output should have expanded and prices (film rentals and admissions) should have fallen. Profits should fall. Under the information and exchange cost hypothesis, the changes should have increased costs and prices and reduced output. Profits should also decline.

According to our argument that the controversial contracting practices were designed to support a steady flow of production, the impact of *Paramount* on contracts is direct: once a distributor was prevented from contracting forward for the exhibition of motion pictures, it could no longer contract backward to maintain a stable of creative talent. The number of creative people under contract would decline and the studio system for organizing motion picture production would unravel. The monopoly hypothesis makes no prediction about the effect of *Paramount* on the contractual status of talent in the industry, but it does predict that employment should increase when production increases. *Paramount* was not a single event, but a whole series of events. Initially, *Paramount* altered contracts and then later forced the studios to sell their theaters. The first change in contracts was in 1946 when the District Court required pictures to be licensed individually (rather than in groups) and theater-by-theater (rather than a whole chain at once).[106] The Supreme Court decision in late 1948 was the landmark decision in the *Paramount* series.[107] The Supreme Court reaffirmed the District Court's opinion as to contracting, rejected bidding as a method for allocating films, and instructed the District Court to reconsider its earlier decision on vertical integration. The District Court followed these instructions and consent decrees eventually were signed by all the defendants in which they agreed to halt the objectionable licensing practices and also to sell their theaters.[108]

What effect did these changes have? Once again, we raise the concern that lots of things were going on in the late 1940s that could affect the movie industry. Even with this caveat in mind, it looks like prices rose and production declined after *Paramount*. In spite of television, the box-office take at movie theaters increased during the 1940s. Theater gross peaked in 1947 at $1.565 billion and exceeded $1.5 billion from 1946 to 1948. By 1950, theater gross was still some 30 percent above the 1940 level. Major studio production declined throughout the decade. Independent production increased slightly and total domestic production declined. Imports came back from a wartime low to reach their highest level in the decade by 1950.

The decline of major studio production from 1944 to 1950 does not fit with the growth in theater grosses or with the high gross per film that prevailed during the period. The decline in motion picture production was largely confined to the fully integrated studios; Columbia and United Artists, which were not fully integrated, held production at a constant rate over the decade while the studios that were forced to sell their theaters produced significantly fewer films (even though they did not all lose their theaters until as late as 1951).[109]

Paramount focused on clearances and runs. It changed the run designations of theaters by granting rights to all exhibitors to compete for the first-run features of the major distributors. Thousands of theaters were raised to first-run status and clearances were reduced in the number of theaters playing late runs at reduced prices. Real admission prices rose moderately after *Paramount*. Part of the rise in the average admission price came from more theaters showing first-run films and fewer theaters showing late-runs at reduced prices (but, of course, that still means prices went up). The size distribution of theater revenue by theaters was the same in 1958 as it was in 1948, so not enough theaters shifted from late-run houses to first-run houses to account for the increase in the average ticket price.[110]

The decline in production following *Paramount* was concentrated among the major producer-defendants who were the primary producers of first-run films. Hence, the percentage of inexpensive "filler" films increased relative to high-quality pictures, and that should have helped to lower admission prices. Competition from television probably shifted down the demand for motion pictures and that should have reduced the price of admissions and film rentals. The combined effect of opening competition to all theaters for first-runs and declining production by the studios increased the film rental price (rate).[111]

This rough evidence on prices and production is not consistent with the hypothesis that *Paramount* broke up a monopoly of the studios on motion picture exhibition.[112] It is consistent with the hypothesis that *Paramount* increased motion picture production and distribution costs.

The studio system of production using contract stars, directors and writers came to an end in the late 1940s and early 1950s. The numbers of actors and writers under contract to the studios declined. The number of directors and producers under contract declined only somewhat. But the two series for actors and writers make a sharp change in direction from increasing up to 1945 to decreasing after and continuing to decrease to the end of the 1950s. By 1950, the number of actors and directors under contract were about 1/3 the highest level they reached in the decade. The decline in creative talent under contract cannot be explained directly by the economic environment, in which theater grosses and gross per film were high and growing. The decline in contract personnel is greater than what could be explained by the decline in film production as well. The drop in writers under contract from 1944 to 1945 is particularly steep, comes back a bit in 1946 and then begins a longer term decline with a sharp drop in 1949. The decline in actors under contract begins in 1945 and then accelerates after 1949. The Department of Justice filed suit seeking theater divestiture in August of 1944 and the decision was dated June 11, 1946. In May, 1948, the Supreme Court decision was issued. The actors' series mirrors these dates with a one-year lag.

Because clearances and runs were ended and replaced with open competition for motion pictures, the result of *Paramount*'s film licensing requirements was to end tiered runs. These were replaced by simultaneous release in more theaters. The consequences were predictable: features played off faster and the variety of films playing at any given time available to the public declined. Earnings per screen in a first-run booking decline faster and generally are lower under a wide-release pattern, so more widely shown films have shorter runs. To counteract this, distributors included minimum run clauses in film licenses. Because a film had to sustain a wider release pattern, studios began to make more big-budget movies.[113] Television advertising became a means for building an audience for movies to fill the loss of the information provided by the sequential release on movies in tiered runs. Television also became the major outlet for the type of "filler" entertainment that made up the second features on the bill. This strengthened the switch to big-budget movies that was under way because of the loss of sequential runs.

Short-term exhibition contracts replaced franchises and trusted long-term relationships. Individual picture and theater licensing made collecting film rentals and getting honest box-office counts more difficult for distributors, because it became harder for them to withhold product from slow-paying or non-paying exhibitors.

Collection problems became more serious as exhibition declined in the 1950s and many exhibitors left the business.[114] To offset slow payment and non-payment of rentals, distributors began to require advances and rental "guarantees"—pre-play payments against which royalties accrued.

Increasingly, distributors relied on competitive bidding for immunity against exhibitor lawsuits, even though the Supreme Court had rejected this mechanism. The use of bidding to allocate motion pictures contributed to the increased use of minimum guaranteed rentals that could not be renegotiated after the film played.[115] In addition, auctioning motion picture licenses required their terms to be inflexible which increased exhibitor risk. Instead of making motion picture allocation and licensing more acceptable to exhibitors, *Paramount* seems to have escalated the long standing hostility between the distributors and exhibitors into open warfare. During the 1950s, 1/3 of the correspondence received by the Antitrust Division of the Department of Justice was from exhibitors complaining about distributor decisions concerning runs, clearances and criteria for selecting winning bids. Many private antitrust cases were brought by exhibitors over these issues.[116]

Independent theater operators were no longer permitted to use pooling agreements and were thereby deprived of economies available to circuits. They were amalgamated into circuits—a device for conserving on transactions costs that the decrees did not bar—and now most theaters belong to a chain.

The loss of moveovers and other flexible contractual arrangements contributed to the development of "multiscreen" theater complexes.[117] A multiscreen theater can move a film among its theaters without violating *Paramount*'s restriction against moveovers. They allow an operator to pool revenue over a portfolio of pictures, which is precisely what the outlawed franchise contracts and blanket deals did with their clauses that permitted overages and underages in runs to be credited against one another. Between 1981 and 1988, four of the seven major distributors purchased over 2,000 screens in the United States. This followed "years of concern" that reintegration would be blocked by the Department of Justice.[118] In 1986, distributors owned all or part of 4 of the 8 largest theater chains in the United States and Canada.[119] Most of these acquisitions were either by post-*Paramount* distributors who were not parties to the decrees or by minor *Paramount* defendants whose decrees did not prohibit them from entering the exhibition business. *Paramount* (now owned by Gulf + Western) purchased three circuits in the United States.[120]

In 1981, the Department of Justice began a review of all the motion picture decrees to determine whether they should be modified or vacated.[121] In 1986, Warner received an order from the District Court that permitted it to acquire theaters.[122] Warner subsequently purchased a 50 percent interest in a theater chain in the United States that had been solely owned and operated by *Paramount* and announced plans to franchise theaters as exclusive outlets in key markets to guarantee play times for its releases.[123] According to some news articles, the two remaining major defendants might seek releases from their decrees and the Department might not challenge further acquisitions.[124]

7.6 Conclusions

In his Griffith opinion, Judge Vaught said the motion picture industry's structure and institutions were developed to produce and deliver products whose value was unknown and could not be forecasted. He cautioned the government not to view as "sinister" and "suspicious" any practice it could not fully explain.[125] Ronald Coase has made the same point. He cautions that we should not view as monopolistic any practices or institutions that cannot be explained by the model of perfect competition.[126] Competition takes many forms and its supporting institutions and industry structure can be very different from those of any particular model. Competition in motion pictures—where so little is known and so much is at stake—looks very different from competition in industries where more is known about what customers want and how to produce it. One cannot hope to understand this industry through standard competitive reasoning or models. We have attempted to explain how the industry's structure and institutions came to be and why they were superior to those laid down by the courts. We have also argued that the courts had little real understanding of what constitutes competitive and monopolistic behavior in this complex business, and that their experiment in designing institutions for the business failed.[127]

Because so little was known about the industry when the cases were tried, few if any of the remedies had the intended effect. For this reason, *Paramount* is full of irony. It is ironic that competitive bidding became the preferred method by which distributors license motion pictures, especially after the Supreme Court made the studios sell their theaters because it distrusted bidding. It is ironic that the attempt to improve the independent exhibitor's access to first-run films fostered an explosion in exhibitor-initiated litigation against the distributors. It is ironic that the less adaptable contracts that *Paramount* required hastened the growth of theater chains and the demise of the independent exhibitor, and made owning theaters more attractive to distributors than ever before. It is ironic that remedies that were intended to stop price fixing caused prices to increase. The final irony is that the remedies were implemented in a way that reduced competition.[128] We believe the *Paramount* decrees should be vacated in their entirety.

8 Was the antitrust action that broke up the movie studios good for the movies?

Evidence from the stock market

8.1 Introduction

The structure of the motion picture industry and its practices were dramatically altered some 50 years ago in a series of antitrust cases known collectively as the *Paramount* litigation. *Paramount* forced the five major vertically integrated studios—MGM (Loews), RKO, Twentieth Century Fox, Warner Brothers and Paramount—to sell their theaters, and changed the contracts and methods by which these studios and the three other major non-integrated studios—United Artists, Columbia, and Universal—could license motion pictures to theaters. Many of the restrictions laid down during the 1940s on the industry's colorful and often hard-to-understand contracts and practices such as "block booking," "blind selling," "track booking," "clearances," "four walling," "pooling agreements," franchising and owning theaters continue to apply in varying degrees even in today's vastly different motion picture market.

The District Court acknowledged that it was unfamiliar with the motion picture business, but that did not dissuade it from a far-reaching experiment in remodeling this wondrous but poorly understood new medium and its exotic practices:

> It is argued that the steps we have proposed would involve an interference with commercial practices that are generally acceptable and a hazardous attempt on the part of judges—unfamiliar with the details of business—to remodel its delicate adjustments which have hitherto provided the public with what is a new and great art. But we see nothing ruinous in the remedies proposed.[1]

The extent of the changes that the courts made to the industry are difficult to fathom: they forced the dissolution of all the major integrated studios from their theaters. The array of practices employed by the major studios are summarized in Table 8.1. Paramount, Inc. lost some 1,395 theaters; Fox divested 636 theaters; Warner Brothers 501; MGM (Loews) lost 135, and RKO 109. Paramount also lost 993 theaters that it owned jointly with other

Table 8.1 Vertical structures and contracts used by motion picture distributors

Distributor	Producer	Theaters	Joint interests	Block booking	Franchises	Master licenses
Paramount	Yes	1,395	993	Yes	Yes	Yes
MGM Loews	Yes	135	21	Yes	Some	Yes
Twentieth Century Fox	Yes	636	None	Yes	Yes	Yes
Warner Bros	Yes	501	20	Yes	Yes	Yes
RKO	Yes	109	187	Yes	Yes	Yes
Columbia	Yes	None	None	Yes	Yes	Yes
Universal	Yes	None	None	Yes	Yes	Yes
Republic	Yes	None	None	No	No	No
Monogram	Yes	None	None	No	No	No
Pathe (PRC)	Yes	None	None	No	No	No
United Artists	No	None	A few	No	No	No

operators; RKO lost 187 joint theaters. MGM and Fox lost about 20 jointly owned theaters each. In addition, all their franchises with independent theater chains were dissolved along with master licenses that governed the general terms of contracts. Forms of what we would now call forward contracting were also prohibited. Season contracts, where a theater would contract (forward) for the entire season of production from a studio (one of the oldest practices in the business) were prohibited. Restrictions against "blind selling" prevented forward contracting for a movie before it was produced (a theater had to be given the opportunity to see a version of the movie before it could book it). Block booking, which was simply the licensing of several films in a single agreement was prohibited.

In place of this rich array of vertical controls and licenses, the court called for a single license for each individual film, theater and play date; in short, spot contracts were determined to be the only permissible method for licensing motion pictures. The District Court also considered competitive bidding to be a superior way to allocate movies to theaters and reasoned that the studios had to be separated from their theaters to free up a supply of pictures for theaters to bid on. It seems to have stopped short of treating movies like commodities only when the Supreme Court concluded that motion picture exhibition licenses were too complicated to be allocated by bidding. A series of consent decrees and decisions flowed out of the long *Paramount* litigation and they gradually transformed the industry. These decrees and decisions are analyzed here.

The objective of this chapter is to use stock market data to assess the impact of these events. Which of the decisions in the sequence was the most important? Did the Court's self-acknowledged experiment in remodeling this emerging medium harm the motion picture industry? Or did the Court succeed in its intent to halt what it perceived to be a horizontal collusion to fix prices? Does the evidence tell us that the court interfered with practices that were efficiently adapted to the peculiar needs of this unusual business of motion pictures?

In this chapter, we use stock market evidence to evaluate the impact of the major events in the *Paramount* litigation. Stock prices reflect the market's assessment of these events and are useful for measuring their importance (Bhagat and Romano, 2002). We find that the market concluded that most of the events in the sequence of decisions and decrees were unimportant—they had no impact that can be detected in stock prices. One decision stands out: the Supreme Court stunned the studios—and the stock market—by forcing the studios to divest themselves of their theaters, something that had been ruled out by the District Court. The major integrated studios lost from 5 to 12 percent of their value when the decision came down. The distribution of the impact of the Supreme Court decision on defendant and non-defendant firms, however, does not support the view that the courts dismantled a collusion to monopolize the market for first-run films.

8.2   The studio system

During the studio era, a studio's pictures were filmed on its stages, financed by its own sources, distributed by its own distribution arm and played its first run in the studio's theaters. Some studios like MGM, RKO, Warner Brothers and Twentieth Century Fox spanned the whole chain of transactions from raw material to theater, while others like United Artists, Universal or Paramount Pictures, from whom the litigation takes its name, spanned smaller parts of it. United Artists owned no theaters; the artists owned the company which distributed films produced by them and other artists. Warner Brothers owned theaters (which they acquired to equip them for "talkies" which Warner Brothers introduced), and employed artists under long-term contracts. Universal owned no theaters, but it relied on franchising to secure theaters for its productions.

The studio's vertical controls over theatrical outlets for first-run films were broad; they included ownership of many first-run theaters. Equally important were the vertical controls exercised by the studios over theaters through their exotic and controversial licensing practices. Block booking, customary price clauses in theatrical licenses, broad clearances of competitive exhibitions in other theaters, booking films over long-established tracks of theaters, franchises and other forms of contracting were considered by the Department of Justice to be vertical controls whose intent was to fix prices and deny first-run films to competing, independent theaters. Justice argued that the studios used all these vertical controls to form a horizontal price fixing cartel.[2] Contrary arguments that the studio system, with integrated theaters, was an effective form of organization for the industry were made by the defendants and a District Court judge.[3]

8.3   The *Paramount* litigation

In 1938, the Department of Justice brought suit against all the major motion picture companies to begin a long series of cases lasting more than 10 years.

This series of cases is collectively referred to as the *Paramount* Litigation or just *Paramount*. The cases and outcomes of *Paramount* are listed and summarized in Table 8.2 (case summaries and references are in the Appendix).

The four major cases we examine challenged studio film licensing practices and theater ownership. In the first decision, reached in June 1946,

Table 8.2 Chronology of events by year in the *United States* v. *Paramount* litigation

1938	Eight major studios charged with restraint of trade and attempted monopoly.
1940	Consent Decree signed only by Integrated Five limiting 1 block booking, 2 blind licensing and 3 theater acquisitions.
June 1946	District Court decision and consent decree. Declared minimum price agreements and fixed runs and clearances illegal. Found no monopolization or attempt to monopolize exhibition. Found illegal joint theater ownership and pools, block booking, and franchises. Refused to force divestment of theaters. Required competitive bidding as remedy.
Dec. 1946	District Court decision and consent decree. Found no monopolization or attempt to monopolize production and distribution.
1948	Supreme Court found all the practices a violation of Sherman Act. Rejected competitive bidding as a remedy. Remanded to District Court for reconsideration of theater divestment.
1948	RKO signs decree agreeing to divest theaters.
1949	Paramount signs decree agreeing to divest theaters.
1949	District Court reconsiders divestment on remand from Supreme Court. Vertical integration found illegal within a conspiracy to fix prices. Loew's, Twentieth Century Fox, Warner Brothers must agree to divest theaters. All eight majors prohibited from 1 fixing admission prices, 2 agreeing on clearances, 3 granting clearances between theaters not in competition, 4 granting unreasonably long clearances, 5 entering franchise agreements with exceptions, 6 performing formula deals or master agreements, 7 conditioned licensing (block booking), 8 licensing except theater by theater solely on the merits.
1950	Columbia, United Artists, Universal sign Consent Decree.
1951	Warner Brothers, Twentieth Century Fox sign Consent Decree.
1952	Loew's signs Consent Decree.

Sources: Cases cited in Tables, and Federal Trade Commission, *Paramount Pictures Inc. et al. Consent Judgments and Decrees Investigation.*

the Department of Justice failed to convince the court that the studios had monopolized exhibition, but the court found fault with film licensing. In the second decision, reached in December 1946, Justice sought to prove the studios had monopolized motion picture production and distribution. The District Court rejected this contention, but it continued to find fault with film licensing. It proposed competitive bidding as a remedy. The court refused to order the integrated studios to sell their theaters "at least until the efficiency of that system has been tried and found wanting."

In the third decision, reached in May 1948, the Supreme Court upheld the District Court's 1946 decision as to licensing, rejected its bidding remedy, and instructed the District Court to reconsider theater divestment. The Supreme Court rejected the Department of Justice's argument that vertical integration of production, distribution and exhibition was illegal per se, but it concluded that a market for exhibition rights that was unrestrained by vertical controls required more than competitive bidding. The Supreme Court objected to competitive bidding for motion pictures on three grounds: (1) films had to be freed from vertical controls to create an adequate supply of films for competitive bidding, (2) the bids would be so complex that an auction would create controversy and invite manipulation that would involve the courts too deeply in supervising the industry; and (3) the studio's own theaters would continue to get the pictures they wanted by outbidding competitors.

In the fourth decision, reached in July 1949, the District Court followed its instructions from the Supreme Court and ruled that divestment was necessary because its bidding remedy had been denied. It found that integration gave the power to exclude competition in the rights to exhibit motion pictures and had been used to do so. Integration of distribution and exhibition became illegal within what the court saw as a conspiracy to fix prices and geographically divide the market for the exhibition of motion pictures. The actual divestment process took several years, beginning before the 1949 District Court trial and ending in 1952.

The courts inferred a geographical division of the market from film booking patterns. Films played on a "track" of theaters where they opened their first run at one or a few centrally located theaters (often owned by the studio), moved into neighborhood theaters and ended their second and later runs in low-priced theaters. Explicit or implicit contracts were made between the theaters on the track and the distributor so that each theater had a customary place in the sequence of runs and showed the same distributor's films year after year. For example, if a particular theater was franchised to Warner Brothers it would receive the entire line of Warner Brothers films to exhibit as exclusive first-run engagements in its area. Competing runs of the same films would be "cleared" from this theater's protected area. Because they were the product of long-term arrangements, runs and clearances were stable over time. The court ordered that fixed runs and clearances were illegal because it saw them as a geographical division of the market.[4]

8.4 Stock prices

Markets use information to value assets. According to the efficient market hypothesis, the price of an asset incorporates all the known information. Once this information is embedded in a price, only the arrival of new information causes the price to change. There are three versions of the efficient market hypothesis which posit the information that is contained in a price. In the weak form, price incorporates all the information contained in the history of preceding price and volume data. In the semi-strong form, price incorporates the weak form information and all publically available information. In the strong form, price incorporates all the semi-strong form information as well as all relevant private information. We are agnostic as to which of these forms of informational efficiency the market might exhibit in evaluating the information revealed in the public announcements made during the litigation. This allows for the possibility that investors may have anticipated the decisions before they were announced publically or that they may not have absorbed the full impact of the decisions until days later. By looking at price behavior in the pre- and post-decision periods, as well as in the decision period, we can gain insight into the efficiency with which the market absorbed the information.

We place a two day window at the day of the event; this covers the day of the decision and the day after. We use price information from a 70 day period preceding each event, the 10 days of the event window, and 10 days following the event. Following capital asset pricing theory, we express the return in each of the studio stocks as a deviation from the risk-adjusted market return.

For each studio and each event we estimated

$$R_{jt} = a_j + b_j R_{mt} + c_{j1} \, TWODAY_t + c_{j2} \, PRE_t + c_{j3} \, POST_t + d_j R_{jt-1} + e_{jt}$$

over an 82 day sample period beginning 70 days before the event, including the two-day event period and ending 10 days after the event, hence $t = -70, \dots, +12$.

The dependent variable $R_{jt} = 100 Ln[(P_{jt} + Div_{jt})/P_{it-1}]$ is the return of firm j on date t calculated as the log of the ratio of i's share price plus dividend on day t to its share price on day $t - 1$. The market return variable R_{mt} is the return on the Dow Jones Industrial Composite Index calculated in the same way. Each event has a pre-event window of 10 days, and a post-event window of 10 days. Each of these time windows is represented by a dummy variable: $TWODAY_t = 1$ for $t = 0, 1$; $PRE_t = 1$ for $t = -10, \dots, -1$; and $POST_t = 1$ for $t = +2, \dots, 11$. The estimates of the coefficients C_{j1} measure the average daily effect of the event on security returns.[5] The estimates of the coefficient C_{j2} of the variable PRE_t measure the extent to which the event was anticipated prior during the 10 days preceding its announcement. The estimates of the coefficient C_{j3} of the variable $POST_t$ measure the amount of adjustment that occurs in the 10 days following the event. Estimates of the coefficient d_j of the variable R_{jt-l} measure the effect of returns during the 70 days preceding the events on returns in the event window. A value different from zero of this coefficient indicates some departure from the efficient market hypothesis.

Research shows that correlated industry-specific effects can bias OLS estimates of abnormal returns. In the decade from 1940 to 1950, when the events happened, the industry experienced a wartime boom that was shared by all the studios. We, therefore, used a generalized least-squares estimation procedure to correct for contemporaneous correlation in the error terms between firms (Thompson 1985; Smith *et al.* 1986; Binder 1985).

Of the eight defendant studios, we have stock price information for: Columbia, Loews (MGM), RKO, Twentieth Century Fox, Universal and Warner Brothers. The other two defendants, Paramount and United Artists, were privately owned companies and no stock prices were available for them. We also collected data on one non-defendant firm Republic Studios. In evaluating the evidence, it is important to recall that only Loews (MGM), Fox, Warner Brothers and RKO among the defendant studios owned theaters. Columbia and Universal did not own theaters. Republic owned no theaters. It produced feature motion pictures, but many of its films were second features on a double bill with a major feature film.

The estimates in Table 8.3 show that the June 1946 decision against block-booking and franchises had no effect on the studios with the exception of a negative effect on RKO. Columbia appears to have gotten a small positive excess return from the decision, though its statistical significance is not high. The market Betas (β_j) for most the studios exceed a value of one, indicating that studios are more risky than the market as a whole. The *Pre* and *Post* coefficients are not significant for any studio. Republic, like the other studios, was largely unaffected by the decision. The efficient market hypothesis test on the value of $d_j = 0$ is not rejected.

The estimates for the decision of December 1946 are reported in Table 8.4. These estimates confirm that the movies are a more risky business than the market portfolio, with all the studios having a market beta over 1.5. Loews experienced a positive two-day excess return in this decision. Columbia had a large positive return, but it lacks statistical significance. The other defendants had normal returns on this event. The market may have seen this decision, which rejected the claim that the studios had monopolized production and distribution, as a confirmation of the legality of MGM's highly developed studio system. The other firm to gain in this decision was Republic, a non-defendant firm who earned a positive excess return of 2 percent. Republic's large gain is not consistent with the position that, as an independent producer, it was harmed by the structure and practices of the integrated defendants.

Estimates for the 1948 Supreme Court decision are reported in Table 8.5. The size and statistical significance of the abnormal returns clearly show that the Supreme Court decision of May 3, 1948 was a landmark event. Columbia, RKO, Universal and Warner Brothers had large negative returns relative to the market as a whole and the estimates are significant statistically. Loews and Fox also had negative excess returns of large magnitude, but these are not as

Table 8.3 Market model regression coefficients for defendants and one non-defendant for the District Court decision *United States* v. *Paramount*: decision dated June 11, 1946

$$R_{jt} = a_j + b_j R_u + c_{j1} TWODAY_t + c_{j2} PRE_t + c_{j3} POST_t + d_j R_{jt-1} + e_j$$

Firm	a_j	b_j	c_{j1}	c_{j2}	c_{j3}	d_j	R^2	Durbin Watson
Columbia[a]	0.31	1.53	0.77	−1.55	−0.70	−0.14	0.22	1.89
	(1.05)	(4.24)	(0.48)	(2.03)	(0.91)	(1.50)		
Loews	−0.02	0.92	0.33	−0.59	−0.45	−0.05	0.31	2.22
	(0.13)	(5.37)	(0.42)	(1.62)	(1.20)	(0.63)		
RKO	0.39	1.20	−1.00	−0.73	−1.38	−0.18	0.16	1.73
	(1.25)	(3.14)	(0.59)	(0.91)	(1.68)	(2.11)		
Fox	0.21	0.92	−1.75	−1.18	−0.63	−0.21	0.32	2.08
	(1.27)	(4.61)	(1.85)	(2.75)	(1.46)	(2.27)		
Universal[a]	0.05	0.84	0.05	−0.34	−0.43	−0.06	0.13	2.02
	(0.25)	(3.19)	(0.05)	(0.62)	(0.76)	(0.62)		
Warner	0.29	0.79	0.13	−0.79	−0.67	0.20	0.17	1.86
	(1.14)	(2.52)	(0.10)	(1.19)	(0.99)	(2.12)		
Republic[b]	−0.10	1.51	−0.27	−1.05	−0.71	−0.23	0.31	1.82
	(0.45)	(5.28)	(0.21)	(1.76)	(1.16)	(2.85)		

Notes
a Did not own theaters.
b Non-defendant who did not own theaters.
Standard errors in parentheses.

Table 8.4 Market model regression coefficients for defendants and one non-defendant for the District Court decision *United States* v. *Paramount*: decision dated December 31, 1946

$$R_{jt} = a_j + b_j R_u + c_{j1} TWODAY_t + c_{j2} PRE_t + c_{j3} POST_t + d_j R_{jt-1} + e_j$$

Firm	a_j	b_j	c_{j1}	c_{j2}	c_{j3}	d_j	R^2	Durbin Watson
Columbia[a]	−0.36	1.47	2.08	−0.65	−0.43	−0.04	0.41	2.42
	(1.45)	(7.27)	(1.52)	(0.99)	(0.65)	(0.49)		
Loews	−0.15	1.19	0.86	−0.28	−0.55	−0.44	0.31	2.34
	(0.28)	(2.71)	(0.29)	(0.19)	(0.39)	(5.02)		
RKO	−0.26	1.70	−0.54	−0.09	−0.58	−0.15	0.67	2.16
	(1.60)	(12.77)	(0.60)	(0.20)	(1.32)	(2.49)		
Fox	−0.26	1.73	0.62	−0.21	−1.21	−0.24	0.63	1.81
	(1.46)	(11.59)	(0.62)	(0.44)	(2.50)	(3.37)		
Universal[a]	−0.41	1.68	−0.13	0.15	−1.15	−0.25	0.52	2.29
	(1.79)	(8.87)	(0.11)	(0.25)	(1.87)	(3.38)		
Warner	−0.22	1.53	0.01	0.27	−0.86	−0.20	0.58	1.87
	(1.24)	(10.36)	(0.01)	(0.57)	(1.78)	(2.85)		
Republic[b]	−0.39	2.30	2.13	−0.22	−0.12	−0.11	0.51	2.37
	(1.23)	(9.01)	(1.23)	(0.26)	(0.15)	(1.47)		

Notes
a Did not own theaters.
b Non-defendant who did not own theaters.
Standard errors in parentheses.

Table 8.5 Market model regression coefficients for defendants and one non-defendant for the Supreme Court decision *United States* v. *Paramount*: decision dated May 3, 1948

$$R_{jt} = a_j + b_j R_u + c_{j1} TWODAY_t + c_{j2} PRE_t + c_{j3} POST_t + d_j R_{jt-1} + e_j$$

Firm	a_j	b_j	c_{j1}	c_{j2}	c_{j3}	d_j	R^2	Durbin Watson
Columbia[a]	0.13	1.53	−3.67	−0.75	0.94	−0.17	0.16	2.03
	(0.37)	(2.99)	(1.86)	(0.80)	(0.98)	(2.06)		
Loews	0.04	1.42	−2.89	−0.08	0.29	−0.04	0.25	1.82
	(0.20)	(4.58)	(2.45)	(0.14)	(0.52)	(0.45)		
RKO	0.08	1.06	−2.16	0.09	1.49	−0.21	0.10	1.88
	(0.24)	(2.35)	(1.24)	(0.11)	(1.78)	(2.87)		
Fox	0.10	1.84	−4.56	0.23	0.001	−0.08	0.25	1.91
	(0.36)	(4.66)	(3.02)	(0.31)	(0.001)	(1.10)		
Universal[a]	−0.04	1.55	−0.39	−0.11	0.08	−0.15	0.08	1.76
	(0.11)	(3.04)	(0.20)	(0.11)	(0.08)	(1.72)		
Warner	0.13	1.73	−5.79	−0.02	0.78	−0.31	0.24	1.93
	(0.42)	(3.97)	(3.48)	(0.02)	(0.97)	(4.40)		
Republic[b]	−0.04	1.60	−3.81	0.18	0.56	−0.32	0.12	1.99
	(0.09)	(2.55)	(1.57)	(0.15)	(0.48)	(3.81)		

Notes
a Did not own theaters.
b Non-defendant who did not own theaters.
Standard errors in parentheses.

statistically significant. The non-defendant Republic also had a large negative and significant excess return. The zero pre- and post-event return coefficients as well as the efficient market hypothesis are not rejected by the estimates. This event was not foreseen and the market did not revise its evaluation in the days following the event; it had a profound and immediate impact on the industry.

The largest loss of value was to Warner Brothers, followed by Universal, Fox, the non-defendant Republic, Columbia, Loews and RKO. Because it upheld the licensing limits of the 1946 decree, the Supreme Court decision would be expected to affect not only studios that owned theaters but also studios who franchised and studios who used any of the offending contracting practices. While Universal had been heavily involved in franchising theaters, it had stopped granting franchises after the 1946 decision. Fox and Loews also franchised to a lesser extent, and their loss on this event reflects the combined loss of their theaters and their franchises.

Republic was not a defendant and did not use the offending practices, yet its stock value also declined. There was no evidence or claim advanced during the litigation that Republic was a participant in the alleged price-fixing cartel and the general view was that Republic was a competitor rather than a conspirator with the major studios. Had the defendants succeeded in using the practices found to be illegal to exclude or restrict Republic from the market, then one would expect its stock value to rise once the offending practices were found to be illegal and Republic gained new access to theaters. That its stock value also fell is some evidence that the court ruling had an industry-wide affect unrelated to those implied by the demise of "cartel" arrangements. This is supported also in the December 1946 decision where Republic gained value, like the defendants, when the court refused to find that the studios had attempted or succeeded in monopolizing either production or distribution of motion pictures. As a competing producer and distributor, Republic should have gained when the defendants lost and lost when they gained. Instead, it shared their fate in the 1946 and 1948 decisions.

The 1949 event was anticlimactic, as shown by the estimates in Table 8.6. Following the Supreme Court decision and prior to the 1949 District Court trial, RKO and Paramount agreed to divest their theaters under separate Consent Decrees. Loews, Fox and Warner Brothers were the only defendants who still had theaters that would be divested under conditions to be determined by the 1949 decision and decrees. Columbia, not owning theaters, had a slightly positive excess return. The market seems to have liked Columbia's

Table 8.6 Market model regression coefficients for defendants and one non-defendant for the District Court decision *United States* v. *Paramount*: decision dated July 5, 1949

$$R_{jt} = a_j + b_j R_u + c_{j1} TWODAY_t + c_{j2} PRE_t + c_{j3} POST_t + d_j R_{jt-1} + e_j$$

Firm	a_j	b_j	c_{j1}	c_{j2}	c_{j3}	d_j	R^2	Durbin Watson
Columbia[a]	0.31	2.39	0.41	−0.67	−0.73	−0.16	0.25	2.11
	(0.97)	(5.06)	(0.23)	(0.80)	(0.87)	(1.83)		
Loews	0.15	1.14	0.25	−0.27	−0.63	−0.14	0.21	2.28
	(0.85)	(4.38)	(0.25)	(0.58)	(1.37)	(1.61)		
RKO	−0.10	1.05	−0.68	0.57	1.05	−0.30	0.17	2.04
	(0.46)	(3.10)	(0.54)	(0.94)	(1.70)	(3.12)		
Fox	0.21	1.79	−1.40	−0.59	−0.99	−0.08	0.39	2.08
	(1.24)	(7.24)	(1.51)	(1.34)	(2.65)	(1.12)		
Universal[a]	0.20	2.82	−1.68	−0.99	−0.75	−0.21	0.30	2.02
	(0.62)	(5.80)	(0.92)	(1.14)	(0.87)	(2.41)		
Warner	0.11	1.16	−2.77	−0.24	−0.47	−0.34	0.17	2.17
	(0.41)	(2.96)	(1.89)	(0.34)	(0.67)	(3.83)		
Republic[b]	0.12	1.91	−1.12	−0.93	0.75	0.02	0.11	2.17
	(0.26)	(2.96)	(0.46)	(0.80)	(0.65)	(0.23)		

Notes
a Did not own theaters.
b Non-defendant who did not own theaters.
Standard errors in parentheses.

prospects once its integrated competitors were divorced from their theaters. But, by 1949, the market had already discounted the previous divestment decision. None of the theater-owning defendants lost value; Loews stock had a zero abnormal return and the abnormal loss to Fox and Warner Brothers are small and not significant. Only Republic experienced a small loss of market value (1 percent). There is no rejection of the efficient market hypothesis and no evidence that the events were anticipated prior to or adjusted to slowly following the announcement of the decision.

The major event of the long *Paramount* antitrust litigation was clearly the 1948 Supreme Court decision. The total loss in value of each studio at the Supreme Court decision is shown in Table 8.7. Warner Brothers led the studios with an abnormal loss of shareholder value of over $10 million, followed by Loews with a loss of value just over $5 million, and then by Fox with a loss slightly exceeding $4 million (these estimates are all in 1948 dollars). Of these, Fox owned the most theaters, followed by Warner Brothers and then Loews. The losses are heaviest to the theater-owning studios which implies that the market did not think they would be able to divest the theaters at a value equal to their value to the integrated studio. The magnitude of these losses is from 4 to 12 percent of firm value. The implied premium for ownership of first-run theaters is about 4 to 12 percent of the total value of the combined studio and theater assets.

8.5 The feasibility of a motion picture cartel

The structural conditions in the industry were not particularly favorable for a cartel during the 1930s and 1940s. There was no dominant firm in

Table 8.7 Two-day abnormal returns and abnormal equity value change on the 1948 Supreme Court decision

Firm	Two-day return[a]	Equity value change[b]
	(Percent)	(1948 Dollars)
Columbia[c]	−7.34	−$2,087,312
Loews	−5.78	−$5,610,464
RKO	−4.32	−$1,140,920
Fox	−7.62	−$3,815,953
Universal[c]	−9.12	−$1,440,193
Warner Brothers	−11.58	−$10,607,509
Republic[d]	−7.62	−$536,768

Notes
a The two-day abnormal return is two times the average daily abnormal return as estimated in the system regressions. The period is the event day and the day following.
b The cumulative abnormal equity change is the cumulative abnormal two-day return times the value of equity on the day preceding the event.
c Columbia and Universal did not own theaters, but were defendants.
d Republic was not a defendant and did not own theaters.

either production, distribution or exhibition. Concentration in motion picture production was low: no studio had a share of film production greater than 10 percent, and the production share of the 8 major companies averaged 54 percent of all motion picture releases (including domestic production and imports) over the period from 1940 to 1950. The courts found that entry into production was open. Concentration in film rental shares was not high and there was great variation in studio rental shares from year to year. The data show that in the 1943–44 season, the five major *Paramount* defendants accounted for 73.3 percent of domestic film rentals, and in 1948 seven of the *Paramount* defendants accounted for 76 percent of all domestic film rentals.[6] Estimates of the variation in studio revenue shares show that the variation is large: the range of each studio's share is from 30 to 100 percent of its mean share. Market shares are highly volatile in the motion picture business since they depend on the success of the films released by each firm.[7]

Concentration in exhibition was low. The five integrated companies owned, had a share interest in, or an operating agreement with just over 17 percent of theaters nationwide. *Paramount*, the largest theater owner, had 1,395 theaters, or 7.72 percent of the national total. The next largest theater owner, Twentieth Century Fox, had an interest in 636 theaters, or 3.52 percent of the national total. Exhibition was characterized by low concentration and easy entry. Similarly, distribution had low concentration and low barriers to entry. That exhibition was open to entry was never challenged; the large increase in the number of theaters during the late 1930s and 1940s belies any such claim. Similarly, distribution was un-concentrated. There were 11 national distributors: the 5 major theater-owning studios, the 3 other major studios, and 3 nonmajor distributors. There also were many other film distributors at the national, regional and state level. The 8 major companies distributed films for themselves and independent producers and their distribution market shares were essentially the same as the production shares, which is to say not concentrated.

Given the structure of production, distribution and exhibition, if the industry had been cartelized by the eight defendants, then some form of explicit or implicit collusion would have been necessary. However, no evidence of explicit collusion was ever developed in the proceedings. The most tenable hypothesis under which the courts could have found the defendants guilty of an attempt to collude is that they implicitly agreed to restrict motion picture output to increase film rentals and admission prices. We know that a successful cartel must restrict output, or (what is the same thing) raise price and absorb the reduction in the amount demanded, agree on a division of output or revenue, discipline cheating by member firms, and close entry. But, in this industry it is difficult to detect cheating; a studio could gain a large increase in market share because it had a "hit" and not because it lowered its price.[8] Because audience response was unpredictable, there was no assurance that when a studio increased production it would affect the film rentals of the other companies.

8.6 Winners and losers

The intended beneficiary of *Paramount* was the independent exhibitor. The independent exhibitors issued a heavy volume of complaints to the Department of Justice regarding the defendants and their actions.[9] Independent exhibitors were highly organized and resolutions adopted at their national meetings were directly communicated to the Department of Justice. Not only is there evidence that the Department of Justice heard from the independent exhibitors, but there is evidence that the Department listened to them. Most Department-instigated changes favored the independent exhibitor over the theater chains and the studios.

Forbidding licensing arrangements that favored theater chains was intended to improve the independent exhibitor's ability to compete with the chains. Justice's insistence that the major studios should sell their theaters and be barred from reentering exhibition initially benefited independent exhibitors by removing the studio theaters as a major source of competition to them. After the Decrees, the Department of Justice acting through the courts regulated entry into exhibition in a manner that protected existing theater operators and probably harmed the public.[10] Banning block booking was primarily intended to benefit the independent exhibitors who were its primary opponents.[11] In sum, the Department of Justice pursued a policy more favorable to the independent exhibitor than to any other interest group.[12]

Paramount's remedies did not have their intended effect, however, and they eventually harmed the independent exhibitor. They promoted wider competition for film exhibition rights and reduced production, which raised motion picture rental rates and admission prices. The higher film rental rates lowered exhibitor net profits. Distributors stiffened the terms of exhibition licenses as the less flexible licenses and licensing methods called for by *Paramount* increased their vulnerability to the decline of exhibition caused by the increased competition from television.[13] The broader release patterns called for by *Paramount* and increasing competition from television induced the industry to produce fewer films.

Perhaps the definitive test of whether *Paramount* improved the lot of the independent exhibitor is to look at their own actions. Just a few years after the studios sold their theaters, exhibitors, fearing television, called on the Department of Justice to let the studios reenter exhibition so that they might have stronger incentives to produce motion pictures.[14] Exhibitors contended that competitive bidding did not solve any problems and increased film rentals (FTC 1997, p. 33). The methods for allocating motion pictures that were devised in the *Paramount* decrees fostered an explosion of exhibitor antitrust cases against the distributors (Cassady and Cassady 1974). Since it is unlikely, though possible, that it became more profitable to sue distributors than to show their movies, we consider the high volume of exhibitor law suits to be an indicator that *Paramount* was a failure even in the eyes of those it was intended to help.

8.7 Conclusions

The integrated studios—Loews, RKO, Fox, Warner Brothers—lost market value in the range of 4–1r percent when the Supreme Court handed down its 1948 decision. But, Columbia and Universal, non-theater owning defendants, also lost market value on the order of 7–9 percent. This suggests the decision was more far-reaching than merely banning producer/distributors from owning theaters; it also barred franchising and other film licensing practices. These licensing practices would have been more valuable to the nonintegrated studio/distributors than to the integrated ones for they were the means through which they maintained vertical controls over theaters.

The revealing evidence is Republic. Why should Republic lose market value when it was not a defendant in the litigation or a participant in the defendant's practices? When eight of the major firms in an industry jointly exercise market power government antitrust action should reduce the market price of their shares and increase the price of their competitors shares (Bhagat and Romano 2002, Bittlingmayer and Hazlett 2000). Should not a competitor to the studios gain market value when it is no longer foreclosed from the market as alleged in the litigation? Republic's outcome is contrary to the antitrust theory of the *Paramount* case.

And then there is the magnitude of the fall in market value. Was the horizontal conspiracy alleged in the antitrust action to include eight of the largest firms in the industry so ineffective that it managed to increase asset values only between 5 and 12 percent for only a few of its participants?

The evidence does not suggest that the District Court's self-acknowledged "experiment" in altering the long-standing and highly developed structure and practices in this new information industry was "ruinous." A fall in value of 5–12 percent is damaging, but not ruinous. But, it did not help and, in the opinion of the stock market, the Court appears to have done harm to the motion picture industry and the value of its assets. The pattern of harm is broad: the integrated and nonintegrated studios were harmed and so was their major competitor who was not a defendant. Nor did the litigation seem to end the long-standing hostilities between the studios and independent exhibitors that had been the source of the litigation. The number of lawsuits brought by exhibitors over distribution practices grew rapidly after *Paramount*. Other evidence shows that neither the independent exhibitors nor the film-going public benefited from the fall in feature film production and rise in theater prices and film rental rates that followed *Paramount*.

9 Stochastic market structure

Concentration measures and motion picture antitrust

9.1 Introduction

Concentration indices are widely used to characterize industry structure. In its 1948 *Paramount* antitrust decision, the Supreme Court relied on simple measures of box-office revenue concentration among the major motion picture distributors to make its finding that the studios, acting together, monopolized the market for "first run" motion pictures. The Court's decision indisputably altered the studio system, forcing the divestment of studio-owned theaters and altering the contracts through which motion pictures were licensed to theaters (De Vany and Eckert 1991).

There are plenty of reasons to be cautious about applying concentration measures to an industry as tumultuous as motion pictures. The volatility of motion picture market shares is driven by the release of new films and audience reactions: together, these create an expansive dynamical process that can "go anywhere" (De Vany and Walls 1996, 1999; De Vany and Lee 2001). Concentration measures of the sort used in the *Paramount* decision, and in antitrust law in general, mask the volatility of market shares and give a false sense of stability and lack of competition where there may be vigorous competition. Antitrust cases and even the Department of Justice Merger Guidelines do not consider the volatility of shares; most decisions turn on concentration measures such as the Four Firm concentration ratio or the Herfindahl–Hirschman Index (HHI) that are non-stochastic averages or stationary values. This practice replaces a random variable with one that is certain and fails to take the volatility of market shares into consideration.

In this chapter, we take a different view. We argue that even if one believes that market shares provide information about the extent of competition, it is essential to take a stochastic point of view when one sees strong evidence that shares are volatile. This means that one must record the variation in market shares and estimate the probabilities that firms will realize different market shares. If there is evidence that market structure is stochastic then it must be explicitly modeled.

When market share is volatile, it makes no sense to suppress its variation in static or average concentration measures. The stochastic structure must be explicitly modeled through the probability distribution of market share. This is not so easy to do in motion pictures because the mean, variance and other moments of the distribution may not exist. Since the expected HHI is the second moment of the distribution of market share, it is troubling that it may not exist. Even though the sample HHI exists, it is an unstable and unreliable estimate of the expectation. Antitrust policy does not acknowledge this troubling problem; we show that standard models of market power that rely on concentration or HHI indices fail completely when the indices are volatile or do not exist. We think many other industries with skew-stable distributions of firm size and shares may share these difficulties.

We show that market shares follow a Pareto–Lévy motion where there are reversals among the ranks of the leading firms and deep plunges and extreme rises in shares over time. Even a firm's average market share is volatile. Motion pictures is a "winner take all" kind of business because small differences among films and information are propelled by nonlinear processes that create extreme differences in revenues among films and the studios that release them. The nonlinear dynamic carries hit movies far into the upper tail of the probability distribution and it is the "heavy tail" of the distribution that is the cause of the infinite variance and the nonexistence of the HHI.

Because they follow nonlinear dynamics, box-office revenues are highly concentrated (skew-stable distributions). Yet, in spite of high concentration, the leading movies and firms change places dramatically and often. Nor can anyone predict which movie or studio will be the winner at any point in time or how long their domination will last. The nonlinear Pareto–Lévy motion of box-office revenues implies that motion picture revenues will be both highly concentrated and extremely volatile. Thus, the industry is stochastically concentrated but vigorously competitive.

There is another reason to be cautious about applying conventional models of market structure to motion pictures. The market for motion pictures is organized and operates in a manner that is completely different from what is assumed in the models that link pricing with concentration. Its organization is adapted to its volatility. We conclude that the use of concentration measures to assess the degree of competitiveness of the motion picture industry is lacking theoretical and empirical justification.

9.2 Market power: theory and policy

One measure of market power is the Lerner Index (L), defined as:

$$L = \frac{(\text{Price} - \text{Marginal Cost})}{\text{Price}}. \tag{9.1}$$

The model assumes that firms with monopoly power can charge prices above marginal cost.[1] The index can be derived from a profit-maximizing single-product monopoly model or a one-stage Cournot oligopoly model.[2] Because of the difficulties associated with obtaining marginal cost data, the index has not been used very often to measure market power. This notwithstanding, the Lerner Index's influence in the antitrust literature remains substantial chiefly through other measures of market power that rely on it indirectly to link pricing to market concentration. Two such indices are the M-firm concentration ratio (CRM) and the Herfindahl–Hirschman Index (HHI).

The M-firm concentration ratio (CRM) is defined as the cumulative market share of the number of firms, M, with the largest market shares:

$$\text{CRM} = \sum_{i=1}^{M} s_i. \tag{9.2}$$

The variable s_i is the market share of the ith firm. The commonly used M-firm concentration ratios are the CR4 and CR8—which measure the cumulative market shares of the four and eight largest firms in the industry, respectively.

Saving's (1970) model exhibits a relationship between CRM and market pricing.[3] All firms produce a homogeneous product and are divided into two groups—a collusive dominant group (consisting of M firms) and a price-taking fringe group. The dominant group jointly maximizes their profit given a conjectural derivative ($\lambda_M = (\Delta Q_F / \Delta Q_M)$), which represents the fringe group's output (Q_F) response to the dominant group's output (Q_M).[4] The Lerner Index (L_M) for the dominant group is related to the M-firm concentration ratio (CRM):

$$L_M = \frac{p - c_M'}{p} = \frac{(1 + \lambda_{FM})(\text{CRM})}{\epsilon_{Qp}}, \tag{9.3}$$

where p is price, c_M' is the (common) marginal cost of firms in the dominant group, and Q_p is the absolute value of the market price elasticity of demand. Hence, the CRM is linked to market power via the dominant group's Lerner Index (L_M). The model implies that the excess of price over marginal cost as a proportion of price is directly proportional to the CRM. It follows that if CRM is volatile, price will be.

The HHI is the sum of the squared values of the n firm market shares:

$$\text{HHI} = \sum_{i=1}^{n} s_i^2, \tag{9.4}$$

s_i is the market share of the ith firm. As in the case of the concentration ratio, the relationship of market power to the HHI is through the Lerner Index. It can be shown that the industry-average Lerner Index ($\hat{L}$) for a homogeneous product industry is given by:[5]

$$\hat{L} = \frac{(1 + \lambda)(\text{HHI})}{\epsilon_{Qp}}. \tag{9.5}$$

The variable λ is the conjectural derivative (assumed to be identical) for all firms in the industry, and ϵ_{Qp} is the absolute value of the market price elasticity of demand.

Note that the HHI in this equation is not a stochastic variable. Moreover, the Lerner Index is an industry average that depends on a fixed distribution of firm sizes and the associated HHI. Once again, the variation in firm shares is suppressed by using an average. When the HHI is volatile, equation (9.1) implies that price is volatile. If the HHI is infinite, so is the average Lerner Index and price is infinite too! The whole exercise falls apart since the model presumes finite and non-stochastic values of the HHI and Lerner Index.

Stochastic market structure undermines the theory and practice of concentration indices. In practice some allowance is made and the HHI is not blindly applied in antitrust. It is supplemented by other criteria such as: (a) factors influencing collusion or coordination (e.g. availability of information on market conditions, firm and product heterogeneity and marketing practices, contracting practices; and historical precedence of collusion); (b) degree of product differentiation; (c) entry–exit conditions; and (d) efficiency gains. These criteria are especially important when the post-merger HHI falls in the "gray area" (or moderate concentration) between 1,000 and 1,800. It is important, however, to note that none of these additional criteria acknowledge volatility of market shares and the possibility of infinite HHI values.

The HHI has come under other criticism. Schmalensee (1987) says that the empirical link between concentration and collusion is weak. He points out that the HHI has not been proven to be superior to the CR4 in terms of predicting noncompetitive behavior.[6] Salop (1987) suggests that measures of market concentration have no predictive value at all, an argument we support with the previous analysis and our empirical results for motion pictures.[7] Many economists regard the model as weak or wrong and the link as far from exact or certain.[8] We would add that, if both the HHI and the size ranking of firms vary, then empirical studies that fail to account for this are flawed. Moreover, if market shares are volatile, there is evidence that concentration is endogenous so that it cannot be treated as an explanatory variable and will have weak predictive power.

The broad use of the Cournot model of concentration and pricing rests on a claim that there is a universal theory of pricing and concentration that applies to every industry and time frame, in spite of their differing institutions and economic environments. This is a remarkable claim that should not be readily accepted in the absence of compelling evidence. We show here that the claims do not stand up to the evidence in the motion picture business.

9.3 Concentration in *United States* v. *Paramount Pictures*

Concentration indices were already in use in motion picture antitrust before they were formally incorporated in the 1968 Merger Guidelines. A central piece of evidence the courts relied on to infer market power of the defendants was the combined market shares of the defendants:[9]

> If viewed collectively, the major defendants owned in the 1945 at least 70% of the first-run theaters in the 92 largest cities and the Supreme Court has noted that they owned 60% of the first-run theaters in cities with populations between 25,000 and 100,000. As distributors they received approximately 73% of the domestic film rental from the films, except Westerns, distributed in the 1943–44 season. These figures certainly indicate, when coupled with the strategic advantages of vertical integration, a power to exclude competition from these markets when desired.

It is interesting to note that during the half century following the Paramount Consent Decrees, all but one of the distributors-defendants (RKO, who exited in 1957) continue to survive.[10] The current market share of the top five distributors (at 72 percent) is similar to that of the market share of the five major distributors-defendants during the *Paramount* litigation (at 73 percent in 1943–44).[11] The current CR8 at 95 percent is the same as during the period of the litigation.[12] Hence, if arguments similar to those employed in the Paramount case were to be made today, the current level of concentration in the motion picture industry would be a concern to regulators.[13]

The average value of the HHI in motion pictures is 1,268, a value that falls in the category of "moderate concentration" according to the Horizontal Merger Guidelines.[14] A merger of any two of the top six distributors, according to the guidelines, would raise "significant competitive concerns" depending on other additional factors.[15] Such a merger would produce an increase in the HHI of more than 100 points in what is a moderately concentrated market.[16] An increase of 100 points in the HHI occurs whenever each of the two merging parties has a pre-merger market share of at least 10 percent.[17] All the top six distributors in our sample meet this condition (Table 9.1). Furthermore, the Guidelines views previous incidences of collusion—as argued in the *Paramount* case—as conducive to collusion.[18] What follows shows these concentration measures to be as irrelevant today as they were 50 years ago.

9.4   Instability of concentration and shares

It is common practice to compute the values of concentration indices using annual data.[19] Motion industry concentration measures have fluctuated in recent years. Between 1990 and 1995 the cumulative market shares of the six major distributors (CR6) ranged between 78 and 94 percent. During these years the HHI fluctuated between 1,200 and 1,577. Moreover, there has been turnover at the top ranks.

Table 9.1 Summary statistics of market shares

Percentiles				
		Smallest		
1%	0	0		
5%	0	0		
10%	0	0	Obs.	2,915
25%	0	0	Sum of wgt	2,915
50%	0		Mean	0.0188679
		Largest	Std dev.	0.0567178
75%	0.0016428	0.4360919		
90%	0.0652009	0.4537629	Variance	0.0032169
95%	0.136359	0.4627493	Skewness	4.168305
99%	0.2881856	0.6245648	Kurtosis	23.92456

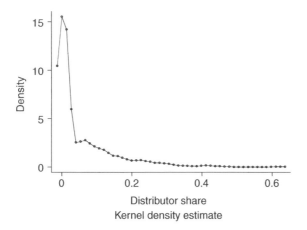

Figure 9.1 Kernel density of market shares.

Here are some details: The summary statistics of distributor market shares are shown in Table 9.1. The high kurtosis is evidence of the highly non-Gaussian form of the distribution. The density plot of Figure 9.1 makes it quite evident that the distribution is skewed to the right and has a long, heavy upper tail. Corresponding to these motion picture market shares are the annual shares of their distributors which is presented in Table 9.2. A longer time frame of market shares is given in Table 9.3. It is interesting to note how the shares change over time and the positions of the top firms vary. Buena Vista has the top share four times out of the 7 years from 1990 through 1996–97 with shares between 15.5 and 19.4. Warner Bros. is number one twice with shares between 13.0 and 19.8 percent. Sony is number one once

Table 9.2 Distributor market shares in 1996–97

Rank	Distributor	Share	Cumulative
1	Buena Vista	18.37	18.37
2	Twentieth Century Fox	14.12	32.49
3	Sony	13.75	46.24
4	Warner Bros	13.00	59.24
5	Paramount	12.93	72.17
6	Universal	11.69	83.86
7	Newline	6.08	89.94
8	Miramax	5.29	95.23
9	53	4.77	100.00

Table 9.3 Market shares of major distributors, 1990–95

Year	1990	1991	1992	1993	1994	1995
Buena Vista	15.5	13.9	19.4	16.3	19.3	19.0
Warner Bros	13.1	13.7	19.8	18.5	16.1	16.3
Universal	13.1	11.0	11.7	13.9	12.5	12.5
Sony	13.9	20.0	19.1	17.5	9.2	12.8
Paramount	14.9	12.0	9.9	9.3	13.9	10.0
Twentieth Century Fox	13.1	11.6	14.2	10.7	9.4	7.6
Others	16.4	17.8	5.9	12.7	19.6	21.8
CR4	57.4	59.6	72.5	66.2	61.8	60.6
CR6	83.6	82.2	94.1	87.3	80.4	78.2
CR8	90.8	88.5	97.4	92.5	90.4	91.0
Herfindahl	1,200	1,204	1,577	1,333	1,367	1,335

Source: Litman (1998), p. 23.

with the highest share of all of 20 and 3 years later hits its lowest share of 9.2 percent. 'Others' take in from 5.9 to 21.8 percent. Fox hit a high share of 14.12 the year after it had its lowest share of 7.6 percent. The concentration indices are relatively stable but they hide the turmoil beneath.

These annual data speak only a bit to the underlying volatility in this market. Weekly data correspond more closely with the natural timescale of the theatrical market; that is the period when films go head to head with one another. Competition between firms is competition between movies. New films open each week and supply is adjusted by extending runs week by week. The statistics on weekly market shares is contained in Table 9.4. Quite dramatic changes are now evident; Universal had a maximum share of 62 percent of all revenues earned that week and a low of less than 1 percent in another week. Buena Vista's share varied between a low of less than 1 percent to a high of 45 percent. Some of the fluctuation in shares is due to the release schedules of the distributors. But, all the major distributors had films playing every

Table 9.4 Market shares and tail weights (α) by distributor

Distributor	Average	Min.	Max.	α
Twentieth Century Fox	0.1308	0.0043	0.4170	1.4957
Buena Vista	0.1778	0.0220	0.4537	1.7137
Miramax	0.0544	0.0020	0.1637	1.4298
NewLine	0.0065	0.0012	0.1755	1.7144
Paramount	0.1349	0.0184	0.4360	1.6977
Sony	0.1441	0.0022	0.3630	1.6090
Universal	0.1162	0.0011	0.6245	1.5447
Warner Bros	0.1292	0.0033	0.3277	1.7627

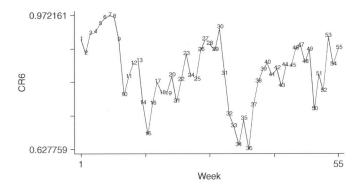

Figure 9.2 Aggregate market share of the top six distributors (CR6) by week.

week, so the volatility of their shares corresponds to how their product is faring against competitors. The average weekly shares correspond to the annual shares, as they must. But the annual figures mask a tumultuous volatility.

The concentration indices also vary in their annual and weekly fluctuations. The annual CR6 fluctuates between 78.2 and 94.1 percent; the weekly CR6 fluctuates from 62 to 97 percent (see Figure 9.2). Instability is not just confined to the CR6. The values of the HHI also exhibit large fluctuations (Figure 9.3). The weekly HHI fluctuates dramatically between 1,305 and 4,166 whereas the annual HHI ranges more timidly between 1,200 and 1,577.

Changes in the HHI's are driven by the market share of the highest grossing film. The fluctuations of the HHI are similar to those of the CR1. The HHI and CR1 are highly correlated (the coefficient of correlation equals 0.84). But, this high correlation masks the important fact that the identity of the top firm constantly changes. The coefficient of correlation between CR6 and CR1 at 0.51 is much smaller.[20,21]

Concentration is clearly evident in Figures 9.4 and 9.5 where the top-ranked films take the largest share of screens and revenues. Here is another

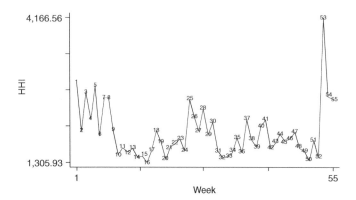

Figure 9.3 The HHI by week.

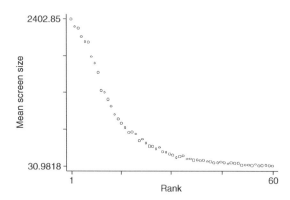

Figure 9.4 Mean number of screens versus rank.

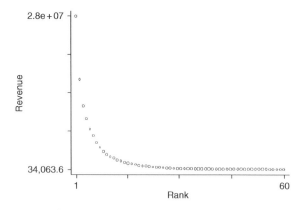

Figure 9.5 Mean revenue versus rank.

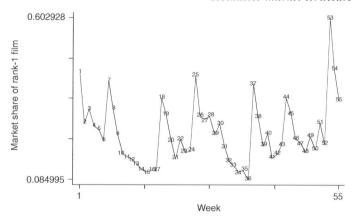

Figure 9.6 Market share of rank-1 films by week.

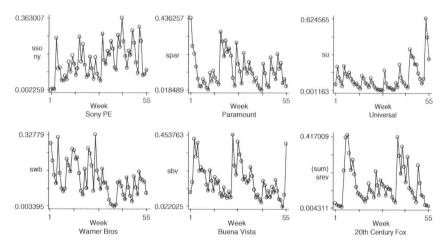

Figure 9.7 Weekly market shares of top firms.

piece of evidence that reveals that concentration is high but shares are volatile (Figure 9.6). We can document the volatility of market shares among the major distributors by examining the weekly shares of the top distributors (Figure 9.7). Just a few weeks before Universal achieved its 62 percent share, its market share was less than 2 percent. There is a negative correlation between the leading distributor's share and the shares of the other distributors when the leading distributor has a huge hit (successful collusion requires a positive correlation of firm shares).

Finally, a dramatic insight into the turmoil in the motion picture theatrical market is revealed in the rich and complex dynamics shown in Figure 9.8 of

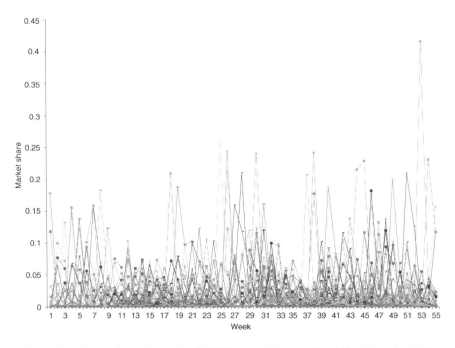

Figure 9.8 Market shares by week of 115 films on *Variety's Top-50* list (May 24, 1996 to June 12, 1997).

the market shares of all the 115 films that made it into *Variety's Top-50* films during the year.

9.5   Turnover of the leading firms

The existence of competition in an industry can be gauged by instability in the ranking of dominant firms. Curry and George (1983, p. 213) express this elegantly:

> One criticism of conventional indices is that they are 'static', in the sense that they simply record the characteristics of a size distribution at some particular point instant. If however, the identity of the dominant firms were to change over time then even persistent high levels of concentration would not imply the absence of competition.

To see how closely volatility in market shares is tied to turnover among the leading firms we track in the identities, week by week, of the first- and second-ranked distributors on the *Variety Top-60* list (blanks indicate a rank below 2).[22] Only a handful of the distributors in the motion picture industry occupy the top two ranks of the *Variety* box-office list. There are 53 distributors in our sample, but only 8 occupied one of the top two ranks during the sample period. They are Buena Vista, Twentieth Century Fox, Paramount, Universal, Sony, Warner Bros., Miramax and Newline. Note that a rank of 1 or 2 is most often attained by a film in its first week but there is some persistence in retaining a top ranking. Every distributor had the top-ranked movie at least one week; Fox managed to hold the top spot 12 times during the 55 weeks; Paramount 11 times and Sony 9 times. Buena Vista held the top spot 7 times. Fox held the number 1 rank five weeks in a row for the longest streak; it also had a four-week winning streak matched only by Paramount.

How important are rank-1 and rank-2 positions? The significance of these positions is revealed by the relationship between rank and the number of screens and box-office revenues shown in Figures 9.4 and 9.5. The convex-to-origin plots show that higher-ranking distributors (e.g. ranks 1, 2) have disproportionately large shares of theater screens and box-office revenues.

The high turnover in firms that occupy the rank-1 and rank-2 positions and the brief occupancy in top rank indicate that competition between major distributors in the motion picture industry is very intense.[23]

We can examine the turnover of distributors by computing the probability of a transition in rank. The probability that a distributor who is ranked number 1 will remain there or make a transition to rank 2, 3 and so on gives a measure of the stability of market shares among the top firms. We use the data to calculate the transition probabilities for 23 of the top distributors. We define a rank = 24 to all ranks outside the top 23. This gives us 24 ranks in all. Hence the dimension of the rank transition matrix is 24×24 (including the 24 rank). In Table 9.5 we report the transition probabilities only among the top 10 firms (the full matrix is available on request).

Reading across a row of the matrix, the entry in each cell is the probability that a distributor in a row-rank will make a transition to a column-rank. For example, the entries indicate that the probability a distributor ranked

Table 9.5 Transition matrix: probability of transition in rank by rank

	1	2	3	4	5	6	7	8	9	10
1	0.4074	0.3519	0.2037	0.037	0	0	0	0	0	0
2	0.1667	0.2037	0.2778	0.2037	0.0926	0.037	0.0185	0	0	0
3	0.0556	0.0926	0.0741	0.3519	0.3519	0.037	0.037	0	0	0
4	0.1111	0.037	0.0741	0.2407	0.2037	0.2778	0.037	0.0185	0	0
5	0.0926	0.037	0.1667	0.0556	0.1481	0.2593	0.2037	0.037	0	0
6	0.0926	0.0741	0.0556	0.0556	0.1111	0.2222	0.2037	0.1667	0.0185	0
7	0.0185	0.0926	0.0926	0	0.0556	0.0741	0.2963	0.2222	0.0741	0.037
8	0.0556	0.0556	0.037	0	0	0.0185	0.1481	0.3148	0.1852	0.1481
9	0	0	0	0.0185	0	0.037	0.0185	0.0926	0.3148	0.2407
10	0	0	0.0185	0.037	0	0.037	0	0.0741	0.2222	0.1852

number 1 remains there is 0.4074 and the probability of falling to rank 2 is 0.2037.

Several features of the transition matrix are of interest:

1   There is some persistence: the probability of remaining in a rank usually exceeds the probability of moving up or down in rank. This is evident by the fact that the probabilities in the on-diagonal cells exceed the probabilities in the off-diagonal cells. The highest probability of remaining in rank is at the rank 24 which corresponds to the state of being a distributor with no movie among the Top-60. A distributor who is in that position has a 0.95 probability of staying there. Nonetheless, there is no barrier to entering the Top 23 distributors as is shown by the positive transition probabilities along the 24 row. The highest rank to which a distributor off the list ascended was rank 7.

2   There is upward and downward mobility. A distributor as low as rank 8 can ascend to rank 1 in one week (with probability 0.056). A 2-ranked distributor can, with probability 0.167, ascend to rank 1. A distributor at rank 1 can descend to rank 2 (probability 0.35), to rank 3 (probability 0.204) and to rank 4 (probability 0.037). The lowest rank to which a distributor can move in one week from rank 1 is to rank 8.

3   The transition probabilities are higher for moving down (e.g. rank 1 to rank 2) than up (e.g. rank 2 to rank 1); this holds for all ranks, but especially for the higher ranks.

4   The probability of remaining off the top list, given that one is off the list, is 0.9569.

There clearly is a good deal of upward and downward mobility even though shares are concentrated; but, shares are concentrated by rank not by the identity of the temporary occupants of the ranks.

9.6   Standard models do not apply to motion pictures

Kwoka (1985) points out that choosing a concentration index is equivalent to choosing the oligopoly theory that links concentration to market power. If the indices do not measure competition, then the conclusions of these models are flawed. But, the second issue is this. The very volatility that makes these indices fail to capture competition also requires the industry to organize in a fundamentally different way from what the oligopoly model assumes.

First, it is important to recognize that, in the motion picture industry, supply is not static. Supply adapts to demand dynamically over time as moviegoers make their decisions based on a variety of information sources, both global (e.g. advertisement) and local (e.g. word-of-mouth).[24] Since the number of theater engagements and the length of the run are adjusted in response to demand, supply cannot be used strategically to affect the admission or rental price.

Second, the Cournot hypothesis that firms are able to make stable conjectures about the responses of their rivals to their supply decisions is implausible. No one knows what a movie is worth before it is released; demand is not known and the number of theater screens and the length of the run are adjusted as the film runs. No firm is able to predict its market share nor can it perceive a change in its market share as originating from the actions of it rivals. The Cournot model assumes fixed demand that allows firms to decode changes in price or market share to read the actions of their rivals. None of this is true when demand is not known until the film is released and when demand is volatile. Motion picture distributors have only vague (and often unrealized) ideas about the level and elasticity of demand for their movies before they begin to play.

Third, the price–cost margin, which is the crucial element in the theory linking market power to concentration, is different in the motion picture industry. There are three "prices" in the industry and they are not determined by the studio/distributor—they adapt to demand and supply conditions. The first price is the exhibition ticket price that moviegoers pay. This price is set by the exhibitor and is usually constant during the run in order to mute an incentive problem and to preserve a pure demand signal.[25] The second price is the film rental rate that sets the payment from the theater to the distributor. The rental rate is a percentage royalty that applies to box-office revenues. The rate is adjusted biweekly and is contingent on the state of demand. Hence, it too, is an adaptive form of pricing, the specifics of which are determined by the reaction of filmgoers to the movie. Thus the rental price is contractually delegated to the motion picture audience whose response to the film determines its rental rate and total rentals. The rental rate increases when demand is high through a contingent pricing arrangement that requires the exhibitor to pay the higher of either[26]

1 a minimum percentage starting at 70 percent that declines biweekly to a minimum of 40 percent of box-office revenue; or
2 90 percent of box-office revenue above a fixed amount known as the "house nut." Should demand be very high, this state will trigger the 90 percent rate. There may also be a guaranteed rental payment against which rentals accrue.

This means that the rental price adapts to demand after it is revealed as a film plays in the theater; it is not set before demand is revealed.

The third price is the price of the distributor's services. The distribution fee is the negotiated percentage (commonly 30 percent) of gross film rentals.[27] The distribution fee thus varies with the state of demand and high fees will be earned on highly successful movies. The rentals and distribution fees that a movie will earn are unknown until it plays to an audience and their demand for it is finally revealed. Hence, neither the producer nor the distributor can determine the price (distribution fee) of a motion picture before if plays. In effect, the pricing of distribution is delegated to the audience.

After a movie is released, the number of prints or copies is fixed and the opening distribution is made, so the marginal cost of accommodating higher demand is small. Essentially, the marginal cost of extending the run is the opportunity cost of the revenue foregone by the distributor and the theater on another film that could be playing on the same screen. Thus, a higher film rental implies an increase in the distributor price–cost margins. While this is true, it does not validate the Lerner Index as a measure of market power in the industry. The potential for a price increase is contracted a priori before the movie is released and whether it occurs or not depends on an unpredictable demand response. Hence, the rental price increases not because distributors have the market power to do so but because demand dynamics trigger the relevant pricing clause in the rental agreement.

Fourth, the application of the concept of conjectural variation to the motion picture industry runs into severe problems. The supply response of distributors and exhibitors occurs chiefly through length-of-run adjustments and this is determined adaptively in response to demand that is revealed during the run. At the time of release, the length of run of a film is unknown. Hence, distributors will not be able to form conjectures regarding how films from other distributors are affected by their own. The best they may be able to do is to try to move the opening date to avoid movies they fear will be "blockbusters" that might reduce the revenues of their own films. This is largely a guessing game because it is not possible to predict which films will be blockbusters or how long competitive films will play or if other distributors will also change their opening dates.

9.7   The probability distribution of the HHI

These key features, and the empirical evidence presented here, require a probabilistic approach to modeling motion picture demand and supply dynamics. Recognizing that a static and certain view of the industry is inappropriate requires one to treat the HHI for what it is—a probabilistic measure rather than a static one. The expectation of the HHI is the more relevant measure than a static calculation of the HHI at a point in time or for a given number of firms.

The market share of a distributor is its box-office sales divided by total box-office revenue for the time period. The expected market share is the first moment:

$$E[s] = \int_0^1 s p(\tau) \mathrm{d}\tau, \tag{9.6}$$

where $p(\tau = s)$ is the probability a firm realizes market share s. The HHI is the expected squared market shares or second moment of the distribution.

$$E[\mathrm{HHI}] = E[s_i^2] = \int_0^1 s^2 p(\tau) \mathrm{d}\tau. \tag{9.7}$$

The HHI is usually estimated by the sample moment $\sum_i^n s_i^2(t)$ of observations at time t. Because firm shares are volatile, the sample moment yields unstable estimates that will depend on sample size. Moreover, the expectation of the moment may not exist. We must estimate the probability distribution to determine if the HHI exists. This is the point Mandelbrot (1997, p. 215) emphasized:

> This (Herfindahl) Index has no independent motivation, and we shall see that its behavior is very peculiar. It is odd that it should ever be mentioned in the literature, even solely to be criticized because it is an example of inconsiderate injection of a sample second moment in a context where even the existence of expectation is controversial.

Following Mandelbrot, the preferred procedure is to estimate the tail index of the probability distribution of box-office revenues. Research has established that the probability distribution of box-office revenues is a Pareto–Lévy distribution with finite mean and infinite variance.[28] Pareto–Lévy distributions have heavy tails and this may prevent the integral in the expectation of the HHI from converging.

The tail weight can be estimated by estimating α in the following probability distribution:

$$P[X > x] \sim kx^{-\alpha} \quad \text{for } x > k, \tag{9.8}$$

where x is the box office revenue, $P[X > x]$ is the probability that X exceeds x, and x is greater than k. The method of Mandelbrot is used to estimate the value of α. This involves a regression of the log of the survival probability $P[X > x]$ against the log of market share. When $\alpha < 2$, the expected HHI is infinite.

Performing this estimation with market shares shows the existence of a linear region in the upper tail of the distribution of market shares where it becomes Paretian. The estimated absolute value of $\alpha(= 1.62)$ is less than 2.00 (see the results in Table 9.6). The kernel density plot shown previously in Figure 9.1 reveals the skew and long upper tail of the distribution that are characteristic of Pareto–Lévy distributions. Since market share is bounded above at 1, there is a bit of truncation in the upper tail. Nonetheless, the dense probability mass at large market share values is enough to cause the integral in equation (9.7) to fail to converge, implying an infinite variance and expected HHI. The estimated value of α for each distributor is given in Table 9.4. Generally, there is more share volatility when α is small as this is an indication that more probability density is located in the upper tail of the distribution. Universal, Miramax, and Twentieth Century Fox clearly have the most volatile shares as evidenced by their low α values.

The volatile nature of the motion picture industry is connected to the non-Gaussian form of the distribution of revenues and market shares. Unevenness

Table 9.6 Estimates of tail weight α for the probability distribution of market shares

Source	SS	df	MS	
Model	276.220253	1	276.220253	Number of obs = 341
Residual	42.6769415	339	0.125890683	F(1, 339) = 2194.13
				Prob > F = 0.0000
				R-squared = 0.8662
Total	318.897194	340	0.937932924	Root MSE = 0.35481

lpsh	Coeff.	Std err.	t	P > |t|	[95% Conf. interval]	
lsh	−1.620027	0.0345853	−46.84	0.000	−1.688055	−1.551998
_cons	−5.464048	0.0745236	−73.32	0.000	−5.610635	−5.317461

in market shares is the consequence of a complex demand process that produces large variations in film revenues; in this business, averages depend on extremes. This means that a snap shot picture of market shares will show high concentration even though there is competition among films. In other words, it is large variations in the demand for the products of firms that produce the high concentration one finds in motion pictures. That variation is the consequence of a vital competitive tournament among films and the cause of the infinite value of the HHI.

9.8 Conclusions

The structure of motion pictures is stochastic; there is high turnover among the market leaders and market shares Pareto–Lévy distributed. The expectation of the HHI may fail to exist and its sample estimate is unstable. We conclude that the HHI is an invalid measure of competitiveness in this volatile industry, that it is usually incorrectly calculated, and that models of pricing that are based on it are deeply flawed.

Part IV
A business of extremes

The movie business is right up there with earthquakes and hurricanes when it comes to the influence of extreme events. They are about as predictable too. Movies, hurricanes, earthquakes, floods, stock market returns, innovations and patents, and just about any really important phenomenon in human affairs are almost completely unpredictable. They are unpredictable because they all follow the "wild" statistics of a Pareto Lévy stable distribution. This distribution seems to capture the complex dynamics of processes that are driven by extreme events and that lack any characteristic scale or size. Just as there is no typical hurricane or earthquake, there is no typical movie. And, just as the damage from floods or returns in the stock market are driven by big events, so are revenues and returns in motion pictures.

Here is a little thought experiment I do with my movie class to demonstrate the influence of extreme events. There are, say, 50 students in the class and their average weight is about 148 pounds (it has risen in recent years and not because my weight has changed). Their total weight is 7,400 pounds. The next person to walk into the class weighs 200 and the average rises to 149. Then a 100 pounder enters and the weight falls back to 148. Not much change even though the weights of the new additions differ by a factor of two.

So far we have been talking about a Gaussian process where people are rather typical and extreme variations in weight are unusual. So the average tends to settle down and converge to a stationary value. The movies are not like that. You have to imagine that the next entrant into the class is a 9,000-pound gorilla. Now the average more than doubles to 315 pounds. And that student may be followed by a 20,000-pound lizard who brings the average up to 680 pounds. Then a ten pounder walks in and the average drops to 667. We are up to 55 students and two of them, a gorilla and giant lizard, account for about 80 percent of the total weight in the room; they completely dominate the average. That is the essence of the movies.

As you will see in Chapter 10 of this next group of chapters, about 5 percent of movies earn close to 80 percent of motion picture profit. The Pareto law of revenues has a grip on the movies where a bit less than 20 percent earn 80 percent of revenues (Chapters 2, 4 and 5). This is a consequence of the skew of the Pareto Lévy stable distribution; there are a lot of small movies and a few big ones, but the big ones are far more probable than the Gaussian distribution would say they are. While the Gaussian distribution is symmetrical and smoothly peaked, with a mean at the center, the Pareto stable distribution is sharply peaked and asymmetric. Its mean is usually far from the center. It's heavy tails, sharp peak and skewed, asymmetric shape make it a statistical monstrosity; it suffers from a condition called leptokurtosis.

But, leptokurtosis is not a condition requiring mouth wash or antideordorant. It is a thing of beauty. It is what makes the movies what they are. Without kurtosis, there would be no *Gone with the Wind* and there would be no aristocracy in Hollywood. The business would be more like dentistry or warehousing.

What produces the exotic form of kurtosis we see in the movies is the nonlinear dynamic that shapes audiences; this, and the fact that a movie can be almost indefinitely repeated. Additional moviegoers are readily accommodated and each viewer leaves no less of the film for others to see. This infinite reproducibility permits the nonlinear dynamic to take revenues to extreme values which gives the distribution its leptokurtotic shape.

The nonlinear dynamics of preferences and movie revenues is apparent in other areas beside revenues and profit. Consider the nonlinear co-evolution of tastes and movies. When you see a violent movie, this conditions you; you begin to tolerate and expect violence. Small differences among movies are amplified by nonlinearity so that a more violent movie that exceeds expectations sets itself apart and may earn large amounts. This leads producers to new violence in order to exceed expectations. Again, these small differences in violence among movies are magnified at the box-office and a new tolerance forms. The audience, its expectations and the motives of the producers to exceed them spirals the movies into the depths of depravity and bloody, gratuitous violence. Nobody planned it and people wonder how it ended up like this; nonlinear processes can produce grotesque outcomes and they are hard to get out of.

It takes a shock or a nonlinear process to get out of these traps. Happily, the nonlinear dynamic can go the other way. Suppose a nonviolent, charming movie with interesting dialogue is a hit. Producers may think this is the next big thing and make more of these movies. Moviegoers begin to expect and appreciate this type of movie. They become more adept at dialogue and small differences among films translate into large differences in revenues. Producers race to find writers who can do better dialogue. The audience comes to expect snappy dialogue and rewards the film that exceeds their expectations. With each round, the dialogue gets better. Suddenly movies now are all about dialogue and interesting characters rather than violent, inarticulate aliens and killers. It could happen; nonlinear feedback produces extreme outcomes. If only.

Is it chaos that makes the movies a business of extremes? Chaos is the result of nonlinear and possibly deterministic dynamics that takes points near one another apart exponentially rapidly. Chaos magnifies small differences into extreme differences. It says that, even if we could write down the equations that determine how the week-to-week changes in box-office revenues behave during a film's run (we can't, nobody can in spite of the forecasts you read in the paper and the studio's belief that it can be done), we will never know how it will come out.

Why this is so lies in the sensitivity to initial conditions. Small differences in the box-office take in the first weekend or week among films can be rapidly magnified by the nonlinearity of the information flow. Film A grosses more than Film B. This is reported on the evening news. People choose A over B. The difference grows. And so on. Pretty soon A can be a hit and B a dog.

But, it could have gone the other way. Due to random circumstances B might have outgrossed A. The audience then chooses B over A and their separation grows with each media report. Now B is the hit and A is the dog.

Fortunately, it is not quite like that, but this is how it would be if the process were deterministic chaos. The funny thing about chaos is that it is completely deterministic—chance plays no role—even though chaos leads to outcomes so bizarre that they appear to be completely random. In the movies, the source of wild outcomes is the nonlinear, expansive dynamic information processing of the movie audience.

So, Hollywood is not in chaos, though it may often look that way. It is a nonlinear business that is driven to extremes. There is a kind of order in everything though and demonstrating that somewhat hidden order is the burden of the chapters that follow in this last Part of the book.

Chapter 10 is another work by David Walls and myself on the statistical attractors to which the complex dynamics of movie revenue and production converge. Chapter 11 is a paper I wrote just for this book about actors and directors. Chapter 12 is also original to this book. In it I go beyond chaos to complexity to show that all we have so-far learned about the movies and all we may love about them fall into a framework called complex systems analysis.

In Chapter 10, David and I propose, test and estimate the general asymmetric stable distribution as a model of motion picture profit. We show that the skew of this distribution and its Paretian tails capture with great fidelity the statistics of the movies. The model implies that the movie business is fundamentally more risky than a Gaussian world. Features of the business such as the "nobody knows principle," the "curse of the superstar," the "angel's nightmare," the instability of profit, and the form of artist contracts are shown to follow from the asymmetry of the stable distribution.

In the discussion on directors and actors in Chapter 11, I show that one can distinguish between luck and talent, but the difference is smaller than one might imagine. A hit–flop coin toss explains many of the features of artists' careers. It is only among the most productive artists that one begins to see a distinction between pure luck and talent. A remarkable self-similarity is uncovered in three measures of director performance: the number of films they make, their revenues and their pay. Self-similarity is another form of the "nobody knows" principle that has figured so prominently in earlier chapters.

If you don't know anything, then how do you pay a director or an actor? I examine a bidding process for signing top actors to show that several problems lead producers to overpay top stars. They face a "winners curse" that is a well-known problem in auctions. They face the curse of the superstar that we will see in Chapter 10. They also face the problem of estimating the probabilities of outcomes for their movie with each potential leading star, an almost unsurmountable task. Producers are also up against the sure thing principle.

But, there is a better way to do it. To solve the problem of what to pay a star when you don't know, the producer should wait until he does know. The producer should pay when he does know, after the film has run. The truth of this proposition is established by Darlene Chishom's research and by new evidence that I develop in Chapter 11. Using a sample of director contracts, I show that contracts condition pay on outcomes far into the tail of the distribution. I call this contracting in the tails and show that the conditioning events are of vanishing Gaussian probability. The contracts are, therefore, completely at odds with the standard theory of contracts. Directors and their agents seem to know that they operate in a stable Paretian world and design their contracts accordingly.

In Chapter 12 of this Part, I draw together the results of all this research and look for a more general model that can reveal the structure that lies just out of sight from an ordinary view. I use complex systems theory to show the deep order of the movies; here too, self-similarity plays a large role.

10 Motion picture profit, the stable Paretian hypothesis and the curse of the superstar

10.1 Introduction

The movie business is a business of extremes. Its participants seem to occupy a third world country where the winners live in palaces and the losers live in slums. Virtually all the statistical measures of the business are dominated by extreme events. It is not a "normal" business, but we all knew that. What we may not recognize is that it is not Gaussian; too many events occur regularly on a scale of magnitude that are of vanishing Gaussian probability. In this chapter, we show this to be true of motion picture profit and we uncover the non-Gaussian probability distribution that captures the extreme nature of this business.

One of the things that makes the movies a business of extremes is the way moviegoers dynamically influence one another. If a movie has a big opening, then lots of people will tell other people about it. It will be reported on the evening news as a top-grossing picture. The news or a good comment from someone who liked the movie might influence someone else to go see it. In a crowded market, just getting that kind of attention can separate a movie from the crowd and get it off to a good start. But, it is the way that the information spreads dynamically that leads to extreme differences among movies—the big hits don't always open big but they do seem to be propelled by a recursive and nonlinear demand dynamic.

The influence of early viewers goes under different names in the economics literature: contagion, network effects, bandwagons, path-dependence, momentum and information cascades are some of the descriptive names attached to these recursive dynamic processes. These models differ in the details, but they are all dynamical processes in which the change in demand depends on demand already revealed just as the spread of a disease depends on the number of carriers who have it. When demand has this dynamical property, initial advantages can lead to extreme differences in outcomes, a property that is at the heart of this business. In an earlier paper we (De Vany and Walls 1996) showed that box-office revenues have a contagion-like property where the week-to-week change in demand is stochastically dependent on previous demand. But, we also showed that there are no guarantees. A big opening of a bad movie can be like a virulent disease that kills off its hosts so quickly that it fails to spread and dies out. A big opening with a good movie can spread quickly to extremely large revenues, such as when a single movie (*Titanic*) out of more than 600 releases grosses about one seventh of total world box-office revenues.

What happens when the participants realize they are in this kind of game and attempt to influence the outcome? They will rationally try to influence the initial exposure of the audience to the movie as though they are attempting to infect a large number of initial carriers with a disease. If a movie is to "open big" it must contain a cast of name stars, or feature expensive and spectacular special effects, and it must open on many screens at once. This is expensive, but if it seeds a contagion, the movie may spread to capture a very large audience. It is also risky because all the money is spent "up front" and can't be recouped if the movie fails to find an audience. Producers and distributors justify these expenses as necessary, though not always sufficient, to give momentum or contagion a chance. But, if contagion doesn't catch on, much money is spent to no effect. What are the consequences for profits of the strategy of pursuing a "big opening"? What is the nature of risk and return in a business where momentum or contagion effects may be powerful and the players try to influence them? How are stars compensated and how are contracts structured to deal with these issues? How far off is the Gaussian distribution as a model of profit? What sorts of mistakes would an executive make if she used a Gaussian distribution instead of the right one? Can any model capture the extremes of this business?

These are some of the questions we address in this chapter. We know from our earlier work on the industry that we require a general statistical model to capture the important properties of revenue and cost behavior. We require a statistical distribution as a model of profit that lies in the basin of attraction of dynamics that have the characteristic that the growth in demand depends on demand in previous periods and we know that the model must accommodate the probability of extreme events in its "heavy tails" and infinite variance because we find these in motion picture revenues (De Vany and Walls 1999, 2002). The distribution cannot impose symmetry because cost has different limits and dynamical features than revenue. And, we know that we require a model that is robust to different initial conditions (openings).

The statistical model that accommodates all these features and which is known to be the most general form of the central limit theorem is the stable distribution.[1] If the finite variance assumption of the Central Limit Theorem is dropped, one obtains the Generalized Central Limit Theorem, which states that the limiting distribution of the sum of a large number of iid random variables must be in the stable class of distributions. A stable distribution $S(\alpha, \beta, \gamma, \delta)$ is a 4-parameter distribution. With its characteristic high peak, "heavy" Paretian tails and skew, the stable distribution captures the winner-take-all nature of the movie business as well as the influence of extreme events.

As it is the limiting distribution of the dynamical process, it is relatively insensitive to initial conditions.

This chapter establishes that the stable distribution is the correct statistical model of motion picture profit and of its underlying components of cost, revenue and returns. The chapter then goes on to examine what this new model says about the movie business and shows clearly that the business is fundamentally more risky than a Gaussian model is capable of portraying. The distribution's properties explain why contracts are the way they are and so different from optimal contract theory. The "curse of the superstar," the "angel's nightmare," and the Goldman's "nobody knows" principle are shown to be implied by the stable distribution and its sometimes counterintuitive properties. We also have an interest in exposing this powerful distribution to a wider audience and revealing some of the diagnostics that are useful in identifying it. We expect the stable distribution to be a powerful and more general model of dynamical, nonlinear phenomena that are dominated by extreme events whose influence is often overlooked when one holds a Gaussian view of the world.

10.2   The stable hypothesis

The stable hypothesis was proposed by Mandelbrot (1963a,b) as an explanation for cotton futures prices and as a general model for natural and social systems. Fama (1963) employed the Lévy flight and stable distribution as a model of stock prices. McCulloch (1978) analyzed dynamic processes with stable increments and showed in his review of the use of the stable distribution in finance that earlier rejections of the stable distribution may have been premature (McCulloch 1996). Subsequent research seems to bear him out as both the S&P 500 stock index and NYSE composite index returns are well-fitted by a (truncated) Lévy distribution (Mantegna and Stanley 1995; Levy and Soloman 1998). Uchaikin and Zolotarev (1999) is a thorough development of stable distribution theory and estimation; it also contains many examples of the use of the stable distribution in finance, biology, geology, physiology and other sciences. McCulloch (1997) and Nolan (1998) have contributed to the methods for estimating the parameters of stable distributions in published papers and in computer code.

The stable distribution's ability to capture the empirical regularities found in motion picture data and the distribution's statistical foundation on the most general form of central limit theorem make it a natural model of motion picture profit.[2] The theoretical reason to think that a stable distribution might apply to motion pictures is because Mandelbrot (1963a) showed that a dynamic process that is stable under choice, mixture and aggregation converges in distribution to the stable distribution. If motion picture revenues and costs are discrete time processes with stable increments then profit will converge to a stable distribution. The limits of a stable process form a generalized central limit theorem which states that the probability distribution of sums of independent and identically distributed terms is stable.[3] The Gaussian distribution is stable, but it is the only stable distribution with finite variance. The stable distribution nests the Gaussian, Cauchy and Lévy distributions as special cases which permits it to be tested against them as a model of the data.

10.3 Motion picture profit

We begin with a brief overview of motion picture profit. It is important to note that our profit series covers all the films released in North American theaters over an eleven-year period. Our source is the industry standard EDI/Neilson data. Profit is constructed using EDI data for North American film revenues and EDI's estimates of production cost. We adjust domestic revenue to obtain world gross and estimated film rental and then deduct production cost to obtain an estimate of profit for each film. All variables are adjusted for price level to obtain real values. Profit is a dark secret in the movie business that is subject to studio overhead, distribution charges, and cost allocations and many other considerations for which data are not available. What we have is complete data from an unbiased source of gross profit, excluding capital, interest, overhead, advertising and other costs. It is unlikely that further detail would fundamentally alter the shape of the distribution in a systematic way, though the location of the distribution might be affected. As a check on our method, we calculated the percentage of films that break even or better and compared this to Vogel's estimate; the estimates agree precisely.

The estimated profit data show that, in spite of the high profit that certain movies earn, most movies are unprofitable. Seventy-eight percent of movies lose money and only 22 percent are profitable. Profit is unevenly distributed among those movies that are profitable: just 35 percent of *profitable* movies earn 80 percent of total profit.[4] Losses are more evenly distributed. Among *unprofitable* movies, 50 percent accounted for 80 percent of total losses.[5] The most dramatic statistic is this: just 6.3 percent of movies earned 80 percent of Hollywood's total profit over the past decade.[6] This concentration of profit is not unlike stock market returns, where 6 percent of trading periods earn 80 percent of returns. By these measures, the movies is a "winner-take-all" business.

Table 10.1 reports attributes and quantiles of the distribution of profit.[7] The sample mean profit is negative for all movies. Profit is negative whether a movie features a star or not. Non-star movies lose slightly more than star movies (−3.595 million average loss versus −2.083 million average loss). The standard deviation is large but it is twice as large among star movies than among non-star movies—this is a sign of the heavy tails of the distribution of star movie profit which we shall soon examine. A movie must land above the 75th percentile to earn a positive profit. Until somewhere between the 50th and 75th percentile, star movies show larger losses at each percentile than non-star movies. From the 75th percentile and higher, star movies show

Table 10.1 Statistical properties of motion picture profit

Attribute	All movies	Without stars	With stars
Mean	−3.351	−3.595	−2.083
Std Dev	11.938	9.601	20.062
Skewness	1.855	2.488	0.860
Kurtosis	15.384	22.744	5.599
Percentile			
1st	−30.328	−25.582	−38.865
5th	−19.976	−16.562	−28.715
10th	−13.796	−12.375	−22.643
25th	−8.299	−7.461	−13.117
50th	−3.787	−3.692	−4.983
75th	−0.466	−0.863	7.012
90th	7.012	3.655	23.661
95th	14.546	9.988	34.487
99th	47.558	32.922	56.693
Observations	2,015	1,689	326

Note
All dollar magnitudes are reported in millions of 1982–84 US dollars.

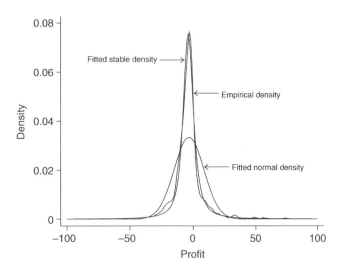

Figure 10.1 Empirical and fitted density functions of absolute profit.

higher profit at each percentile than non-star movies—more evidence of the importance of extreme outcomes to this segment of movies.

The skewness and kurtosis statistics indicate that the profit distribution is asymmetric and sharply peaked. The data depart from Normality and this is plain in Figure 10.1 where the empirical density is overlaid with the fitted

Gaussian density. In addition to this visual evidence, statistical tests clearly reject the Gaussian distribution as a model of profit.[8]

10.4 The stable distribution

There are many parameterizations of the stable distribution. We shall use the parameterization: $X \sim \mathbf{S}(\alpha, \beta, \gamma, \delta)$ where the characteristic function of X is given by

$$\mathrm{E} \exp(itX)$$

$$= \begin{cases} \exp\left\{-\gamma^{\alpha}|t|^{\alpha}\left[1 + i\beta\left(\tan\frac{\pi\alpha}{2}\right)(\mathrm{sign}\ t)((\gamma|t|)^{1-\alpha} - 1)\right] + i\delta t\right\} & \text{if } \alpha \neq 1 \\ \exp\left\{-\gamma|t|\left[1 + i\beta\frac{2}{\pi}(\mathrm{sign}\ t)(\ln|t| + \ln\gamma)\right] + i\delta t\right\} & \text{if } \alpha = 1 \end{cases}$$

$$(10.1)$$

This parameterization is a variation of that given by Zolotarev (1986) and it is convenient in several respects: (1) the interpretation of the parameters is clear; (2) the parameterization is a location and scale family so that if $X \sim \mathbf{S}(\alpha, \beta, \gamma, \delta)$ then for $a \neq 0$ it follows that $aX + b \sim \mathbf{S}(\alpha, (\mathrm{sign}\ a)\beta, |a|\gamma, a\delta + b; 0)$; and (3) the characteristic functions are jointly continuous in all four parameters.

The characteristic exponent α is a measure of the probability weight in the upper and lower tails of the distribution; it has a range of $0 < \alpha \leq 2$ and the variance of the stable distribution is infinite when $\alpha < 2$. The skewness coefficient β is a measure of the asymmetry of the distribution; it has a range of $-1 \leq \beta \leq 1$, where the sign indicates the direction of skewness. The scale parameter γ must be positive. It expands or contracts the distribution in a nonlinear way about the location parameter δ which indicates the center of the distribution; the scale parameter is, roughly, the "width" of the central part of the distribution.

The stable distribution is the limiting distribution of all stable processes so that it contains the other well-known stable distributions as special cases. Lévy showed that there is a class of distribution functions which follow the asymptotic form of the law of Pareto which Mandelbrot defined as

$$1 - F_X(x) \sim \left(\frac{x}{k}\right)^{-\alpha} \qquad x \to \infty \tag{10.2}$$

Such distributions are characterized by the fact that $0 < \alpha < 2$ and they have infinite variance. The Lévy is a generalization of the Gaussian distribution when the variance is infinite. Its tails are Paretian and the variance is infinite when $\alpha < 2$. Its mean need not exist for values of $\alpha < 1$. The Normal (Gaussian) distribution is a special case of the stable distribution when $\alpha = 2$. The stable distribution becomes the Cauchy distribution when $\alpha = 1$ and $\beta = 0$, and the Lévy distribution when $\alpha = 0.5$ and $\beta = \pm 1$. As the characteristic exponent α approaches 2, the skewness coefficient β has less impact on the shape of the distribution and it becomes more symmetric. When $\alpha = 2$ the skew parameter ceases to have any impact and the distribution has only two parameters, location and scale, which correspond to the familiar mean and variance of the (symmetric) Gaussian distribution.

10.5   Stable estimation and diagnostics

The stable distribution parameters are estimated by the method of maximum likelihood.[9] Estimates of the stable parameters for all the movies in the sample are reported in the upper panel of Table 10.2. Log-likelihood values for the general stable distribution as well as for the symmetric stable ($\beta = 0$) and the Gaussian ($\alpha = 2, \beta = 0$) distributions are shown. The test clearly rejects the null hypothesis of Normality at the 1 percent marginal significance level in favor of the symmetric stable distribution with a likelihood ratio test statistic of $\chi^2_{df=1} = 1151$.

Diagnostic tests suggested by Nolan (1999) were used to detect a departure from stability. First, Figure 10.1 shows the fitted stable density function corresponding to the maximum likelihood estimates. For comparison, the fitted Gaussian density function where α is restricted to equal 2 and β is restricted to equal 0 is also shown in the plot. The empirical density function corresponding to the data is also shown in the figure.[10] It is clear that the stable distribution is a better approximation to the empirical distribution than is the Gaussian distribution. Second, the cumulative distribution functions of the fitted stable distributions are plotted with the data in the probability plot shown in Figure 10.2. The fit is good, even for the extreme values.[11] Third, the quantile–quantile plots shown in Figure 10.3 show that the data are consistent with the random variation of a stable distribution and fall with the corresponding 95 percent confidence bands.[12] The PP-plot, QQ-plot and

Table 10.2 Maximum likelihood parameter estimates

	α Index	δ Location	γ Scale	β Skewness	Log-likelihood
All movies					
Normal	2	−3.351	8.442	0	−7855.37
Symmetric α-stable	1.268	−4.079	4.032	0	−7279.87
α-stable	1.259	−4.042	4.020	0.043	−7279.46
Movies with stars					
Normal	2	−2.083	14.186	0	−1439.69
Symmetric α-stable	1.582	−4.568	10.555	0	−1419.16
α-stable	1.624	−6.385	10.805	0.768	−1410.82
Movies without stars					
Normal	2	−3.595	6.789	0	−6216.46
Symmetric α-stable	1.358	−3.932	3.507	0	−5739.47
α-stable	1.335	−3.827	3.441	−0.122	−5737.95

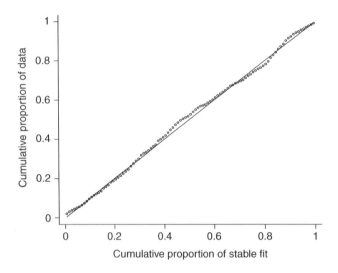

Figure 10.2 Probability plot of stable distribution fit of the data.

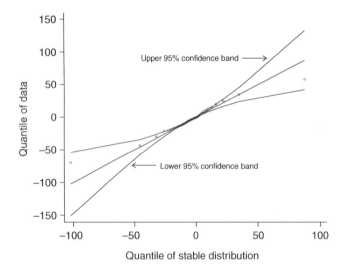

Figure 10.3 QQ-plot of stable distribution fit of the data.

comparison of empirical and fitted densities support the stable distribution over the Gaussian and other stable distributions. Indeed, the fits are remarkable and even the deviations at the extreme values are predicted by the model and fall within its confidence bands.

10.6 Probability under the stable model

As we said, the movies are fundamentally more risky in a stable world than in a Gaussian one. At first glance, the data in Figure 10.1 appear to be Gaussian. Motion picture analysts and studio executives might well use a Gaussian model of profit. But, the potential error of using the wrong statistical model is large. For example, the probability that a movie is profitable at all is 0.30 according to the stable estimates, but 0.39 according to the Gaussian distribution; this is a sizable difference.

Values of the tail parameter α shown in Table 10.2 indicate that stable distributions place far more probability mass in the tails than the Gaussian distribution. When the tail index α of the stable distribution is less than 2, profit has infinite variance. The Gaussian variance is finite since in that case α has a value of 2. Thus, an investor would have more confidence than is justified in profit forecasts were she to use a Gaussian model instead of the correct stable model.

To illustrate this point, consider the probabilities of observing the extreme outcomes in the data. *Home Alone* was the most profitable movie in the sample at nearly $93 million (1982–84 dollars) and *Waterworld* was the least profitable at −$85 million in real dollars. According to the stable distribution estimates, the chance of observing a hit as profitable as or more profitable than *Home Alone* is nearly 0.83 percent and the probability of observing a bomb as unprofitable as or more unprofitable than *Waterworld* is 0.45 percent. According to the Gaussian distribution one would incorrectly calculate the respective probabilities to be essentially zero (2.817×10^{-16} and 3.413×10^{-12}). The Gaussian model fails to predict the probabilities of extreme outcomes which account for the overwhelming share of loss and profit.

To see this more clearly, consider the shape of the tails of the Gaussian and stable distributions. An expanded view of the upper tails of each distribution is shown in Figure 10.4. For profit in excess of $30 million, the stable density has far more probability mass than the Gaussian density. The Gaussian distribution is a poor model of the extreme events that dominate profit and loss.

The stable distribution is also a better model of the central part of the distribution. First, consider the location parameter δ. Table 10.2 contains the estimates of δ for the Gaussian distribution and the symmetric and general stable distributions. The location parameter of the Gaussian distribution is larger than it is for either of the symmetric stable models. In all cases, the distribution is located at a negative value. The central or typical outcome for movies is a loss of about $4 million. The Gaussian mis-estimates this typical outcome as a $3.35 million loss.

The probability peak of the stable distribution is higher and thinner than the peak of the Gaussian distribution. The peak of the distribution corresponds to the most probable event. Thus, the stable distribution shows that a loss is more probable than the Gaussian distribution suggests. The central probability mass

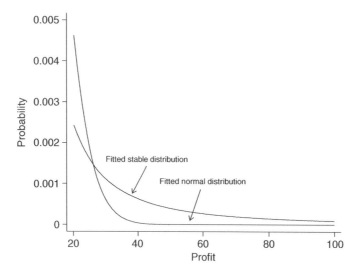

Figure 10.4 Expanded upper tails of profit density functions.

Note
Expanded upper tails of density functions shown in Figure 10.1.

of the stable distribution falls off far more rapidly from the peak than the Gaussian distribution and then shoots into the far upper and lower tails. This is an indication of the fragility of the most probable event in estimating the expected outcome.

We can conclude that estimating the profit distribution under the constraint that it is Gaussian (namely that $\alpha = 2$) leads to significant changes in location and scale parameters and seriously biased estimates of probabilities at all points in the support of the distribution.

10.7   "Nobody knows"

Screenwriter William Goldman (1983) famously said that, when it comes to predicting the success of a movie, "Nobody knows anything." This inability follows directly from the form of the stable distribution. Given the infinite variance, mean-squared error and other second moment-based measures of the precision of forecasts would necessarily be zero. Given the skew, the most likely event is not the expected value. The instability of the sample average makes it a poor estimator of the expected value. The expectation is dominated by rare events. And these events are far more probable than a Gaussian model would suggest. The sample standard deviation and variance are finite for any realization of the process, but they are poor estimates of the theoretical values, which are infinite, and they are unstable. The asymmetry of the

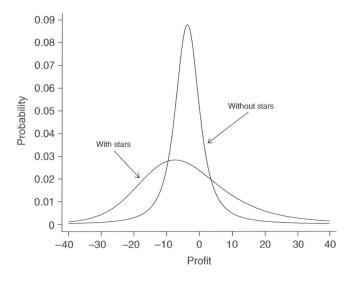

Figure 10.5 Stable density functions of profit of non-star and star movies.

distribution and the extreme variations of profit confound studio regression models. The narrow central region of the distribution means that forecasts of the most probable event are fragile. All these features give support to the "nobody knows" principle which is at the core of the organization of the industry and its contracting practices (De Vany and Eckert 1991; De Vany and Walls 1996; Caves 2000).

10.8   Conditional stable analysis and superstars

Is there evidence to suggest that movies with superstars have a different distribution of profit than other movies? To address this issue, estimates of the parameters of the Gaussian, symmetric stable, and general stable models were made for movies that featured a superstar from the "A-list" of the 100 top stars in Hollywood and for movies that did not feature one of these stars. These estimates are reported in the two lower panels of Table 10.2.

The tests reject the hypothesis that the parameters are equal between these groups of movies. For superstar movies, the tests reject Normality in favor of the stable distribution. They also reject symmetric stability in favor of asymmetric stability. Thus, the profit distribution for superstar movies is not Gaussian and it is not symmetric; it is a general, asymmetric stable distribution.

The profit distribution for movies that do not feature an A-list star is also stable. The tests reject Normality in favor of stability. The profit distribution for non-star movies is skewed toward losses, but we cannot reject symmetric stability in favor of the general stable distribution.[13] We conclude that the

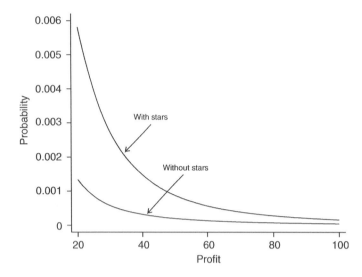

Figure 10.6 Expanded upper tails of stable density functions of profit of non-star and star movies.

Note

Expanded upper tails of density functions shown in Figure 10.5.

profit distribution of non-star movies is stable, symmetric and has infinite variance.

The subtle distinctions between the distributions of star and non-star movies are important. Figure 10.6 shows the extreme upper tails of the probability density functions of superstar and other movies. These tails are Paretian (power laws). Movies with stars have much higher probability mass in the upper Paretian tail. The probability of a movie without a star earning a profit greater than $20 million is 0.02 while for a movie with a star the probability is 0.10. Superstar movies show a higher probability of earning a positive profit, $p = 0.37$. For movies without stars, the probability of earning a positive profit is $p = 0.23$.

Table 10.3 tabulates the upper and lower tail probabilities of profit for superstar and other movies in the interval $[-100, 100]$. As is evident from the fitted density functions, stars place much more mass in the upper tail of the profit distribution. The tables reveal what is not clear in the figures, which is that the probability of extreme catastrophes, say losses in excess of $95 million, is higher for movies *without* stars. Putting a star in a movie places more mass on the upper tail and less on the lower tail.

Calculations of expected profit from the stable estimates are in Table 10.4. Expected profit is positive for star movies and negative for non-star movies. These values are consistent with the fact that probability is skewed to the

Table 10.3 Upper and lower tail probabilities

Profit	Lower tail probilities Prob($\pi \leq -Profit$)		Upper tail probabilities Prob($\pi \geq Profit$)	
	Without star	*With star*	*Without star*	*With star*
0	0.7693248	0.6288536	0.2306752	0.3711464
5	0.3875867	0.4990274	0.0910419	0.2668165
10	0.1285646	0.3582706	0.0498598	0.1906549
15	0.0596434	0.2300214	0.0328644	0.1380565
20	0.0362870	0.1339655	0.0239798	0.1024961
25	0.0253501	0.0746570	0.0186244	0.0783693
30	0.0191672	0.0431966	0.0150847	0.0616906
35	0.0152458	0.0275664	0.0125905	0.0498569
40	0.0125605	0.0194864	0.0107485	0.0412221
45	0.0106185	0.0148503	0.0093385	0.0347486
50	0.0091556	0.0118935	0.0082283	0.0297738
55	0.0080181	0.0098512	0.0073339	0.0258656
60	0.0071110	0.0083590	0.0065997	0.0227354
65	0.0063726	0.0072236	0.0059873	0.0201855
70	0.0057610	0.0063328	0.0054697	0.0180774
75	0.0052471	0.0056168	0.0050269	0.0163119
80	0.0048098	0.0050302	0.0046445	0.0148161
85	0.0044337	0.0045417	0.0043111	0.0135360
90	0.0041072	0.0041293	0.0040182	0.0124305
95	0.0038214	0.0037772	0.0037591	0.0114682
100	0.0035693	0.0034735	0.0035285	0.0106244

Note
Profit is in millions of dollars.

Table 10.4 Expected, average and most probable profit

Measure	Stars	No stars
Average	−2.083	−3.595
Expected	7.684	−4.477
Most probable	−7.500	−3.750

positive tail in superstar movies and to the negative tail for others. The conclusion is that superstar movies are more profitable and less risky than other movies.[14]

10.9 Sample average and expected value

When a stable process generates profit, the sample average profit is not stationary because extreme events dominate the average. To illustrate this feature of

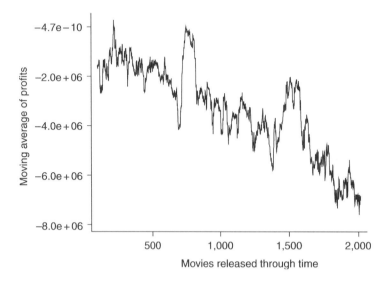

Figure 10.7 Time series of average cumulative profit.

motion picture profit Figure 10.7 shows the sample cumulative average industry profit as movies are released each week (the x-axis is ordered by release date so the graph also demonstrates the falling profitability of the industry from 1984 to 1996). The series is volatile and does not converge. When profit is stable-distributed the scale of profit changes $\Delta X/\Delta t$ is $(\Delta t)^{1/(\alpha-1)}$ (McCulloch 1978). In this series Δt is on the order of one week. The industry average profit can change dramatically because it follows a stable Lévy motion. This means that some decreases may be so large that the bottom may appear to fall out of the industry from time to time as McCulloch (1996) suggests of financial markets. When there are large positive increments the industry may appear to be in a rapid expansion. But, these surges and falls are fully consistent with the stable motion and neither a cause for alarm nor joy. Dramatic changes just "happen" and do not signal a trend or structural change. The tendencies for the industry press to overextrapolate current events (the decline and fall of the industry has been predicted many times before) and for studios to curtail production excessively in response to revenue swings might owe their origin to the large changes inherent in the stable Lévy motion which industry profit seems to follow.

The volatile difference between the average and expected values can be a trap for decision makers. If they base expectations on recent events they will often be seriously in error. There is some evidence and much folklore to suggest this error often happens in Hollywood. The error could well account for the rush to imitate successful movies. The potential for error is greatest for superstar movies because the difference between their sample average and expected value is large.

10.10   The curse of the superstar

The curse of the superstar has its origin in the positive skew of the distribution which causes the expected value to be substantially greater than the most likely outcome. The tails of superstar movies are so long and skewed that their expected profit is positive ($7.684 million) even though their sample average profit is negative (−$2.083 million).

Given the shape of the stable distribution there is a big difference between the most probable event—the small interval of profit in the support of the profit distribution that is of maximum likelihood—and the expected value. If the distribution of profit were Gaussian, these values would not differ from one another. The shape of the stable distribution—skewed, asymmetric, and "heavy tailed"—makes the sample average, expected, and most likely superstar profits different from one another.

Table 10.3 contains the values of the average, expected and most likely profit for superstar and other movies. Because the distribution is negatively skewed, the theoretical expected profit of non-star movies is negative and less than the sample average and most probable profit, which are nearly equal and also negative. The distribution of superstar movie profit is positively skewed so the theoretical expected profit is positive and far exceeds (by $9 million) the sample average profit, which is negative.

Superstar profit comes from the upper tail of the distribution, above $100 million.[15] The moderately unprofitable movies in the central part of the distribution between −$10 and −$30 million bring an expected loss of about −$6 million.[16] These findings agree with and further verify Ravid's conclusions that star movies are not more profitable; we both find that the sample average profit of superstar movies is negative. But, our results add a new dimension in that we examine both the sample mean and the theoretical expectation and find that they are unequal. The expected value exceeds the average profit for superstar movies and this is the source of the curse.

An executive faced with a decision to "greenlight" a non-superstar movie is not apt to be misled when comparing the sample average, expected, and most likely outcomes since they are close to one another. The expected profit is negative and slightly less than the average profit, and this may add a measure of caution to the executive's reliance on the sample average as a forecast. As most movies lose money, if a movie is to be greenlighted it must be because the studio believes it has something that might carry it into the upper tail, which, being Paretian, contains non-trivial probabilities of extremely high profit. The symmetry of the non-star profit distribution prevents the decision maker from systematic error, though the harsh reality that most movies lose money is not easily overlooked.

A studio looking at a superstar movie is more apt to be misled by the differences between the sample average, expected, and most likely profit. In small samples, which is all most studios can afford to make, the most likely outcome (mode) will dominate and this is a loss of $7.5 million. In larger samples, the sample average will dominate and this is a loss of $2.083 million. Only in large samples will the average approach the theoretical expected profit of $7.684 million. Most studios and certainly most studio executives don't have the resources or time to produce a large enough sample of movies in order to realize the expectation. They would be broke or fired before that. So, they have to be lucky. They have to hit in the tail of the distribution on their first or first few tries or their losses will be too great. And the odds are against them because the most likely outcome is that the movie will lose money.

The difference between the mode and expectation leads directly to the curse of the superstar. If a superstar is paid the expected profit of her movie, she would receive about $7.5 million. But, the most probable outcome (mode) for the movie would be a loss of $7.5 million. So, the studio is looking at a most probable outcome of a loss of $15 million when the superstar has extracted the expected rent.[17] In order to earn a profit when the star is paid the expected profit, the movie has to hit a profit of $10 million or greater, an event whose probability, according to Table 10.3, is only about 0.19. So, a superstar movie for which the star is paid her expected profit has an 81 percent chance of losing money. This is the superstar curse: if a superstar extracts a fee equal to the expected profit then the movie almost surely will lose money.[18]

The studio that bids closest to the superstar's expected profit is most vulnerable to the curse of the superstar. A rational studio bidding for a script or a superstar should shade the bid well below expected profit. Better yet, the bidder should guard against the curse by offering the superstar some form of contingent compensation in exchange for a reduced fixed fee.[19]

10.11   Success breeds success

An interesting property of the stable distribution is that conditional expectation does not converge. The tails of stable distributions are Paretian distributions and the conditional probability that $x \geq x_0$ is $P[x > x_0] = (x_0/x)^{\alpha}$. The conditional mean, given that $x > x_0$ equals $\bar{x}_{x_0} = x_0\alpha/(\alpha - 1)$. Since the estimated value of $\alpha = 1.624$ is greater than 1 the conditional mean exists. Since α is a constant, the conditional expected value of profit depends linearly on x_0.

Thus, in a sense, a movie does not exhaust its profit potential. Conditional on having earned a profit, the expected profit continues to rise with current profit and this does not end as the movie earns more profit. For example, the expected profit of a superstar movie is only $7.6 million, but a movie that has already earned $10 million has an expected profit of $26 million and a movie that has earned this much has an expected profit of over $50 million.

This property holds true of non-superstar movies as well. Expected profit of non-star movies is −$4.47 million, but a movie that has earned $10 million has an expected profit of $40 million.

There is nothing paradoxical about this: movies that make it into the upper tail of the profit distribution have been strongly selected from their competitors. The "heavy tails" of the stable distribution imply that probability does not decline rapidly enough for the conditional expectation to converge in the value of the conditioning event. For the Gaussian or log Gaussian distributions, the conditional expectation converges to a constant as the conditioning event increases. The linear conditional expectation of the Paretian distribution means that blockbuster movies that have already attained high profit have an expectation of even higher profit and this prospect does not diminish as profit grows. This property seems to capture the intuition behind a statistical form of momentum.

10.12   Cost momentum and the angel's nightmare

We showed that something like momentum can occur on the revenue side. What about the cost side? There is a form of momentum here too and we call it the "angel's nightmare" because it accurately captures the rational fear of investors that a movie's budget is capable of growing out of control. This fear is rational and derives from the Paretian lower tail of the profit distribution, where the budget is the dominant influence.

The budget distribution has a Paretian tail whose exponent is $d = -2.64$. Thus, the mean and variance of budget are finite, but the variance is very large. The upper (high) budget tail is long and heavy because a few massively large budgets strongly influence the average budget. This is consistent with the way studios budget movies and with the way movies sometimes run far over budget. The Paretian distribution implies that a movie's expected production cost grows in proportion to the costs already expended.[20]

Consider a movie whose budget stands at y_0. The expected budget conditional on y_0 is $y_0(d/(d-1)) = y_0 \times 1.609$. Now suppose $20 million has already been spent on a movie that was expected to cost only the sample mean of $16 million. What is the expected cost of the movie, conditional on $20 million having been expended? It is $20 \times 1.609 = 32$ an amount that exceeds the original expectation. At this point, the movie is already $4 million over budget and yet its expected cost is, $32 million, twice the original planned budget! This is the "angel's nightmare."

The angel's nightmare is inherent in the self-similarity of the Paretian probability distribution, a form that typically is found in complex iterative processes (Shroeder 1990). The nightmare haunts movies and other complex, sequential production processes. It is not a hazard of morals (though that may further complicate matters); it is a rational expectation about the behavior of cost as the project unfolds. The investor rationally expects that the expected cost is increasing in cost expended. Thus, the decision to halt production can be taken on the basis of sunk cost. A rational investor ignores costs that are sunk, but in this case sunk cost predicts expected cost. This gives some relevance to sunk cost even though the decision criterion is expected cost relative to expected revenue.

As the angel's nightmare affects complex film projects, motion picture investors retain the right to withhold funds beyond an agreed limit. But, that leaves the funds already expended sunk. To salvage value from sunk cost, investors require a completion guarantee before they put money in a film. But the completion guarantee does more than that; it hands authority to an independent party once the original budget is exhausted (or exceeded by a specified amount) and this party's sole responsibility is to finish the film. The completion guarantee thus brings the potentially infinite process to an end, though the final product that results may not be artistically or financially successful.[21] The stable Paretian model explains the need for and existence of the completion guarantee.[22]

10.13   Stable distribution and contracting

Optimal contract theory plainly does not fit the stable distribution environment of motion pictures. Expected values are dominated by extraordinary events, which are rare and unpredictable but so large in magnitude that little else really matters. The incentive clauses of optimal contract theory are designed to alter the probabilities of favorable outcomes and raise expected values. But to apply standard incentive models in this setting, one would have to determine the impact of effort on the probabilities of tail events that are of zero Gaussian probability. It is hard to imagine that the relationship between effort and the odds of extremely rare events could be measured or predicted when motion picture outcomes are stable-distributed. Nor would such a relationship be verifiable in a court of law were the contract to be challenged. The asymmetric information often emphasized by optimal contract theory is not a factor either because both principle and agent are in a state of symmetric ignorance about the prospects of a movie owing to the "nobody knows" property.[23]

What does the stable model imply about contracts? Since "nobody knows" you contract to pay compensation after events have materialized and you do know. This is the option principle: delay decision until you have more information. Contingent compensation delays the decision of how much to pay until more information is available and everyone sees how well the movie performs. But, profit is private information, so the contract meets the requirement of verification by conditioning pay on gross revenue, which is accurately and openly reported by third party sources.

Extreme events drive the business, so contracts condition pay on rare events. This is "contracting in the tail" of the distribution, meaning pay is conditioned on rare events in the Paretian tail. The stable hypothesis states that the option-like pay clauses will be condition on events that may have vanishing or zero Gaussian probability.

How do the studios avoid the curse of the superstar? They do so by avoiding the payment of a fixed, upfront fee equal to the expected profit. This is done by conditioning compensation on the outcome of the movie.

Many Hollywood superstar movie contracts contain some form of participation in profit. Actors, screenwriters and directors on the A+ list are the only players whose fees are high enough to warrant profit participation. In a profit participation deal, the artist takes a smaller, fixed upfront fee in exchange for compensation which is contingent on the outcome. Contingent compensation avoids the superstar curse of an almost-sure-loss by paying a fee less than the expected profit. But it permits the star to capture part of the upper tail of the distribution. The positive skew and long tail of the distribution make it clear that this is where the action is. These contracts condition on events so far into the tail that they are of vanishing Gaussian probability, so they cannot be explained with that model. No one writes contracts with clauses that stipulate actions on events of zero probability.

In the movie business contingent compensation takes many forms, but there are some common features. Only outcomes on the positive tail are shared, the events must be well out on the tail to be "in the money," and compensation is roughly linear in profit. Because profit is not observable, many profit participation contracts are written on theatrical box-office revenues, which are readily monitored. In this case, the share of gross revenue paid often is nonlinear, with the share rising at higher outcomes, to reflect the nonlinear dependence of profit on revenue. In a complex contract, there may be several breakpoints where the star's percentage share increases, this nonlinearity reflecting the nonlinearity of profit in revenue.

More sophisticated yet is a contract, usually between studios cofinancing a movie, that allocates all or a different portion of revenue to different participants at different revenue points. These contracts allocate portions of the upper tail of the profit distribution. Thus, after the film has attained a revenue at which both parties have agreed that their costs are covered in a 50/50 split, or a split that approximates each's contribution to cost, they may "take turns" in receiving revenue at different revenue points along the upper tail. One studio might take all the revenue from $100 to $200 million, and the other all the revenue above $200 million. The alternation may continue at higher revenue. In their selection of portions of the upper tail, each studio can tailor its risk and profit potential. *Titanic* was co-financed by Twentieth Century Fox and Paramount who divided the upper tail this way. In this case, the holder of the far upper tail received the larger return.

Theatrical exhibition licenses are also written to share events in the center and the tail of the distribution. Events in the tail are the high-revenue weeks during a movie's theatrical run; these weeks can occur at any time during the run. High revenues trigger an escalation of the rental rate from a minimum percentage that may range during the run between 70 percent and 40 percent to a high rate of 90 percent of revenues in excess of the "house nut." During these high-revenue weeks, the rental clause allows the exhibitor to retain his (negotiated) cost per week of operation plus 10 percent while allocating 90 percent of the box-office revenue in excess of the nut to the distributor. Alternatively, the contract may contain a "best week" clause to the same effect.[24]

10.14   Conclusions

The stable Paretian model has the ability to capture many aspects of this exotic business where information dynamics are important but highly diffuse and potentially chaotic. Motion picture profit data obey an asymmetric stable distribution that has a finite mean and an infinite variance. The heavy-tail property of this distribution implies that events of extreme magnitude have a probability that is much larger than a Gaussian distribution would indicate. Thus, the world of the movies is fundamentally riskier than a Gaussian world. The stable model explains a number of puzzling aspects of the motion picture business: The infinite variance of profit supports the "nobody knows" principle. The skew and asymmetric tails of the distribution give rise to the "curse of the superstar"—to pay a star the expected profit of her movie almost surely guarantees a loss. A curious property of the stable distribution's Paretian negative tail implies the "angel's nightmare"—the cost of a movie that is over its budget is proportional to the cost already expended. A property of the profit distribution's Paretian positive tail is that success breeds success. The paradox that motion picture profit sharing contracts are conditioned on events that are of zero Gaussian probability is easily resolved by the stable distribution model. These and related problems might be found in other industries where the probability distributions of outcomes follow the Generalized Central Limit Theorem, but are not Gaussian.

11 Contracting with stars when "nobody knows anything"

11.1 Introduction

What do directors and actors have to do in order to become a top star? There is a surprisingly simple answer that has nothing to do with relationships or casting couches. If you want to be a star, direct or appear as lead in a movie that grosses $50 million or more in the North American market. One movie can do it, but it has to gross $50 million or more. That criteria alone is sufficient to identify the top 100 actors and directors. To become a super star, an artist must be associated with a movie that grossed at least $100 million. Their names will be familiar to nearly everyone.

Of course, this rule of stardom is not very helpful. How do you manage to direct or act in a movie that grosses $50 million or $100 million? One way is to already be a star, that is to already have at least one blockbuster on your resume. Once you are a star, better things come your way and the probability that you will show up in another hit increases. That, again, is not very helpful because it doesn't say how to get that first break. It really is true that being lucky and paying your dues is the way to stardom. Paying your dues by working steadily gives you more chances of appearing in a blockbuster or directing it.

In this chapter, I look at stardom and develop a simple model of the process of becoming a star. Then I look at a number of questions: How do producers place a value on a star and how do they bid for their services? What errors are they prone to in these tasks? Is anyone really worth $20 million? Or, do producers error in estimating the worth of a star? How do producers pay stars when no one knows how much a movie will gross? Who are the top directors and actors and by what measures do they come out on top? How durable is their status? How fragile are their careers? Can we distinguish between luck and talent? And if this distinction can be made, are artists paid in accordance with their talent?

Surprisingly, a good part of their success and productivity can be attributed to luck rather than talent. Throughout the analysis, we shall see how risk and uncertainty touch every aspect of an artist's career, from longevity to opportunities. As we should by now expect, there is evidence of statistical self-similarity on the creative side of the business, just as we have seen it on the revenue and cost sides. Extreme events and risk participate in all aspects of an artist's career.

It turns out that productivity and wealth are unevenly distributed in Hollywood. This uneven distribution of wealth and influence makes Hollywood a kurtocracy ruled by a handful of people whose wealth and influence are immense. Their influence is magnified by the way Hollywood is organized. Why wealth and influence are so highly concentrated in Hollywood is part of the mystery this chapter investigates.

11.2   On becoming a superstar

I am going to focus on the super stars, those who made a movie that grossed $100 million or more. This moves us up the ladder of stardom to the top 30 actors and directors. The rule for super stardom is to make a movie that grosses $100 million or more; mere stardom (top 100 status) can be achieved with a single film grossing $50 million or more.

Tables 11.1 and 11.2 rank the top 30 directors and actors by the accumulated North American theatrical gross revenues of their movies from 1982 through 2001. The tables list by artist the number of movies (n), the cumulative sum of their movie revenues (sbo), the largest grossing film (mbo) and the average box-office revenues of their films (abo).

Both lists contain the familiar names. The qualification shared by every one on it is that they made a movie that grossed at least $100 million. The top 30 artists are busy, each made no fewer than 6 movies as an actor or 4 as a director. George Lucas is an exception because his two Star Wars movies grossed enough to put him among the top grossing directors.

In keeping with the dominance of extreme events in the movies, cumulative grosses depend heavily on a single movie. This is more true of directors than actors. *Forrest Gump*, one of the 23 movies Tom Hanks made, accounts for 14 percent of his cumulative gross. The same movie is one of 10 directed by Robert Zemeckis and it accounts for 22 percent of his cumulative gross. *Titanic* earned 53 percent of the gross revenues earned by James Cameron's 7 movies and 80 percent of Leonardo DiCaprio's cumulative gross from 6 movies. For actors, the portion of cumulative gross that is due to one film averages 22 percent and is no less than 10 percent; its high value is 80 percent. For directors, the average portion of their cumulative gross that is due to one film is 32 percent and this varies from a low of 16 percent to a high of 75 percent.

The lesson is clear. If you want to be a super star direct or act in a movie that is a monster hit. There are no exceptions to this rule. Every director or actor who is in the top 30 had a movie that earned at least $100 million. This point, be lucky enough to make a hit, holds as one goes lower in the rankings too.

Here is another pattern: nearly everybody starts out small, with minor films that earn modest revenues. All the super star actors and directors experience

Table 11.1 Top grossing actors

	cst1	n	sbo	mbo	abo
1	Hanks, Tom	23	2.44e+09	3.30e+08	1.06e+08
2	Gibson, Mel	23	1.70e+09	1.83e+08	7.37e+07
3	Murphy, Eddie	19	1.66e+09	2.35e+08	8.75e+07
4	Ford, Harrison	19	1.62e+09	1.97e+08	8.54e+07
5	Cruise, Tom	16	1.58e+09	2.15e+08	9.84e+07
6	Williams, Robin	23	1.55e+09	2.19e+08	6.74e+07
7	Schwarzenegger, Arnold	20	1.42e+09	2.05e+08	7.12e+07
8	Willis, Bruce	21	1.40e+09	2.94e+08	6.67e+07
9	Travolta, John	23	1.21e+09	1.40e+08	5.26e+07
10	Carrey, Jim	10	1.18e+09	2.60e+08	1.18e+08
11	Stallone, Sylvester	22	1.16e+09	1.50e+08	5.27e+07
12	Douglas, Michael	19	1.13e+09	1.57e+08	5.93e+07
13	Costner, Kevin	19	1.12e+09	1.84e+08	5.92e+07
14	Roberts, Julia	10	9.05e+08	1.52e+08	9.05e+07
15	Martin, Steve	21	8.38e+08	1.00e+08	3.99e+07
16	Eastwood, Clint	17	7.96e+08	1.02e+08	4.68e+07
17	Neeson, Liam	14	7.89e+08	4.31e+08	5.63e+07
18	Fox, Michael J.	15	7.77e+08	2.08e+08	5.18e+07
19	DiCaprio, Leonardo	6	7.46e+08	6.01e+08	1.24e+08
20	Keaton, Michael	16	7.42e+08	2.51e+08	4.64e+07
21	Myers, Mike	6	7.04e+08	2.64e+08	1.17e+08
22	Connery, Sean	15	6.99e+08	1.34e+08	4.66e+07
23	Washington, Denzel	17	6.58e+08	1.16e+08	3.87e+07
24	Hoffman, Dustin	10	6.53e+08	1.77e+08	6.53e+07
25	Gere, Richard	18	6.41e+08	1.78e+08	3.56e+07
26	De Niro, Robert	24	6.31e+08	1.66e+08	2.63e+07
27	Cage, Nicolas	18	6.25e+08	1.02e+08	3.47e+07
28	Bullock, Sandra	10	6.07e+08	1.09e+08	6.07e+07
29	Reeves, Keanu	14	6.05e+08	1.71e+08	4.32e+07
30	Murray, Bill	9	5.90e+08	2.39e+08	6.55e+07

n, number of movies; *sbo*, cumulative sum of movie revenues; *mbo*, largest grossing film; *abo*, average box-office revenue.

a turning point in their career when they make a $50 million movie. From then on, projects just seem to flow and they tend to become busy and get opportunities to make many movies. Since it is hard to direct more than a movie per year or act in more than three per year, the super stars tend to have long careers. If you don't make a hit, you can still have a career, but it is less lucrative and more uncertain. There is such a thing as a big break in Hollywood—no one can become a top star without a break-out movie that makes at least $100 million.

You can identify the real super stars in Hollywood if you raise the bar to three or more movies that gross at least $100 million. Only 16 stars have accomplished this feat in the past two decades. If you raise the standard to

Table 11.2 Top grossing directors

	dir	*n*	*sbo*	*mbo*	*abo*
1	Spielberg, Steven	13	2.08e+09	4.00e+08	1.60e+08
2	Zemeckis, Robert	10	1.52e+09	3.30e+08	1.52e+08
3	Cameron, James	7	1.13e+09	6.01e+08	1.61e+08
4	Howard, Ron	13	1.13e+09	2.60e+08	8.67e+07
5	Columbus, Chris	9	9.59e+08	2.86e+08	1.07e+08
6	Donner, Richard	12	8.87e+08	1.47e+08	7.39e+07
7	Reitman, Ivan	10	8.46e+08	2.39e+08	8.46e+07
8	Schumacher, Joel	14	7.78e+08	1.84e+08	5.56e+07
9	Burton, Tim	8	7.29e+08	2.51e+08	9.12e+07
10	Scott, Tony	10	7.28e+08	1.77e+08	7.28e+07
11	Levinson, Barry	16	7.16e+08	1.73e+08	4.48e+07
12	Emmerich, Roland	6	6.64e+08	3.06e+08	1.11e+08
13	Eastwood, Clint	15	6.15e+08	1.01e+08	4.10e+07
14	Sonnenfeld, Barry	6	6.10e+08	2.51e+08	1.02e+08
15	Lasseter, John	3	6.00e+08	2.46e+08	2.00e+08
16	Bay, Michael	4	5.99e+08	2.02e+08	1.50e+08
17	Petersen, Wolfgang	9	5.76e+08	1.83e+08	6.40e+07
18	Lucas, George	2	5.69e+08	4.31e+08	2.85e+08
19	Scott, Ridley	11	5.67e+08	1.88e+08	5.16e+07
20	Marshall, Garry	11	5.66e+08	1.78e+08	5.14e+07
21	McTiernan, John	9	5.63e+08	1.21e+08	6.26e+07
22	De Palma, Brian	11	5.28e+08	1.81e+08	4.80e+07
23	Shadyac, Tom	4	5.17e+08	1.81e+08	1.29e+08
24	Pollack, Sydney	6	5.17e+08	1.77e+08	8.62e+07
25	Stone, Oliver	12	5.14e+08	1.38e+08	4.28e+07
26	Reiner, Rob	12	5.09e+08	1.41e+08	4.24e+07
27	De Bont, Jan	4	5.03e+08	2.42e+08	1.26e+08
28	Nichols, Mike	11	4.85e+08	1.24e+08	4.41e+07
29	Musker, John	5	4.52e+08	2.17e+08	9.03e+07
30	Craven, Wes	11	4.41e+08	1.03e+08	4.01e+07

six hits, you are left with just nine actors: Jim Carrey (6), Tom Cruise (7), Harrison Ford (6), Mel Gibson (8), Tom Hanks (11), Eddie Murphy (7), Julia Roberts (6), Arnold Schwarzenegger (6) and Robin Williams (6).

It is easier to remain a star than it is to become one. When you become a star, you gain access to more opportunities. You are offered the best roles in better movies with higher production and advertising budgets. And your movies open on more screens. So, the problem is easy; find that great role in a movie that will gross more than $100 million. Then make careful choices in selecting your new projects and you are a super star.

But, how do you get that first big role in a hit? It is the movie that makes someone a star. Yet, no one knows how much a movie is going to make. You just have to be lucky. Only about 3 percent of movies are big hits; just 196 movies out of 6,289 movies released in North America during the two decades from 1982 through 2001 grossed at least $100 million. So, at the most, there is just a 3 percent chance of becoming a star. Just about half of the hits were first time break-out movies for actors and about half of the roles in hits went to stars who already had a big movie under their belt. There are about 200 hit movies per two decades, according to the sample data. About 100 of these roles will go to established stars and 100 to new ones. This leaves room for about 50 actors per decade to break out in a $100 grossing movie.

According to the model, Sylvester Stallone broke out in 1982, Harrison Ford and Eddie Murphy in 1984, Tom Cruise in 1986, Robin Williams in 1987, Tom Hanks and Arnold Schwarzenegger in 1988, Mel Gibson in 1989. Sean Connery and Bruce Willis broke out in 1990 and Julia Roberts in 1991. Brad Pitt hit in 1995. Nicolas Cage, Matt Damon and Leonardo DiCaprio all broke into the $100 million plus ranks in 1997. They remain top stars today if we measure by the number of roles they get or what they are paid. It is a bit surprising how recently these actors came on the scene, so familiar are their names. Only Stallone has been around for 20 years. The average length of time on the list (at this date, they may stay longer) is 12.33 years. The differences among them are not that large; the average deviation from the mean is 3.9 and the median deviation is just 3.

11.3 Artists and kurtosis

Kurtosis measures the skew and unevenness of a distribution. In Hollywood, the distributions of work, income and status are highly skewed; their mass is piled high on the left on the low outcomes and their upper tail to the right is long and heavy; this is where the extraordinary events are to be found. The kurtosis of a normally distributed random variable is just 3. Low kurtosis makes things similar; no one is far above the mean and extraordinary individuals are vanishingly rare. Low kurtosis makes a mediocracy. Hollywood is high on kurtosis; it is a kurtocracy rather than a mediocracy. The kurtosis of box-office revenue is 45 and the kurtosis of the cumulative box-office revenues of an actor's body of work is an extraordinary 104. These are levels many times the kurtosis of a normal distribution.

Even though most movie fans haven't heard of kurtosis, they intuitively know that the movie business is a kurtocracy. This is evident in the fact that their favorite actors or directors are many times more successful than less-favored talents; these are the kurtocrats. High kurtosis is what gives Hollywood its pecking order and hierarchical structure. The differences between the talents of kurtocrats and others may be small, but, as we have seen in other chapters, nonlinearity can magnify small differences in talent (or luck) into extreme differences in outcomes.

Kurtosis is evident in the degree to which the industry is dominated by a few directors and actors. Consider Figures 11.1 and 11.2. These are histograms of the number of movies made and released in North America by all active directors and actors over the 20 year period from 1982 through

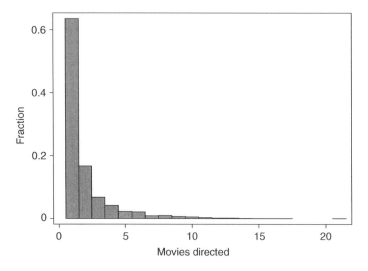

Figure 11.1 Number of movies made by directors.

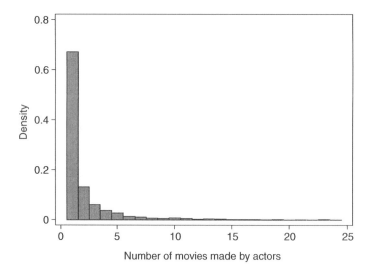

Figure 11.2 Number of movies made by actors.

2001. They are remarkably similar (statistical self-similarity again) and have the characteristic shape of high kurtosis.

This similarity shows in the statistics of productivity, which are remarkably similar. The mean number of movies made by directors and actors

Table 11.3 Summary statistics of movies acted

	Percentiles	Smallest		
1%	1	1		
5%	1	1		
10%	1	1	Obs.	2,818
25%	1	1	Sum of wgt.	2,818
50%	1		Mean	2.12704
		Largest	Std dev.	2.697013
75%	2	23		
90%	4	23	Variance	7.27388
95%	7	23	Skewness	3.941963
99%	15	24	Kurtosis	21.90554

Table 11.4 Summary statistics of movies directed

	Percentiles	Smallest		
1%	1	1		
5%	1	1		
10%	1	1	Obs.	3129
25%	1	1	Sum of wgt.	3129
50%	1		Mean	1.991051
		Largest	Std dev.	2.01717
75%	2	16		
90%	4	16	Variance	4.068974
95%	6	17	Skewness	3.211839
99%	11	21	Kurtosis	16.07401

are remarkably similar, 1.99 for directors and 2.12 for actors (Table 11.3 and 11.4). Both medians are 1. The mean is located at the 75th percentile in each case. The skewness coefficients are similar. Actors have a bit more kurtosis than directors, 22 for actors and 16 for directors. That is to say, the productivity differences among actors is greater than among directors. The most active actor (De Niro) made 24 movies; the most active director (Woody Allen) made 21.

What about revenues? If we sum the revenues the movies each director and actor made from 1982 through 2001 we obtain a measure of the gross revenues earned by their movies.[1] The mean cumulative actor gross is $32.8 million and the kurtosis is very large 103.99. The median is $1.01 million, far less than the mean, which shows the influence of a few high grossing movies. Indeed, the mean is far up into the tail at the 85th percentile, showing that a few stars have cumulative grosses that are massive. In fact, the largest cumulative gross goes to Tom Hanks at $2.44 billion.

According to the stardom model, an actor who gets exposure in a hit movie becomes well-known. Once that top star status is earned, it doesn't go away easily because it is hard to disentangle performance from movie. Causality is hard to nail down in a complex system, so when a movie is a big hit the most readily identified element, the artist, gets much of the credit. Once a reputation is so constructed it becomes an advantage in securing future roles in more ambitious movies. As more opportunities come, the actor becomes more famous, and still more opportunities open up. This is nonlinear, positive feedback and it can carry actors or directors to heights of frenzied activity. But, fans can tire of an actor. If her grosses fall off, or if she chooses a bad project, the nonlinearity can go the other way. Low grosses translate into fewer opportunities which feeds the downward spiral until the artist no longer gets any calls. We can see this process at work by examining career statistics.

11.4   Career expectations and opportunities

What are the career expectations of actors and directors? As we saw, a director can expect to make two movies. But, the usual career, experienced by about 75 percent of directors, is to direct just one movie. The situation is similar for actors.

One of our extreme distributions, the stable Paretian distribution, best describes the situation artists face. For directors and actors I have estimated the tail exponent of the Paretian distribution of movies made (seen in Figures 11.1 and 11.2). The values of the tail weight coefficients are 1.5 for directors and 1.8 for actors. This means that the expectation exists for both distributions, but the variance is infinite since $\alpha < 2$. The infinite variance of the Paretian distribution is a warning to remember that the longest career and sample variance will continue to change and their largest values will never diminish as the sample grows. Hence, our estimate of the expected career of an artist will always be subject to revision and an ever increasing error.

A special property of the Paretian distribution gives a surprising twist to the career expectations of artists. A director who has made n movies can expect to make many more. This follows from a property of the Paretian distribution; its conditional mean rises with the conditioning event. The conditional mean of a random variable x, given that $x > x_0$, equals $\bar{x}_{x_0} = x_0 \alpha / (\alpha - 1)$. Since the estimated value of $\alpha = 1.5$ is greater than one, the conditional mean exists. Hence, the expected number of movies a director will make, conditional on having made n is $n \times \alpha / (\alpha - 1) = n \times 1.5/0.5 = n \times 3$. The comparable value for actors is $n \times 1.8/0.8 = n \times 2.25$.

Note what these values mean. A director who has made n movies can expect at that point forward to make $3 \times n$ movies in all. An actor, having made n movies, can expect to make $2.25 \times n$ movies in all. A director who made, say, 3 movies has a career expectation of making 9. Having made 3 movies, an actor can expect to make 6.75. Sadly, this leads to the law of the artist's cemetery: the artists buried there died with one half to two thirds of their careers ahead of them. When a director's career ends, he leaves two thirds of his promise unfulfilled; a dying actor leaves just less than half or her career work unfinished.[2]

The expansiveness of the conditional expectation is best summarized by saying that success breeds success. The odds of making another film depend positively on the number already made.[3] When an artist becomes famous, more opportunities come his way, he makes more movies, his fame increases, and he gets more opportunities, and so on. This positive feedback is responsible for the vastly different careers and grosses among artists. Small differences in talent or plain luck are magnified enormously by positive nonlinear feedback.

Another implication of these results is a variant of the law of box-office champions. In the movies, as in sports, super stars have the longest careers (see the law of box-office champions in Chapter 1). When it comes to artists the law means that the longer you have been on top, the longer your future career is likely to be.

Why do these nonlinear expectations hold? In terms of the dynamics of box-office revenues, we know that small differences among movies are magnified by nonlinear feedback to produce a highly kurtotic distribution of revenues. This makes it possible for an actor or director who is only slightly more talented to earn many times what a less talented artist earns.

There is another nonlinear effect stemming the way production is organized. Movies are produced by temporary organizations of skilled workers and artists assembled by producers. Producers are the hubs of personal networks of contacts among a pool of freelance artists. Reputation, trust and work flow over this network according to the number of connections. When an artist makes a hit, she becomes more densely connected in this network (ask Nicole Kidman what happened to her opportunities after *Moulin Rouge*). The network of contacts becomes more densely connected with each hit the artist makes. This increasing connectedness expands opportunities for the artist, probably as a power law. Network effects produce a nonlinear flow of opportunities amplifying small differences in talent into vastly different outcomes.

Because they have established reputations and are densely connected to the network of work and opportunity, the top creative artists are far more productive than their peers, whether productivity is measured by the number of movies they make or by their grosses. The less connected function along the edges of the industry, working sporadically and hanging on as they wait for their break.

11.5 Luck or talent?

How much of the hierarchy of stardom is due to talent? And how much is just pure luck? Suppose we start off by assuming everyone is equally talented.

Let a purely random coin flip determine whether a director's or actor's movie is a hit or a flop. Heads is a hit and tails is a dud. When the artist makes a hit, she gets another chance. When he makes a flop, his career is over.

If we start with a thousand directors or actors and each makes a movie with a fifty/fifty chance that it succeeds or fails, then after one year, a half would be left. The next year, only a quarter would survive, and then one-eighth. Only 1.5 percent of them would make 10 movies in a row and only 0.7 percent would make 15 movies without a failure. And it would be entirely up to the luck of the coin tosses.

How far is this randomly generated survival distribution from the actual distribution of movies made by actors and directors active in the past two decades in Hollywood? Let's compare the statistical distributions to chance. The Pareto function that fits the director data is $f(n) = kn^{-1.5}$ where k is a normalization constant (r^2 is 0.91 and the coefficient is highly significant). The same function fits the actor data, but with a value of -1.8 for the exponent. These functions are shown in Figure 11.3 along with what would happen under pure chance. The figures are statistically self-similar and strikingly close to one another.

Using the pure luck distribution and the Pareto distribution, that actually fits the data, to calculate the probability that actors and directors will make n movies gives the odds in Table 11.5. As is evident, until an actor or director has made 7 movies the odds are no different than what we would have from pure chance. So, there is no telling whether it is luck or talent at that point. Beyond 7 movies the odds depart from pure chance and we

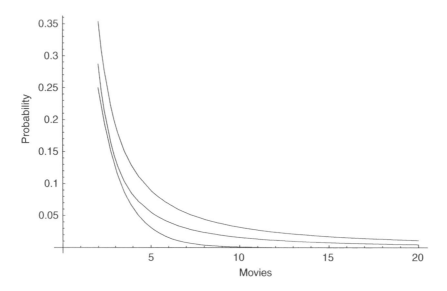

Figure 11.3 Probability comparison among actors, directors and chance.

Table 11.5 Probability of making n movies

Movies	Actors	Directors	Luck
2	0.2584	0.2651	0.2500
3	0.0830	0.0962	0.1250
4	0.0371	0.0468	0.0625
5	0.0198	0.0268	0.0312
6	0.0119	0.0170	0.0156
7	0.0077	0.0115	0.0078
8	0.0053	0.0083	0.0039
9	0.0038	0.0062	0.0019
10	0.0028	0.0047	0.0009
11	0.0022	0.0037	0.0004
12	0.0017	0.0030	0.0002
13	0.0013	0.0024	0.0001
14	0.0011	0.0020	0.00006
15	0.0092	0.0017	0.00003
16	0.0007	0.0015	0.00001
17	0.0006	0.0012	$7.629e - 06$
18	0.0005	0.0011	$3.814e - 06$
19	0.0004	0.0009	$1.907e - 06$
20	0.0004	0.0008	$9.536e - 07$

can see something like reputation or talent begin to show. The odds that a director will make 15 movies is very small, only 0.0092, but this is 56 times higher than pure chance. And it is almost 500 times more likely that a director will direct 19 movies than pure chance. The odds for actors are similar.

The high odds ratios for the most prolific directors suggests there is something beyond luck in determining how many movies a director will make. In seeking to further draw a line between luck and talent, we rely on a remarkable property of the Paretian distribution. A merely lucky director would find that the probability of succeeding with her next film is 0.5. And this would be the same for each film, no matter how many the director made. That is to say, the probability of success is not altered with experience as measured by the number of successful films made.

If talent, skill, or learning have anything to do with success, then the probability of success should not remain constant; it ought to increase with the number of successes realized. And that is just what the Pareto distribution implies. Under the Paretian probability law, the probability that the random variable $x \geq x_0$ is $P[x > x_0] = (x_0/x)^\alpha$. This is increasing in the conditioning variable x_0 (Feller 1957; Samorodnitsky and Taqqu 1994).

Given this property, suppose a director has already made 5 movies. What is the probability she will make 6 or more movies? It is $(5/6)^{1.5} = 0.76$. We can generate the conditional probability that a director will make more movies, given that she has made n of them, from the Paretian distribution.

Table 11.6 Probability of making more movies conditional on having made *n* movies

n	*P[movies > n|n]*
1	0.35
2	0.54
3	0.64
4	0.71
5	0.76
6	0.79
7	0.81
8	0.83
9	0.85
10	0.86
11	0.87
12	0.88
13	0.89
14	0.90
15	0.90
16	0.91
17	0.91
18	0.92
19	0.92
20	0.92

Table 11.6 lists the conditional probabilities that a director who has made *n* movies will make more for various numbers of movies. Notice that the probability of making more movies rises with the number of movies a director has already made. Having made 2 movies, the probability of directing another one is 0.64, which is well above the random value of 0.5. Having made 3, the probability of directing more rises to 0.71 and this value continues to rise with the number of movies directed. This feature captures our intuition about experience or reputation.

On the other hand, we might get a similar pattern if studios failed to recognize survivor bias in selecting actors or directors for their movies. Suppose an actor's first three movies are successful when the process is purely random. Suppose a studio were to compute the conditional probability according to the proportion of successes when the true probability is just 0.50. The director would have to make three failures in a row before the studio would drop him. The probability of this event is only 0.125, so the director still has a reasonable probability of continuing her career for some time to come. The point here is that a decision model based on relative frequencies could generate statistics virtually identical with the data. In other words, pure luck combined with a failure on the part of studios to properly recognize survivor effects would produce almost precisely what we find in the data.

11.6   Competing for stars

Since 1995, from 25 to 47 movies per year have been released with budgets of at least $50 million for an average of 33 big budget movies per year. These movies compete for a small group of established stars. Only about 11 of these will earn $100 million or more. About 5 low budget movies per year will gross more than $100 million; they are the source of new stars. If every movie with a large budget tries to hire an established star, one with 3 previous hits, there will be 33 movies per year chasing 18 actors. If they look for an actor with just 2 or more previous hits, there will be 36 of them.

Suppose the process of signing stars to pictures goes like this: producers rank stars by placing a value on each star they would like to cast as the lead in their next movie. They make their first offer (a script and fee) to the star at the top of their list. If they are refused, they go to the next actor on the list. They continue in this fashion until they get a leading actor signed to their movie. This is a sequential bidding process for lead actors. The producer tells his agent, "Get me Hanks, get me Cruise if Hanks won't do it, and then try Ford, followed by Gibson..." and so on. Since there are 33 big budget projects this year, Hanks gets 33 offers and takes the best, Cruise gets 32 offers and takes the best of them, Ford agrees to the best of his 31 offers, and so on. We are sorting actors and projects sequentially. The actors at the top receive the most bids and the competition is greatest for their services. Bids and competition diminish down the list as the actors above pick their projects.

The producer works down the list getting refusals, an acceptance, or finds that the actor is already committed. Eventually, some 33 actors will be hired to star in the big budget movies scheduled for the year. To secure all these actors, the producers of these movies will have to go more than 33 actors down the list, allowing for refusals, commitments, and for the fact that there must be some match between the character and the actor. To fill all the roles, some producer is likely to have to go down the list some 50 to 100 slots.

This is a private value auction because each producer bidding places a personal value, unknown to other producers, on each actor on the list. The price of each actor's services is determined by the second price principle. The price of the last actor hired, will be less than the producer's value but more than the actor's opportunity cost. The price of the second to last actor hired is between the highest and second highest values of the two producers bidding for her.

As we go up the sequence of actors ranked according to producer values, the price will move toward the second highest value among the active bidders. That is, no producer will pay more than his personal value for the actor, but he must bid above the value the producer just below him places on the same actor. This pushes the actor's price to the second highest value among the producers actively bidding for the actor. As a result of this competitive bidding, actors at the top ranks can earn far more than those below them.

This is a private value auction because no producer discloses his personal values of the actors. Nor are the values revealed when the deal is made, because the terms of contracts remain private. Because values are not known or revealed, the winner's curse raises its head. The winner's curse occurs if you win the actress against many competing bids. This tells you that you had the highest estimate of value among many bidders and you probably are wrong; at least a lot of other people think your bid is too high. The problem is more severe in bidding for the top stars because there are more bidders. The winner's curse implies that the top stars are much more likely to be paid more than they are worth unless producers shade their bids downward whenever they face a large number of competing bidders.

There is a further problem, pointed out in Chapter 10. When the producer estimates the value he places on a star, how is he to do this? He has to think about probabilities and form expectations. But, we have seen, that is hard to do. And few producers will be aware of the exotic properties of the probability distribution they are up against. The curse of the super star makes its appearance. The producer must be aware that the most probable outcome for her movie is far less than the expected outcome. This means her bid will almost surely be too high if it is based on her expectation. This is completely different from the situation when the distribution is Gaussian; in that case, the expected and most likely values of motion picture revenue are equal. But, the distribution of probabilities is not Gaussian, so it is irrational to bid an amount close to expected value. The problem is that the producer is dealing with a small sample, one movie, where the most likely event dominates and the more likely outcome (the mode) is a value far below his expectation. This means the producer must bid less than the expected value she puts on the artist.

So far we have two reasons why top artists are likely to be paid more than their worth—the winners curse and the curse of the super star. Here is another problem for our producer. She has to form a probability function of outcomes for each actor she sets a value on. This is simply an impossible task and the evidence tells us there is really no foundation for doing it. The producer must engage in a mental exercise something like this: what will the movie gross if Hanks does it, or if Cruise does it, or if Ford does it, and so on? But, he must go further because we know that the expectation is higher than the most likely outcome and, after all, the forecast has no precision because the variance is infinite.

What the producer must do is to determine the probabilities of different outcomes with each of the prospective actors. He is unlikely to have a staff sophisticated enough to do this. Chapters 4–6 show that stochastic dominance is the tool to be used for this task. But, we showed there that few actors show any association with the probabilities of outcomes. There is some evidence that only three or four stars are associated with a higher least outcome, that is they may affect the worst outcome favorably. And this is primarily related to having more screens at the opening. But, we know that the opening conveys almost no information about the eventual revenue of a movie that turns out to be a hit.

A further point should be made about the task of forming reasoned conjectures of the possible outcomes: we cannot say that a star is the cause of better outcomes even when his movie has a high gross. There is only a statistical association, weak and poorly understood, between a handful of stars and the revenues of their movies. Correlation is not causation. The star may have had better movies to select from and the effect could be this rather than the star's presence. And there is the real possibility that the movie makes the star; after all, that is the simplest and, at this point, best explanation of stardom. In the epilogue, I add another problem to this long list of things that tend to drive super star fees above their true worth. In that chapter I explain the sure thing principle that biases the producer's assessment of probabilities in favor of established stars over other artists.

What is a producer to do in the face of all these problems that seem to lead him to pay super stars too much? What the producer must do is what producers have always done: use judgment and take reasonable steps to reduce the risk. Or, structure the contract in such a way that the amount the actor receives is in part determined by how much money the movie makes. This requires paying the actor in two parts; one part is a fixed fee that is less than the actor's expected value, and another part that depends on the movie's box-office revenue or profit. We will look at this two-part compensation scheme.

11.7 Artist contracts

Given that stars are made by hits, why don't producers capture the rents that come to the actors they make into stars? After all, they gave the actor her chance to break out. In the days of the studios, that is what happened. Artists were under long-term contracts. The studios developed stars, probably paying them more than they were worth to start, and then retaining some of the rent for themselves when the artist becomes a star. This practice recognized that the source of stardom is the chance to appear in movies and it directed some of the rents of stardom to the studio in light of the opportunity they granted. Once the studios were no longer able to hold stars to long-term contracts, they were no longer able to retain the rents of stardom. Star salaries escalated in the "free agency" conditions created by the demise of long-term studio contracts.

A hold-over of the studio era is the conditional contract that gives the producer the right to use an actor in subsequent movies. When Christopher Reeve signed on as the first Superman, both he and the producer knew that he would be in demand if the movie was successful. To anticipate that the movie would make Reeve a star and that his persona would be associated with Superman, the producer contracted to have Reeve appear in sequels at a salary below his market value if *Superman* turned out to be a hit. Recognizing that Reeves' commercial value would also be increased in other movies, the producer could have gone farther; his contract could have been written so that Reeve had to pay the producer a percentage of what he earned from working on other movies. That, in effect, was the nature of the long-term studio contract. It would be difficult to enforce such a contract today and it may even be illegal. Many would think it unethical, but it makes a good deal of sense.

11.8   Pay when you do know

There is really only one answer to the problem of contracting with top artists. If you don't know what a movie will earn or how much a star contributes, then wait until you do know before you pay. This is the option principle. It cuts right through the uncertainty. The artist, director or actor, is paid a conservative fixed fee that is less than her expected value along with a portion of net revenues. This is the two-part scheme we mentioned earlier that solves the bidding problem.

The pay-when-you-do-know contract is an option. The artist supplies services in exchange for a fixed fee and a portion of the film's revenue. The option comes into the money and pays off if the film's revenue hits certain target values. I call this contracting in the tail of the distribution, since the option contract pays off only on events far out into the upper tail of the distribution.

We have anecdotal evidence these kinds of two-part pricing schemes are used: Dannie De Vito and Arnold Schwarzenegger made a fortune on *Twins* because they took a percentage of revenue in lieu of their customary fixed fee. No one would take the risk of making *Twins* otherwise. Tom Hanks did the same thing on *Forrest Gump* and made a killing off it. Jimmie Stewart was not a rich man until he made *Winchester 45* for a percentage of the gross. We also have direct evidence from Darlene Chisholm's research. She analyzed 118 contracts between actors and producers and found that actors who are compensated through a share of revenues differ from those who receive fixed payments. They have more experience, they have at least one large grossing movie, and they have some Oscar recognition either as nominees or winners. Supporting the network property, they also have a history of prior collaboration with the producer on the other end of the contract. As I argued above in the discussion of *Superman*, share arrangements are more common for sequels than for first off movies.

Both theory and evidence indicate that, as film budgets escalate, producers will seek established artists they have previously worked with and offer them a share of revenues in exchange for a lowered fixed fee. This is borne out by Professor Chisholm's analysis, but the details of the sharing arrangements remain obscure. In this section and the one following, I test the implications of the Paretian contracting model using a sample of director contracts.

The questions I am interested in are these. Are director fees conditional on the outcomes of the films they direct? Is contingent compensation roughly

Table 11.7 Statistics of director pay

	Percentiles	Smallest		
1%	278,441	278,441		
5%	278,441	527,374		
10%	527,374	1,726,319	Obs.	13
25%	1,837,100	1,837,100	Sum of wgt	13
50%	1.08e+07		Mean	1.16e+07
		Largest	Std dev.	1.42e+07
75%	1.39e+07	1.39e+07		
90%	1.70e+07	1.55e+07	Variance	2.03e+14
95%	5.49e+07	1.70e+07	Skewness	2.285522
99%	5.49e+07	5.49e+07	Kurtosis	7.809765

linear in revenue? Do the contracts condition on extreme events that are of negligible Gaussian probability? Are cost overages penalized?

I begin with the statistical evidence in Table 11.7 which presents summary statistics of a sample of director contracts. The mean directing pay from all sources is about $11.6 million for the prestigious directors in the sample and the standard deviation is $14.2 million. The range of the data is large; the highest pay is nearly 200 times the least pay. Pay is skewed to the right and the average is between the median and the 75th percentile. Skew and kurtosis are large. These are all earmarks of the stable Paretian distribution.

The option principle is at work as there is a high correlation between what directors are paid and the gross revenues of the movies they direct. The correlation between pay and gross is 0.8186 and between cumulative pay and cumulative gross is 0.9743. This latter correlation is larger because these directors gained stature over the sample time period and, thus, began to receive participation fees. The correlation between pay and profit is a bit weaker at 0.7443. The correlation of cumulative pay with cumulative profit is 0.9593; a bit less than the cumulative pay/cumulative gross correlation. The graph in Figure 11.4 shows the data for pay and gross. The pattern of cumulative pay and cumulative gross suggests that pay shifts from a flat fee to contingent pay after the director's movies achieve a certain level of cumulative gross revenues, or (and this comes to the same thing) after a movie they direct grosses $100 million or more.

Director pay follows a Pareto distribution. The estimated value of the tail coefficient, α, of the upper tail of the distribution of the gross revenues is reported in Table 11.8. Figure 11.5 shows the fit in this upper tail. The tail coefficient for the upper tail of the distribution of pay is 1.53, see Table 11.9. When the tail coefficient of the distribution of grosses of these director's movies is estimated the value is 1.49. So, the α values for both movie grosses and director pay are approximately equal, 1.49 versus 1.53. The distributions are self-similar in appearance and in their tail weights. This strongly suggests that director pay is a linear transformation of gross, since a transformation

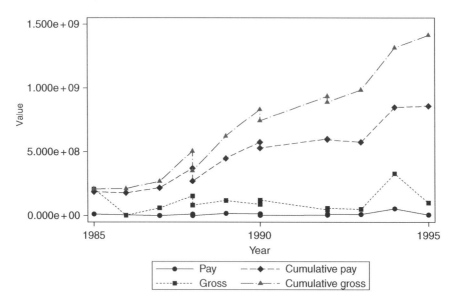

Figure 11.4 Gross, cumulative gross, pay and cumulative pay by year.

Table 11.8 Estimate of α coefficient value for gross revenue

Source	SS	df	MS			Number of obs.	=	11
						$F(1, 9)$	=	134.38
Model	5.20340873	1	5.20340873			Prob > F	=	0.0000
Residual	0.348484712	9	0.038720524			R-squared	=	0.9372
Total	5.55189344	10	0.555189344			Adj R-squared	=	0.9303
						Root MSE	=	0.19678

lpg	Coeff.	Std err.	t	P > |t|	[95% Conf. interval]	
lgr	−1.492448	0.1287437	−11.59	0.000	−1.783687	−1.20121
_cons	26.33478	2.356481	11.18	0.000	21.00405	31.66551

of a stable-distributed variable is also stable-distributed with the same *alpha* coefficient. Earlier, we estimated survival probability of a director's career. This too is Paretian with a value of the $\alpha = 1.5$. All three distributions have a tail coefficient of $\alpha \sim 1.5$.

It seems remarkable that the tails of these distributions should be so similar. But, it is not so remarkable when you recognize that they represent different pictures of the same thing: the shape of probability in the extreme upper tail of events in three measures—movies directed, grosses and compensation. Such a finding of self-similarity is not surprising in view of the considerable research reported in other chapters of this book, and by other authors, that the stable

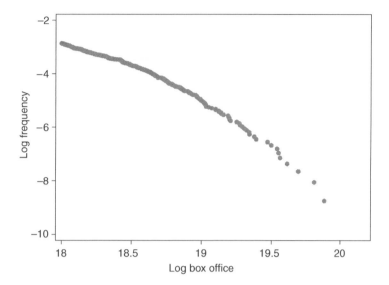

Figure 11.5 Paretian upper tail of gross revenue.

Table 11.9 Estimate of α coefficient value for director pay

Source	SS	df	MS			
				Number of obs. =	8	
				F(1, 6) =	15.46	
Model	2.49504557	1	2.49504557	Prob> F =	0.0077	
Residual	0.968304532	6	0.161384089	R-squared =	0.7204	
Total	3.4633501	7	0.4947643	Adj R-squared =	0.6738	
				Root MSE =	0.40173	

lpayf	Coeff.	Std err.	t	P > t	[95% Conf. interval]	
lpay	−1.532411	0.3897323	−3.93	0.008	−2.486052	−0.5787705
_cons	23.57605	6.29246	3.75	0.010	8.178959	38.97315

Paretian model captures the statistical details of the movie business with great fidelity and insight. If the probability of success is Paretian distributed, we should expect the number of films directed and the pay of directors to be similarly distributed.

As these results and the Paretian contracting model suggest, contingent pay is linear in gross revenue. The estimates of the linear equation are reported in Table 11.10. A nonlinear function does not fit as well. This is consistent with the option principle which says that contingent pay is conditioned on observable outcomes in the upper tail. Neither a linear nor a nonlinear model of pay as a function of profit fit the data as well. Nor does pay fit profit as well as it fits gross revenue. The estimates indicate that the total compensation received by the successful and well-established directors in the sample is equal

Table 11.10 Regression of director pay and domestic box-office gross revenue

Source	SS	df	MS		
				Number of obs. =	12
				$F(2, 9)$ =	19.38
Model	1.8861e+15	2	9.4303e+14	Prob > F =	0.0005
Residual	4.3784e+14	9	4.8649e+13	R-squared =	0.8116
Total	2.3239e+15	11	2.1126e+14	Adj R-squared =	0.7697
				Root MSE =	7.0e+06

Pay	Coeff.	Std err.	t	P > t	[95% Conf. interval]	
year	1811101	696631.9	2.60	0.029	235209.7	3386991
gross	0.1448944	0.0256821	5.64	0.000	0.0867974	0.2029914
_cons	−3.61e+09	1.39e+09	−2.60	0.029	−6.75e+09	−4.73e+08

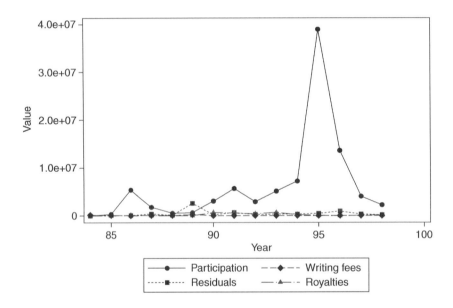

Figure 11.6 Fees and participation pay by year.

to 0.14 of the gross North American theatrical revenue of their movies (adding in gross theatrical revenues from other markets and other sources of revenue would cut this to about 7 percent of world gross and video royalties).

There is no doubt that contingent income is the dominant source of director pay as Figure 11.6 clearly indicates. Recall that the option principle says that the fixed directing fee is the fixed component of a two part compensation scheme that depends on the outcome. Participation far exceeds directing fees in almost all years, the early years before their reputations were established are the only exceptions. Moreover, contingent and total pay are dominated by extreme events, the very essence of the stable Paretian model. Table 11.7

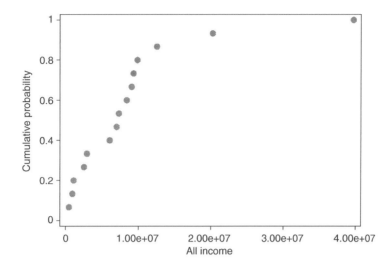

Figure 11.7 Cumulative probability of pay.

summarizes the data. Average pay is about $12 million per movie, but this is at the 75th percentile, well above the median. The data are skewed to the right. If the top pay is removed, the average falls to $8.14 million and if the top two are removed, the average falls to $7.26 million. The largest single pay event accounts for 38 percent of cumulative pay over the sample. The cumulative probability distribution also reveals the extreme skew and dominance of the largest pay events as can be seen in Figure 11.7.

Clearly, these contracts condition pay on the extreme events. Now we turn to the contracts themselves to see how they do this.

11.9   Director contracts

The following observations are drawn using a sample of over thirty contracts covering major motion picture directors.

All the contracts stipulate a director's fee and often include also a writer's and/or producer's fee. The director's fee is a fixed amount. Contingent pay is a feature of many, but not all, of these contracts. Just over half of the contracts include some form of contingent compensation. As the directors gained in stature and had a body of work, they attained contingent compensation and virtually all their subsequent contracts are of this form.

All the contracts contain a penalty for budget overruns. The penalty usually recoups some of the overage from the director's fee, sometimes from the direct fee and other times from the contingent fee. A typical deal might permit the studio to recoup the budget overage by withholding 50% of the director's contingent compensation.

The contingent compensation clauses are the interesting issue as they are responsible for the overwhelming share compensation and the extreme variation that we saw in Figure 11.6. Here are the terms of a "mature" contract of a director who is well established with a history of high grossing and award winning movies.

The distinctive features of the contingent compensation clauses are these. They condition on net distributor receipts, "gross proceeds." This is often confused with box-office revenue, but it is not. The distributor does not collect gross theater revenues. Distributors collect film rentals from the theaters as payment for the rights to exhibit a film. A distributor may collect from 30 to 60 percent of theatrical gross box-office revenue as rentals. The average rental rate is around 48 percent, but it will usually be higher for a film that earns high theatrical revenues. Even "first dollar" gross proceeds in a contingent contract do not refer to theatrical gross revenues. They refer to the distributor's receipts of theatrical rentals, which, in turn, are about half of theater gross box-office revenues.

Contingent compensation is awarded as a percentage of gross proceeds (adjusted rentals and video royalties). The percentages increase at successively higher break points in gross proceeds. One contract that illustrates this particularly well is the following:

Contingent pay is the higher of the director's fixed fee or

- 16.25 percent of gross proceeds less than or equal to $150 million
- 17.5 percent of gross proceeds between $150 and $165 million
- 18.5 percent of gross proceeds between $165 and $172.5 million
- 20 percent of gross proceeds between $172.5 and $180 million
- 22.5 percent of gross proceeds above $180 million.

I have calculated the break points from the details of the contract. What is notable about the amounts on which compensation is conditioned is the extraordinary box-office revenues that must be taken in for a film to trigger these clauses. These figures correspond to gross rental receipts that are roughly equal to the amount in the break points. Thus, the box-office revenues associated with these break points are roughly twice that and go from $300 to $360 million.

How likely are these numbers? Table 11.11 reports the statistics of box-office revenue. Mean box-office revenue is only $14.7 million and twice this is $29 million. These clauses condition pay on events above the 99th percentile! But, the conditional probabilities for the kind of movies these directors are involved with may be quite different from these probabilities. So, I have calculated these statistics for high budget ($90 million or more) movies in Table 11.12. Mean box-office revenue for these movies is $131 million with a standard deviation of $102 million. Now these clauses condition on events from the 75th and well into the 95th percentiles.

Table 11.11 Motion picture domestic box-office revenue

		Domestic box office		
	Percentiles	*Smallest*		
1%	1,513	211		
5%	5,352	309		
10%	13,136	321	Obs.	6289
25%	101,071	350	Sum of wgt.	6289
50%	1,723,487		Mean	1.47e+07
		Largest	Std dev.	3.25e+07
75%	1.47e+07	3.57e+08		
90%	4.26e+07	4.00e+08	Variance	1.06e+15
95%	7.05e+07	4.31e+08	Skewness	5.018068
99%	1.57e+08	6.01e+08	Kurtosis	45.10717

Table 11.12 Motion picture domestic box-office revenue for high budget movies ≥ $90 million

	Percentiles	*Smallest*		
1%	6,719,973	6719973		
5%	2.86e+07	1.00e+07		
10%	4.37e+07	2.86e+07	Obs.	50
25%	6.38e+07	3.47e+07	Sum of wgt.	50
50%	1.10e+08		Mean	1.31e+08
		Largest	Std dev.	1.02e+08
75%	1.71e+08	2.42e+08		
90%	2.25e+08	2.60e+08	Variance	1.05e+16
95%	2.60e+08	4.31e+08	Skewness	2.438852
99%	6.01e+08	6.01e+08	Kurtosis	11.12553

The 22.5 percent clause would come into play only at gross domestic revenue of $360 million. The probability of this event is 0.012 using a normal distribution. With the Pareto distribution the probability is 0.052, more than four times as probable. The remarkable thing is how the percentages escalate at probabilities that are very small; the least revenue that increases the percentage above the minimum 16.25 percent is an amount with a probability of only 0.035 with a normal distribution and 0.063 with a Pareto distribution. The clauses are so far into the upper tails of either of these distributions that the contingent compensation is hard to connect with the usual reasons given by standard contract theory. Nor is risk a likely explanation for these events distribute proceeds among all parties and involve only the positive upper part of the distributions; so they allocate only extreme, positive outcomes.

The Pareto probabilities are roughly four times greater than the Gaussian probabilities of the gross proceeds written into the contingent compensation. And the escalation occurs in a range where the normal probabilities are rapidly decreasing, whereas the Paretian probabilities decay slowly. Does this mean the parties to the contracts realize the distribution they are dealing with is Paretian? I would guess they do, roughly. The entire content of these contracts is aimed at sharing events in the upper tail of the distribution. These events would seem to be too rare for a Gaussian to care about. But, they are important to a Paretian, not only because they are more probable than a Gaussian would believe. The expected values at these extremes are very different in the Paretian world than in the Gaussian world. I think this is the fundamental reason behind the contingent compensation clauses. The conditional mean grows with the value of the conditioning event. If the parties recognize this they would rationally design contracts to match contingent compensation to conditional expectations in the upper tail.

11.10 Conclusions

Consider what the conditional expectation of gross proceeds is with the Gaussian and Paretian distributions for a high-budget movie. At the first break point of the contract, the expected additional revenue is only $16.7 million under the Gaussian distribution and yet it is a massive $1134.13 million under the Paretian distribution. At the highest break point, the corresponding conditional expectations are $4.8 million under the Gaussian and $1,035 million under the Paretian distribution. A Gaussian model would tell the contract designer that at these break points there is really no revenue left for the movie to earn. But, the Paretian model tells her not to neglect these improbable, but extremely high value expectations.

If the parties thought they were facing a Gaussian distribution they would not write contracts conditional on events that are unlikely ever to occur. But, if they knew, even if only intuitively, that they were facing a Paretian distribution these contracts would make sense. The extreme events specified in the contract are not improbable in a Paretian world and when they occur their expected values are enormous. Thus, a simple principle seems to explain the features of these contracts: contingent compensation should closely match contingent expectation at high outcomes.

An agent maximizing expected value would want a contract whose conditional expected value rises as favorable events unfold. The contract is a dynamic object that alters its terms as the movie plays out its run and revenue comes in. At each break point the contract asks "what is the expected value from this point forward?" The answer is that the conditional expected value is rising as each break point is crossed. The probabilities, which were unknown at the time of the deal, are revealed as the film runs and expectations are revised according to this new information. The director's contract share rises to reflect the increasing conditional expected value of going forward rather than stopping at each break point. In other words, the contract is a dynamic stopping rule[4] when "nobody knows anything" and each week of the run reveals new information.

12 How extreme uncertainty shapes the movie business

12.1 Introduction

Hollywood seems to believe in chaos. A belief in chaos is revealed in the heavy artillery that Hollywood brings to the opening—stars, massive advertising, many screens and big production budgets. These marketing strategies are aimed at controlling the initial conditions in order to give a film a good launch. This reveals a belief that films are sensitive to the initial conditions of their runs. With nonlinear dynamics at work, small opening weekend differences can become very large over time. Believing the opening is critical and that small differences become large ones amounts to a belief in two of the key elements that characterize chaos; nonlinearity and sensitivity to initial conditions.

Where the industry fails to acknowledge chaos is in its belief that a film's revenues can be predicted from its opening. Managing chaos, if that can ever be done, means managing the initial conditions because once the dynamical process is under way nothing else will work. The big budgets, special effects, stars, advertising and opening screens are employed to gain an edge over other movies in the initial conditions of the run.

There is a lot of chaos in the movies; it appears in some guise in nearly every chapter in this volume. But, more than chaos is at work; there is a hidden order beneath the chaos. You just have to know how to look at the business to find the order there. This requires a new point of view, which I will try to lay out in this chapter.

The tools for seeing the order are in the new sciences of chaos and complex systems. We have found a deep order in the business that builds on the concepts of self-organization, feedback and adaptation. This hidden order is expressed in the dynamics of information and the self-similarity of the statistical attractors of the dynamics of nearly all the measures that are interesting about the business. Revenues, profits, pay, productivity, status, careers, cost, screens, run life, the twists and turns of box-office receipts during the run, the angel's nightmare and the curse of the superstar are all scenes shot from the same script. Every shot reveals self-similar chaos and complexity.

12.2 Complex systems

Let me discuss chaos and complexity and then make some distinction between them. To do this, I rely on what is called the complex systems approach. Complex systems analysis is a paradigm shift in the study of social systems. Traditionally, scientists attempted to reduce behavior to elemental causes. Complex systems theory shifts the focus from reductionism to connectionism over the economic, social, artistic and institutional realm. A study of the movies based on this approach touches on the way moviegoers are connected, how they process and exchange information and why this leads to complex dynamical processes. The creative, social and economic processes are interwoven and are at the heart of the institutions that shape how the movie business is organized and does things.

12.2.1 Chaos

Chaos is different from complexity, but both have something of value to add to our understanding of the movies. Chaos is a deterministic, nonlinear dynamics that is so wild that it appears to be completely random. If motion picture dynamics were chaotic, we could write down the equations that movie revenues follow during the run. Even if you could write the equations down, you still will not be able to predict revenues. The reason is that the equations must be nonlinear; evidence to that effect is abundant and is to be found in all the preceding chapters, most notably in Chapter 2 where we show that revenues diverge rapidly after the opening and in Chapter 3 where we show that revenue bifurcates into hit and non-hit branches.

Nonlinearity in motion picture revenues means that small differences are greatly multiplied with each interaction among fans transmitting information. Even very small differences in opening revenues can become enormous in a matter of weeks. This is the famous "butterfly effect" of chaos, which is a colorful though not necessarily true example of the sensitivity to initial conditions typical of nonlinear systems. The story is that, given the expansive, nonlinear dynamics of weather systems, a butterfly flaps its wings in Hong Kong and this disturbance becomes a storm on the coast of South America. Taken literally, sensitivity to initial conditions would mean that slight differences in opening box-office revenues, or even minute errors or rounding their values when they are input into the equations, will lead to enormous differences.

When I tell my motion picture class about the butterfly effect I call it the "popcorn effect." I ask them to imagine that the popcorn at the Roxy is bad the night of the much anticipated opening of "Dickens in Love." The bad popcorn influences the audience to give their friends a more adverse opinion of the movie than they would have had the popcorn been good. Their friends had eagerly anticipated "Dickens," but the reports disappoint their high expectations. Now some people who hear this message go in spite of it but some do not, and this holds true farther down the chain of information.

The domino effect is underway; we have a contagion of eager, but disappointed, expectations, all set off by bad popcorn at the Roxy on opening day. The movie, which had been expected to be number 1, falls to 2 or 3 in its second week. The press reports the fall off as disappointing. Soon "Dickens in Love" is gone and nobody knows why. Blame it on the popcorn at the Roxy and chaos.

Like the butterfly effect, don't take the popcorn effect literally. But, do realize that there is a sensitivity to initial conditions and that movie revenues can go off in strange directions for small reasons that totally escape analysis by industry observers and post mortems at the studio. It can drive a producer nuts because there are so many things that are beyond control that can influence how a movie performs. And small differences with large effects can come from obscure places that simply cannot be identified. Explanations will be attempted after the fact—it was the trailer, it was the advertising, it was a bad movie, etc. They can all be given an air of plausibility because when you have chaos anything is a possible explanation. But, nobody knows.

Motion picture revenues can go anywhere because they are not bounded; they are pure information. Information is unbounded because using or relaying information doesn't destroy it and usually produces new information. Take an unbounded variable and apply nonlinearity and you get increasing returns, one of the most important characteristics of motion pictures documented in this volume's research.

12.2.2 Complexity

Complexity encompasses chaos, but needn't be deterministic. In a complex system, many diverse agents interact dynamically. They don't know everything about the system and their knowledge is primarily local, gained from agents with whom they interact and other sources. Their information is partial, incomplete and noisy. They are imperfect decision makers even when they are fully informed because information processing is hard and costly. They learn and adapt because, in this climate of imperfect information, costly and imperfect processing and changing conditions, optimizing is out of the question.

The agents operate in a dynamic world that is constantly evolving and their actions influence the evolution of the system. The organization that comes of their interactions emerges; it isn't written down, centrally commanded, or even apparent to the agents as they interact and make choices. It isn't deterministic. It may be chaos or it might even be a bit more "wild" so that it is not deterministic at all.

Complex adaptive systems, then, exhibit emergent order. They are never in equilibrium and often in chaos. They are far-from-equilibrium systems that live between order and chaos. Self-organized, complex systems live in a twilight region between order and chaos called the Critical Zone. Rod Serling has shown us that humans live in a twilight zone somewhere between order and chaos; our research suggests the movies live there too.

In a complex system, problems shift and are novel so even if one were to frame a problem well enough and have sufficient information so as to optimize the solution, the solution would not be adapted to the new and constantly changing circumstances. In a complex environment, the agents often form organizations and institutions to structure their interactions and lay down a broad framework for learning and adaptation. Stability of institutions promotes learning because it makes interactions more predictable and narrows the space of possible outcomes. Organizations that are well suited to these problems often are hierarchical and tend to delegate decisions to different levels through a process called decentralization. The old Hollywood studio system was a hierarchical organization well adapted to complexity.

12.3 Universality

The consistent structure that captures the enduring and universal complexity of the business is the stable Paretian distribution. The stable Paretian distribution is the statistical attractor to which the dynamics of motion pictures converge. It is a "strange" attractor and so contains a hint of chaos; it is strange because it has infinite variance and points that begin near one another are propelled apart rapidly by the nonlinear information dynamics of the movies.

The stable Paretian hypothesis says that motion picture dynamics are attracted to the infinite variance stable distribution with Paretian tails. This probability distribution is the source of the "nobody knows" principle. Much of what we know of risk and returns, or discovery and adaptation are grounded on the non-Gaussian stable Paretian probability distributions of outcomes.

It follows that the movies is not a normal business. I think most readers already knew that. But, we are saying something different from a commentary about life among the stars, agents, wannabes and studio heads. We are saying that the movies live in a non-Gaussian world where nonlinearity and chaos lead to extreme events and infinite variability; in this world, nothing is "typical" and every movie is unique.

In the stable Paretian world of the movies narrow, static goals of prediction, optimization and control are replaced by concepts that recognize the importance of contingency, emergence, novelty, self-organization, evolution, adaptation, diversity, hierarchy, network effects and flows. These are the key drivers of the industry's organization and function.

The movies seem to self-organize into a poised or critical state where disturbances lead to events of all scales. Many of the changes are abrupt and some are massive in scale. The evolution to the critical state occurs without design from any outside agent. The state is established solely by the dynamic interactions among elements of the system. The avalanches can only be understood from a holistic description of the properties of the entire system rather than from a reductionist description of individual movies. The movies are a complex system.

A theory of the movies is a theory of process, not an account of accidental details of each unfolding of the process. The theory must be statistical and so cannot produce specific details. The theory then is abstract and statistical, not what a studio executive might want. Most of the attention of motion picture analysts is drawn to forecasting how much money a movie will take in; this is a study of details that fails to see the process. We can predict the statistics of phenomena but not specific outcomes.

The conclusion of complex systems analysis is that one should not try to formulate a theory of specific explanations (why movie X grossed more than movie Y), but rather a general theory that encompasses universal features of movies, large and small. This has been my task throughout this research.

12.4 Deep order

My coauthors and I have discovered features of motion pictures that transcend genre, budgets, stars, studios and decades; they are universal features that will remain durable over many time and cultural perspectives. They have always been true of the movies and will remain so. This is not merely a claim based on the universality of the theory; many other scientists have confirmed and reproduced the discoveries reported in these chapters for other time periods and other countries.

12.4.1 *Increasing returns to information*

What are the deep universal features of the movies? First, all the things that matter are information variables. As I said earlier, information variables are unbounded and have no natural scale; they can take on any value. Information can be shared without losing its value. Information is different from physical goods: two dishwashers cannot differ by much in their output. But the output of movies or movie stars can differ in the extreme, and they do. The top movie or star may receive thousands of times more than other movies or actors. That is because motion picture revenues and salaries are information goods; they are just numbers that have no natural scale.

Motion pictures are pure information. Movies depict other lives and times creatively and symbolically to plant new models and insights in the mind. There is nothing tangible there. All a person has when he leaves the theater is the information the movie planted in his mind. When a movie fan tells a friend about a movie she loses none of this information and may even take additional pleasure or insight from the act of communication. So, no information is lost in the transfer and a new beneficiary is created. Information sharing increases its value and this gives rise to increasing returns.

Increasing returns to information were seen in the Bose–Einstein statistical dynamics of Chapter 2 where it was shown that it becomes more probable that a movie will be chosen if a large number have already seen it. These dynamics lead to an increasing inequality of revenues among movies the longer they run and to a log concavity of the probability distribution. For very long runs and high revenues, the log concavity gives way to a linear relationship of log probability to log revenue; this is the signature of the stable Paretian distribution. As the run progresses, the distribution becomes more skewed and leptokurtotic because the leading movies progressively increase their leads over the lesser-rans. As the tail of the probability distribution becomes heavier—usually by the eighth week of the run—the curve flattens and the relationship converges to linearity. These results were first discovered in my paper with David Walls in Chapter 2 in the US theatrical market. Chapter 3 shows this convergence from another point of view: it shows that the frequency distribution goes from log-normal to Paretian among long-running movies.

These findings have been confirmed for many other countries: in Hong Kong by Walls (1997), in the UK by Collins *et al.* (2002), in Australia by McKensie (2003), and in the UK and Ireland by Ghosh (2000); these studies cover many different time frames as well as different countries. They document increasing returns to information—when a movie gains market share, its share is likely to rise further. The information contained in a movie's box-office revenue expands the probable number of new viewers, if it is good. The information is not used up; it is multiplied as it is communicated. If a movie is bad, it goes the other way, cascading into a death spiral. The non-linear information transmission can take small differences in information and preferences and turn them into very large differences in revenues and returns among movies.

12.4.2   *Nobody knows*

The "nobody knows" principle comes from the stable-Paretian statistics of the business; it is revealed in the infinite variance and scale-free form of the probability distribution. When the probability distribution is scale free it has no characteristic size and there is no typical movie. If variance is infinite, then prediction is impossible; one can only say that the expected revenue of a movie is X plus or minus infinity. Not very precise or satisfying.

The infinite variance is not to be taken literally, as any movie must earn only a finite amount of revenue. It is a warning that the variance is unstable and ever changing, usually increasing with the length of the time period over which observations are made. This implies that the average revenue is unstable as well. This point is documented in several of the chapters of this book. Both the average and the variance of revenues resolutely refuse to settle down or converge to steady-state values. Too bad if you want to use a Hollywood rule of thumb of using the average, a statistic about the past, to predict the future. It won't work.

If you have a time-varying average and ever-growing variance, then you have a real problem forecasting or extrapolating from past results. The Paretian distribution implies that any time series will have a fractal or ragged look, rather like the stock market. There will be plenty of upward and downward moves in, say, average box-office revenue to trap the unwary into seeing trends and patterns. It is just all part of the process, signifying nothing. No one but a fool would forecast a movie's expected revenue from an average of similar movies from the past.

12.4.3   *Dominance of extreme events*

Yet another deep property is the dominance of extreme events; it is a business of extremes after all and virtually all the measured statistics of the business show strong dependence on extreme events. The mode (most frequent outcome) is less than the median (midpoint), which is less than the mean because most movies do not attain high grosses and the average is dominated by extreme events. The mean is up around the 75th percentile or more, a clear indication that it depends heavily on the largest events. It is because the mean is heavily dependent on extreme events that sample statistics are unreliable and forecasting is impossible.

Another implication of the powerful effect of extreme events is the "winner-take-all" form of the distribution of revenue, salaries and productivity. Just less than 20 percent of movies earn 80 percent of gross and less than 5 percent earn about 85 percent of all profit in the business.

It may not be evident that the dominance of extreme events is also evidence of increasing returns and a deep property of the stable Paretian distribution. We found increasing returns in an unusual property of revenues and careers. In both instances, the conditional expectation grows with success—success breeds success. In Chapter 3, we discovered that the expected future earnings of a motion picture rise the more the movie has already earned. This conditional expectation should vanish if movies were Gaussian because, in some sense, the revenue should be all used up after some point. This does not happen because probability does not decay as rapidly as the Gaussian distribution would suggest. There is a lot of probability in the upper tail and it does not diminish much even at a point where the movie has earned 3 or 4 standard deviations above the mean. Chapter 11 demonstrated that the probability that a director will make another movie rises the more movies she directs; the same sort of rule applies to actors too.

Increasing returns to information drives a few movies far into the outer reaches of gross revenue, far beyond the levels attainable in a Gaussian world.

12.4.4   *Self-similarity*

Now for a more esoteric kind of structure, self-similarity. This is evident everywhere in the business. Self-similarity is a statistical property; it says that viewing a process in different perspectives reveals a similar pattern. A self-similar object retains that property on all time and distance scales, with suitable transformations. Such objects are fractals as well. A fractal has a ragged, uneven aspect to it, like a mountain range. Seen from far away or up close, the raggedness is still there. Portions of the range show the same ragged pattern as the whole range. Because they appear similar at all scales (near and far), fractals are self-similar. Self-similar statistical distributions have the same property and, for that reason, they are often called fractal distributions. In order for a statistical distribution to be fractal and self-similar it must be what we call scale free. There must be no natural, typical size or scale of the objects represented in the statistics.

What this means is that the probability density function must remain unchanged in form when the variables are scaled up or down. More technically, a stochastic process, $\{X(t)\}$, is statistically self-similar of degree $H = 1/\alpha$ if

$$\{X(at_1, t \in T\} \simeq \{a^H X(t), t \in T\}. \tag{12.1}$$

$\simeq$ in this context means equality in distribution.[1] What this says is that expanding or contracting the time window by a value such as a expands or contracts the size of the value of the process, $\{X(t)\}$, by $a^{1/\alpha}$ but it does not change the statistical distribution. We know $\alpha = 1.5$. Suppose that the largest grossing movie in a one-year period, t, made \$100 million. Then, by self-similarity, in a two-year period, $2t$, the largest grossing movie is likely to make $2^{2/3}$ \$100 million, or about \$159 million. Stretching the time space scale expands the state space scale.

The density function for which this scale invariance is true is the power law or Pareto distribution. The importance of the Pareto distribution is hard to overstate in the movie business. The distribution characterizes the tails of the statistical distributions of virtually all the variables of interest to the business. And the tails are where the action is because extreme events dominate the business.

If the extreme events fall in Pareto tails then they are self-similar. The fractal distributions of motion picture variables imply that extreme events are self-similar. This finding is supported in many pieces of research, reported here and elsewhere. It is of enormous practical as well as theoretical value. Let me spell out a few instances of self-similarity and what they mean for the movies.

- *Cultural self-similarity.* The size distribution of motion picture revenues in various countries is Paretian. Motion picture revenues in the United States, Hong Kong, the United Kingdom, Australia and Ireland are Pareto-distributed. With allowance for the sizes of these respective markets one finds a Pareto tail weight of about 1.5 in each one. They are self-similar in that they share a similar statistical distribution; if you plot the revenue against the rank, from highest grossing to lowest grossing, the shape of the curve will be the same in the upper region.

This means that the concentration of revenues among movies is universal among cultures and film markets. There are no features of tastes or market organization in these countries that exert an influence on revenue concentration. The movie business is neither more nor less uncertain in one country over another and the specific organizational features of the market in these countries do not alter the concentration of revenues among movies. This does not mean that the top-grossing film in the United States will be the top-grossing movie in the United Kingdom or Ireland. It does mean that the top film will earn about the same proportion of total revenues in each market where it plays.

- *Temporal self-similarity.* One can do a similar exercise using different time periods. In Chapter 4, David Walls and I showed that the US motion picture revenue distributions each year were similar for more than 10 years, beginning in 1984 and continuing up to the present. The shape and tail weight have been stable for more than a decade. The distributions are strikingly similar in shape. They are pictures of the same process at work over a period of a dozen different years.

  The implication of this result is that, in spite of many changes in films and theaters and even likely changes in tastes, audiences still rank movies in such a way as to give the same proportional revenues to the top movies. The winner-take-all nature of the movies has not changed. Seen in another way, the movies have not become more risky; the distribution is still highly skewed, but not more so, and the variance of revenues is still infinite.

  Another striking form of temporal self-similarity is seen in the time series of motion picture revenues. That can be seen in Figure 10.7 of Chapter 10 where average motion picture profit is shown by release dates of motion pictures over more than a decade. The shape of the series is ragged and if you look at the whole series or pieces of it, it still has the same ragged profile. You get the same thing if you plot average motion picture revenues.

  What gives both series their ragged appearance is the way extreme events cause the curve to rise or fall sharply. Thus, the rugged self-similarity we find in the time series reveals the statistical self-similarity we find in the Paretian distribution of extreme events; again, they are pictures of the same process wherein the total and average magnitudes are heavily influenced by extreme events.

- *Budget self-similarity.* You get a similar pattern if you slice up movies according to their production cost. If you group movies into low-, medium- and high-budget categories you still get a Pareto distribution of revenues in each category with the same tail weight. The minimum and average revenues change a bit, but the highest revenues do not. In each category, the average revenue is not the typical outcome; there is no typical outcome because they diverge over all possibilities. The variance of revenue outcomes is still infinite in every budget category.

A similar point can be made about returns on production cost. The ratio of revenue to cost is a Pareto distribution and it does not differ, except by a scale factor, in shape among budget categories. The average return differs a bit among budget categories, but the variance is infinite in all cases. So, risk, as measured by variance, does not differ among high-, medium- or low-budget movies. So much for a strategy that made the rounds in the studios about "getting out of the middle." The idea was to increase returns and reduce risk by making only low- and high-budget movies.

It didn't work and couldn't because movies in all budget categories have a similar distribution of probabilities over outcomes. By another measure of risk, how much you can lose, there are differences. Low-budget movies lose less because they cost less, not because they have less variance in their outcomes.

- A striking pattern of self-similarity was seen for careers, movie gross revenues and artist pay. All these are Pareto distributed with about the same tail weight of $\alpha = 1.5$. Chapter 11 on artists showed that all these measures—career length, gross and pay—are conditioned on extreme events in the tail of the revenue distribution. Therefore, they are all rescaled measures of the same phenomenon and must be statistically self-similar. And they are. This is a powerful confirmation of the stable Paretian model of motion pictures. A suprising and novel result that only seems natural once the model lets you discover and see it.

In short, the movies are statistically self-similar over time, genre, cultural, rating, budget and other windows.

12.4.5   *Runs, legs and chaos*

Finally, there are deep properties of the dynamics of box-office revenue in the motion picture theatrical run. By about the fourth week, film revenues bifurcate onto a hit branch and a bomb branch. From that point onward, they follow different dynamics. The paths rapidly diverge from one another, reflecting the expansive growth of revenues for hit movies and the contracting revenues of movies less fortunate.

There are no hits (revenues exceeding $100 million) that have short runs. Nor is their fate set at the opening. The opening is only a good predictor or revenue for movies that bomb. The expansive dynamic of a hit makes its conditional expectation—what can you expect given that you have grossed X—rise through the later weeks of the run. This means that success breeds success. When consumers behave this way, they are often said to be following a bandwagon or herding. When a hit movie is running, crowds flock to it as though they are simply imitating others. But, we find more evidence for this: when a hit branches off from the pack, a powerful feedback is at work. The feedback is between word-or-mouth or quality information and box-office reports, which speak of mere quantities. When both these sources of information confirm one another, the movie gets its legs and is on its way to becoming a megahit.

12.5 Complexity and the stable Paretian distribution

All these observations of what I have called deep order are consequences of the nonlinear and complex dynamics through which moviegoers select movies. Let's try to pull it all together.

I said the stable distribution is the statistical attractor of motion picture revenue dynamics. This means that the probability and size distribution of revenues converges to the stable distribution with Paretian tails. The stable Paretian distribution is then called the attractor of the probability distribution as the run progresses. As the run progresses, the size of the sample rises; call this n for the number of observations of the distribution. Let the function $F(S_n)$ be the probability distribution of the stochastic process S when n observations have been taken. S is the sum of identically distributed random variables. As n increases, $F(S_n)$ changes form and is drawn to the stable attractor. For large n the probability distribution converges to a stable distribution. The distribution is stable if the function $F(S_n)$ remains of the same form as n increases. The best-known stable distribution is the Gaussian distribution. But, it is the attractor for stochastic processes that have finite variance. It is the only stable distribution all of whose moments are finite.

The issue of finite moments draws attention to the tails because that is where probability mass pulls the statistics out of the Gaussian domain. And we have seen that motion picture distributions have very heavy tails.

Motion picture revenue dynamics turn out to be stochastic processes with infinite variance and power law (Pareto) tails. Hence, motion picture stochastic processes converge to non-Gaussian attractors. The stable attractor of motion picture revenues generally does not have a variance (second moment) and may not have a mean (first moment). Moreover, the probability distribution is scale free; there is no characteristic size of motion picture revenues. Revenues diverge over all possible values and the mean does not correspond to the typical outcome.

Apparently, there is a kind of statistical wildness to the motion picture market. Revenues can follow many paths as a movie makes its run through the theatrical market. Movies that break away accelerate rapidly ahead and their lead grows each week, this is the nonlinear information flow. How can such a process be called stable when it is so wild?

The stochastic process is stable because it converges on a probability distribution that is stable, meaning it is unchanging in form and characterized wholly by its fat tails. To capture the movies we need processes that drive results far into the outer limits to get heavy tails and the stable distributions they characterize. The Bose–Einstein process, where choices become more probable when others make similar choices, converges to a stable Paretian distribution. This nonlinear choice logic propells movies into the outer tails.

Call it imitation or herding or the emergence of a consensus on the best movies, but there has to be some kind of nonlinear feedback at work to produce the heavy tails where the runaway hits are located. Remarkably, when agreement emerges the outcomes become more disordered and divergent; movies differ by so much in their revenues that they may as well come from different worlds. When there is disagreement among movie fans, outcomes are more orderly because no breakaway hits bulge the tail out into a wild probability distribution.

What this says is that, in the movies, agreement leads to disorder and disagreement leads to order. No wonder no one knows anything.

Epilogue
Can you manage a business when "nobody knows anything"?

In the movies, artists and management have to collaborate. The hardest part about their collaboration is that most of the money has to be spent up front to produce a movie before anyone knows how much it is worth. All the best talent and intentions do not assure that the can of celluloid that results from the collaboration of talent and money contains the images and sounds of a vivid and interesting story. Everybody makes bombs and never knows why. No one sets out to make a bad movie, but many are bad. Some are good, but don't catch on or are run over by a blockbuster.

Learning is hard in this business. Every movie is unique, so one can't learn much about what will capture an audience. And just when you think you have it down, tastes change or something new comes along and the old formula comes crashing down. The craft of filmmaking can be learned, but there is no learnable craft when it comes to predicting how a movie will play with audiences. Experience is not reproducible when it comes to greenlighting movies; it may even be a hindrance because experience relies too much on past successes and selective memories that blot out failures.

There really is nothing that is predictable, not costs, not performance value and certainly not revenue. Genre means nothing; rating has only a small effect on probabilities and varies so much that only a portfolio of films is capable of capturing the difference. Stars only change the odds slightly and at high cost. These are the overriding lessons of the exotic statistics developed in the chapters in this volume.

But, there is further cause for pessimism or delight in the movie business. And that is that we can't even trust that the statistics, as wild as they are, will remain stable in the future. We do appear to be able to say that the broad features of the movies—self-similarity, infinite variance, volatility, skew, non-linear dynamics, kurtosis and inquality—will always be true of the business. This complex of features revolve around the stable Paretian model and it is this model that offers some help in managing the movies. This epilogue explores that idea briefly.

Probabilities

First, the stable Paretian model tells us to think in terms of probability distributions, not individual movies. Every movie is to be thought of as a prospect; that is, as a probability distribution of possible outcomes. When you consider movies to be probability distributions, you avoid the sure thing principle and its associated biases (see section "Narrow framing and sure things"). No movie is a sure thing, no matter who makes it or who stars in it or who directs or produces it. Some movies offer better odds than others of high grosses, but they are still probabilities, not sure things. And one must pay for whatever can be identified as the source of the better odds. Stars raise the least expectation of revenues a movie might earn, but only if their presence convinces theater booking agents to allocate a lot of opening screens to the movie.

Because stars get the best material it is impossible to attribute causality; the better odds of higher grosses associated with star movies may come from better material, heavier advertising, more opening screens and so on. In a real sense, a producer is signaling quality to the audience when she pays a big salary to a high-priced star; she is saying my movie is so good I am willing to spend $20 million on actor Z to play in it. And actor Z is saying, I think the material is so good that I will risk my reputation on it.

What makes a star anyway? It is the chance to play a great role in a successful movie. Of course, the performance has to be up to the role. But the star movies that make real money are those that have long runs; so both the movie and the performance have to have the goods. But, again, it is all probabilities we are talking about since nothing is sure. If we ask the question in that way it turns out that the success of a star's last movie does not predict how well the next one will do. And, most stars do not shift the probability toward more success by enough for it to show up in the statistics.

It is hard to think about movies as probability distributions. And the common studio practice of greenlighting movies one at a time makes it even harder. When you look at a single movie you see it in a different perspective than if it were part of a portfolio of movies. You see it as a unique object to which the whole idea of probabilities is hard to apply. When you look at a portfolio you consider how the hits and misses add up to determine a total return. You also are mindful of the opportunity cost; for every movie you greenlight, you give up the opportunity to make something else. You also fail to see the scale effect in looking individually at movies. As a consequence you do not aggregate returns over a portfolio which makes you undervalue several smaller movies whose aggregate prospects may exceed those of a single bigger movie.

Inside and outside the studio

Neglecting to consider movies as probability distributions is engendered by the internal decision making of the studio. Movies get chosen based on internal dialogue and discussions in meetings where the most optimistic and biased forecasts tend to win the argument. The choices inside the studio tend to be driven by playing out alternate movies as scenarios. This is story-telling, not good decision logic and it is prone to errors of all sorts.

We need to distinguish between the inside view of the business and the outside view. The making of a movie is a complicated process; there are choices to be made at every step and many steps from development to a finished answer print. As a movie makes its way along that complex path, supporters must argue their case against other projects under consideration. A film that actually gets made survives many comparisons against other projects and there are always proponents and opponents vying for a studio's limited financial and creative support. In effect, a coalition of supporters of the film must be sustained through the long creative and production process. These coalitions of supporters are fragile, they are hard to hold together and there are many others vying for support.

The inside view of this process is what the decision makers see. Who has the producer lined up, who is supporting the project, who argued against it at the last meeting, how strongly is marketing behind it, and how does distribution think it will do? This kind of inside view is a grave source of error. It anchors expectations on plans and optimistically biased forecasts. The preference for inside views which dominates in the process-oriented production of movies and projects is intuitively preferred by the participants, but it is a source of serious errors of judgment.

The outside view is the one developed in this book; it is anchored on external experience, not inside studio perspective and politics. The way to make Hollywood management right more often and to make better movies is to be constantly testing the inside view against the outside view. Internally generated prospects and odds must be measured against the real, external odds.

A team locked in developing a movie is susceptible to the fallacies of inside thinking: they tend to make scenario projections anchored on present values and extrapolate current trends. These can lead to optimistic forecasts of the movie's potential. Pervasive optimistic bias is based on (1) unrealistically positive self-evaluations; (2) unrealistic optimism about future events and plans; (3) an illusion of control. People exaggerate their control over events and the importance of their own actions in ensuring desirable outcomes. Projects that are forecast to have the highest returns are most likely to fall short of expectations. On the other hand, no movie would be made based on worst case analysis. The worst case for every movie would be that it loses money. But, in insider decision making, worst case analysis tends to focus on only mildly pessimistic cases rather than the worst possible outcome.

Risk and decision bias

"Managers accept risks, in part, because they do not expect that they will have to bear them." Hollywood executives accept risk because they are under the illusion that they can control it. A movie is risky, but if the illusion of control convinces a manager that the risk can be controlled by putting a star in the movie, or adding special effects, or introducing a story line for broader appeal, or spending heavily on advertising, then we will get just the kind of movies we have today. But, the risks cannot be controlled; they are inherent in the very nature of the business. Yet, the illusion of control will profoundly affect the kind of movies that are made and how they are marketed.

In Hollywood, probable gains are underweighted in comparisons with sure ones; this is known as the certainty effect. Vague or ambiguous probabilities seem to enhance the certainty effect and heighten the bias when there are unknowns. Thus, an unknown movie producer or a movie without a major star is seen as a prospect with an ambiguous probability distribution and as a small decision. Managers are excessively risk averse in small decisions (low-budget movies) and when they face ambiguous probabilities (unknown or little known artists). The certainty effect means that small projects by unknown or lesser-known individuals are underweighted in comparison to large projects with known individuals. Everyone trying to get a first break in Hollywood knows this. It is far easier to get a large project done with known individuals than to get a small one done with new faces.

Narrow framing and sure things

There was a time when the great studios planned a season of production at the beginning of each year. The studio heads were able to take the aggregate of projects—the film portfolio—and weigh their prospects. Today, there is no institutional structure to permit choices among portfolios of films. Nor is there a mechanism for aggregating small projects. Movies are judged one at a time and, therefore, subject to aggregation bias. Consequently, smaller movies are undervalued. The narrow framing of decision in studios today, treating projects individually rather than as portfolios or aggregates, produces a severe form of myopia which leads to excessive risk aversion.

The narrow framing of decisions leads to the sure-thing principle. Bold forecasts enable cautious decision makers to take bold risks. Biased forecasts are more likely to come from inside the studio, where projects are treated as unique and where competion among supporters influences their forecasts. Inside forecasts tend to be made using scenarios about the process of bringing the project to completion. It is story-telling about the project and its release, wrapped in obstacles to be overcome and deals to be made, but never anchored on external odds.

It is impossible to avoid bias when you evaluate the accuracy of these scenarios and forecasts when they are made one at a time. People are overconfident in evaluating the accuracy of their beliefs one at a time. Their assessments of their overall accuracy is about right or may even be pessimistic. But, they are too optimistic in evaluating the accuracy of their beliefs regarding individual choices. Thus, someone might well believe that a coin is fair and that the odds are 50/50 that it will come up heads in a sequence of flips, but still believe that they can do better than that at forecasting the next toss. The same bias grips an executive who may know that only a few movies exceed a gross of, say, $50 million, but who believes that his movie is sure to.

Biased forecasts and excess confidence in forecasting single outcomes lead to the sure-thing mistake. The sure-thing mistake also makes single events have a large impact. Thus, when a movie of a certain type is successful, there are always many imitators. The producers, misled by the sure-thing mistake, believe "it can't miss."

Financing movies

Studios are really banks that have a distribution arm for leasing movies to theaters. They finance movies based on what they think they will cost and how much they expect they will make. All the risk is borne by the studio unless they grant points to artists in exchange for a lower fee. To shed some risk they might pre-sell one or another revenue stream. This turns risk into cash. A film has many such streams and a studio can choose which ones to sell for cash and which ones to keep. Turning these unknown streams into cash may seem comforting, but the studio may regret taking cash upfront instead of retaining some part of the uncertain gain. If the movie is wildly successful, they will have sold the stream for a pittance, transferring to the buyer all the uncertain, but potentially very large gains. Remember, the rewards lie in the far upper tail of the probability distribution. The heavy tails tell us that a studio should always retain some part of the upper tail of any revenue stream they pre-sell because that is where the really big gains are.

There are better ways to finance movies. We know that we should take portfolios of projects and that there are extreme statistics to deal with. So, investors in motion pictures must manage their risk through the use of portfolios; these can be created by packaging projects together into one investment package. Or, the projects can be securitized in some manner so that an investor can manage risk by diversiying securities.

Except for the odd portfolio offering, such as the Silver Screen limited partnerships (where the junior partner typically does not fare well), few movies are financed by packaging them as part of a portfolio. Studios implicitly build portfolios of 4–15 films in their annual productions, but they do not use portfolio analysis in picking movies because they greenlight them individually.

So, is there a better way? We can design securities to solve many of the problems in film finance. How should such a security be designed? Design must be grounded in the specifics of motion pictures. What is to be securitized is the revenue or profit of the revenue streams that flow from a movie for the whole of its economic life. It is desirable to securitize a stream that can be monitored by the investor and verified by a court. Revenues, as reported by any of the independent sources of box-office revenues, is an ideal variable on which to condition payoffs. Profits are more controversial and are difficult to allocate over a portofolio (this leads to what has been called the studio accounting problem). Conditioning security payouts on revenue is more likely to draw investor support because studios or producers have less room to influence revenue than profit and independent sources publish information on revenue. By reducing moral hazard, revenue-based securities are more likely to support a market. The importance of a market is that it would bring greater liquidity to motion picture securities and, thereby, increase their value.

Suppose film securities are Arrow–Debreu securities that pay off conditional on states of the world as determined by box-office reports. Because the movies is a business of extremes, these securities would condition on the low-probability, large-magnitude events in the upper tail. For that reason, I call them extremal securities.

Define an extremal security as follows: an extremal security in film A pays an amount X if A's box-office revenue equals or exceeds K and zero otherwise. The payment is some share of revenue; zero at less than K and, say, share σ at any greater amount.

The security could have several break points and percentages. This permits an investor to tailor her exposure to probabilities and returns. Such a security would pay a different share at different revenue levels. Letting R denote revenue and σ the share, an extremal security might pay these shares at different revenue break points:

$$
\begin{array}{lll}
\sigma_1 R & \text{if} & R \geq K_1 \\
\sigma_2 R & \text{if} & R \geq K_2 \\
\cdots & \cdots & \cdots \\
\sigma_n R & \text{if} & R \geq K_n
\end{array}
$$

for revenue shares $\sigma_1 < \sigma_2 \ldots < \sigma_n$.

Such an extremal security is equivalent to several call options that come into the money when the movie hits the revenue break points.

As I said, the studio should always retain some risk so it can capture part of the gains of extremely successful movies. By financing movies with extremal securities, the studio can always retain some portion of the low probability-high gain upper tail events instead of pre-selling them to their later regret. Individual investors can tailor their exposure to the infinite variance of the stable Paretian distribution of revenue by choosing particular pieces of the upper tail of the distribution.

We can value these extremal securities using the probability distribution of the events specified in the payout schedule. We can't use the Black–Scholes formula to calculate the value of the call options in the extremal securities because the formula assumes the distribution of events is a log normal distribution. We know that the motion picture revenue distribution is stable Paretian rather than log normal, so we have to use another method. We have to price what are called stable call options. That is doable, but it is an exotic topic best left for another place.

Causality and Kim Basinger

Kim Basinger's career is a case study of what can happen in the uncertain stable Paretian world of the movies. Recall that she became famous for her *Playboy* spread and her bad behavior in *The Marrying Man*. She was brilliant (I thought) as the money and sex hungry object of Mickey Rourke's manipulation in *Nine 1/2 Weeks*. Her promising career was brought to a dramatic halt when she was sued by the producer of *Boxing Helena* for refusing to take the role. The jury dismembered Kim's career like Helen's lover dismembered her body when it concluded that she dishonored a verbal contract and awarded the producer $8 million in damages. The ground for this award was the jury's belief that, had Ms Basinger appeared in *Boxing Helena*, it would have earned $8 million more in profits. The judgment forced her into bankruptcy, though the award was later tossed out on appeal.

Given all I've said about the uncertainties in the business, it is clear that the jury made many errors in awarding damages against Ms Basinger for refusing to be put in Helena's box (or the attorneys failed to tell them what they needed to know to avoid their errors). The movie was not a sure thing; it was a prospect which we have come to see as a probability distribution over returns. Kim Basinger could only have brought a different probability distribution to *Boxing Helena*'s prospects. By the time of the trial, the movie had appeared, so it no longer was a prospect and that information should have been used to calculate damages. Given that it was a poor performing movie, earning only $1.76 million in North America, the jury should have calculated the conditional probability that she would lift its gross revenues by $25 million. This is roughly the revenue it would have to take in to lift profit by the $8 million in damages the jury awarded.

Even the unconditional probability of this is small. But, since we know that the movie was released and grossed an improbably low amount, we have more information. We should calculate the probability it would earn $25 million with Kim in the lead, given that it only earned $1.76 million with another actress in the role of Helena. Now we are into vanishingly small probabilities. We know the movie was a turkey and there is no way Kim could have saved it.

But, there is more the jury should have been told and it is worth telling because it illustrates so well the lessons we have learned from the research covered in this book:

1   There is no single cause in a complex system. Kim's nonappearance is only one of immeasurably many elements of the system. The jury could no more trace the final outcome of the movie's run back to a single factor than they could trace a particle of water in a wave at the seashore to its location a mile out into the ocean.

2   The butterfly (popcorn) effect is real. The world was a very different place when the movie was released from what it was on the date it was going to be released with Kim in it. The nonlinearity of the information

cascade would take the movies to very different places given even slight differences in initial conditions.

3   Past is not future. What Ms. Basinger's other movies earned is irrelevant because *Boxing Helena*, like every other movie, is unique. Taking averages over unique movies is meaningless. Further, complexity means that average revenue from the past is a poor predictor of tomorrow's outcome. Only a fool would use the average of Ms Basinger's past movies to forecast revenue for *Boxing Helena*. (This is a story movie agents might not want to hear.)

4   Motion picture revenues do not follow a normal (bell-shaped) distribution; they follow power laws. Because of this, the jury should have been told that:

   (a) An average taken over the past, say, over her prior movies, is not an expectation. A power law is scale free; so it has no characteristic scale or central location, like a normal law, and its expectation need not even exist.

   (b) Even if the expectation exists and can be estimated, the variance is infinite. When the error of the estimate is infinite it lacks precision and contains no information. It is completely uninformative as a forecast of the outcome.

5   The information cascade can kill. Ms Basinger's appearance in *Boxing Helena* might have gotten it booked on more screens initially. For two reasons, this higher initial exposure might have propelled the movie to an even earlier death and smaller, not larger, revenues.

   (a) The unfavorable information about the movie would have begun from a larger initial audience. The negative information cascade would have been more powerful and cause the audience to decline more rapidly.

   (b) Booking the movie on many screens dilutes box-office revenue over more theaters, putting more of them below the critical level at which they drop the movie.

6   Last, and the most counterintuitive result of all: because motion pictures are a critically organized system, movies can die for no apparent reason. The market may be in relative stasis for a period of time, but the equilibrium can be punctuated with inexplicable extinctions on every scale. A movie can die just because of the way the system operates, and its death cannot be traced to any intrinsic qualities of the movie itself.

In a bizarre twist that can only occur in a complex system, Kim Basinger went on to appear in *LA Confidential* and win the Academy Award for best supporting actress. In a matter of a few years, her career went from new starlet, to notoriety and a blossoming of real talent, to bankruptcy, to an Academy Award. You don't get that kind of variability selling lawnmowers or doing income taxes.

A skeptical attitude

Anyone who claims to forecast anything about a movie before it is released is a fraud or doesn't know what he is doing. The margin of error is infinite. That does not mean that he won't ever get it right, only that he seldom will and only because of sheer luck.

Even if he gets it right several times in a row, you wouldn't want to bet on it anymore than you would bet heavily on heads after three tails in a row. Most of the success one sees in the movies can be explained as pure luck. So, even if the forecaster (or producer) gets it right several times in a row these are confirmations, not tests. You don't know if the theory is right because it hasn't been tested, just weakly confirmed. If you invest in someone who tells you all swans are white, you can lose it all when a black swan shows up. One black swan falsifies the theory; many white swans don't make it true.

A skeptical attitude is your only defense. Don't take success for skill unless you cannot explain it as pure luck. The probabilities of the past may not be those of the future. No amount of confirmation makes a theory about the movies right. There are black swans out there. There is no formula. Nobody knows anything.

Notes

Part I Box-office champions, chaotic dynamics and herding

1 Cassey Lee and I modeled this process by simulating the exchange of information among movie fans in our "Quality Signals in Information Cascades and the Dynamics of the Distribution of Motion Picture Box Office Revenues" (De Vany and Lee 2001).

1 The market for motion pictures: rank, revenue and survival

1 See Alchian (1950), Nelson (1977) and especially Day (1975) for a discussion of evolutionary models.

2 The theatrical release usually precedes the video market, and we are modeling only the theatrical stage. The video and foreign markets yield about as much revenue now as the theatrical market, though these revenues depend on a successful theatrical run. Consequently, the total revenue of a motion picture will usually exceed the theatrical revenue we are here modeling.

3 Fixed costs with information feedback and limited outlets create "superstars" or "hits" as in Rosen's (1981) model. As this market is organized sequentially, increasing returns come through a lengthening of product life for successful films; the adaptive run is the primary mechanism for capturing hit revenues.

4 See Blumenthal (1988) for an analysis of auctions with blind bidding.

5 The contract "clears" an area around the licensing exhibitor of competing showings of the film.

6 De Vany and Walls (1996) show that the information dynamics of motion pictures can produce bandwagon effects as well as swift death.

7 See De Vany and Eckert (1991) for an analysis of motion picture exhibition contracts.

8 Kenney and Klein (1983) note that the exhibitor may have an incentive to over-search and show that the liquidated damages provision of the block-booking contract often used before it was outlawed limited search. In the present context, over-searching would mean an exhibitor ends the run prematurely, something that is guarded against, in part, by the minimum run and the hold-over clause.

9 Our data are *Variety's Top-50* motion pictures, consisting of the top fifty revenue earning motion pictures playing each week. Hence, without restricting generality, we speak here of 50 slots in the tournament.

10 Some of the films in the sample were censored because they remained on the charts after the terminal date of our sample. The survival model estimated in Section 5 explicitly accounts for the censored data.

11 The hypothesis that births/deaths are Poisson distributed could not be rejected using a χ^2 goodness-of-fit test. The marginal significance level was 0.545.

12 Note that a constant hazard rate is a necessary and sufficient condition for the survival time to follow an exponential distribution. A constant hazard rate reflects the exponential distribution's lack of memory. When the survival times follow an exponential distribution the probability of failure in an interval (t, τ) depends only on the length of the interval $(\tau - t)$.

13 The function $\psi(z)$ may take values greater than or less than 1 corresponding to decreasing or increasing survival times, respectively.

14 An alternative parameterization is to model the effect of z on the hazard function directly: $h(t; z) = \psi(z)h_0(t)$. This form is referred to as the proportional hazards model. Cox and Oates (1984) have shown that the accelerated life model and the proportional model coincide when survival times follow a Weibull distribution.

15 For example, correcting for heterogeneity in the parametric Weibull model can over-parameterize the model and lead to seriously misleading statistical inferences (Heckman and Singer 1984).

16 See Pagano and Gauvreau (1993) for a discussion of the product-limit estimator and the tests for equality of survivor functions. See Kalbfleisch and Prentice (1980) and Cox and Oates (1984) for more mathematical discussions.

17 The Logrank test had a marginal significance level of 0.507, and the Mantel-Haenszel test had a marginal significance level of 0.502.

18 See Kalbfleisch and Prentice (1980, pp. 15–28) on the maximum likelihood estimator of the hazard function.

19 We also estimated a Cox proportional hazards model to provide a check on the assumption of a Weibull distribution. The estimated relative risks were nearly the same as those obtained from the proportional hazards parameterization of the Weibull model (De Vany and Walls 1993).

20 Screens and showcases were very highly correlated, and many observations of screen data were missing, so we used showcases instead to proxy the total number of screens.

21 The highly convex revenue distribution could also result with competitors that are homogeneous with respect to talent. See, for example, Adler (1985) and Chung and Cox (1994).

2 Bose–Einstein dynamics and adaptive contracting in the motion picture industry

1 We are influenced in this endeavor by Brock (1993) who argues that to come to a real understanding of distributions one must model the underlying processes and dynamics that produce them.

2 The sample is described more fully in Section 2.3 and its subsections. In this chapter, we consider only the first run of films playing in domestic theaters; subsequent runs, foreign runs, and video and TV runs are not considered. The domestic theatrical run accounts for about 30 percent of film rentals. It is an important market in its own right and still is important to launching a film.

3 De Vany and Walls (1993) show that movies have hazard rates that are characteristic of highly stressed products; there is a sharp peak in the hazard at about four weeks.

4 See Kenney and Klein (1983) and Blumenthal (1988) for analyses of the auction institutions under which exhibitors bid for distributors' films.

5 See De Vany and Eckert (1991) for details and an analysis of motion picture exhibition contracts.

6 This quality search model is equivalent to Rothschild's (1974) model of searching for the lowest price when the distribution of prices is unknown. Lippman and McCall (1976) deal elegantly with this model. In the price search model, the searcher relies on her previous trials to update; here the distribution of quality

is unknown and a Bayesian searcher relies on the quality reports of the searchers ahead of her to update θ, the location parameter of her distribution.

7 If we assume that the n viewers are distributed among the s screens according to probabilities $(p_1, \ldots, p_s)$, the multinomial distribution is Dirichlet. The distribution of viewers among films $m \leq s$ showing on multiple screens will also be Dirichlet (DeGroot 1968). The movie search problem—a search for quality with an unknown distribution—is similar to the search for price with an unknown distribution. Viewers who do not know the distribution begin with a uniform prior and adapt from there. Rothschild (1974) shows that a searcher facing an unknown price distribution has a Dirichlet prior if and only if his experience can be parameterized in exactly the way we have done it for movie quality searchers. We think this is a completely natural way to characterize information sets for Bayesian film buffs when screen bookings are set to equalize prior expectations of revenues.

8 This model also has a well-known interpretation as a Polya's Urn Model (Ross 1993, p. 117).

9 This diffusion of audiences over movies is similar to the process in Arthur's (1989) model, where the probability of choosing a product depends on the number who previously chose it, as well as the number who chose other products; in fact, it suggests that the Arthur model may generate the Bose–Einstein statistics.

10 To obtain the distribution of each film's revenues, we need to consider the probability that a group of k theaters showing film j will contain a total audience of r. Feller (1957) gives this as: $P\{\sum_i^k X_i = x_j\} = \binom{k+r-1}{k-1}\binom{s-k+n-r-1}{r-k}/\binom{s+n-1}{n}$.

11 Consider an audience of just 1,000 movie-goers and 49 theaters plus 1 "did not go" element to give a 1×50 outcome vector. There are 2.03×10^{210} possible outcome vectors.

12 Notable examples are *Howard the Duck* and *Last Action Hero* two highly promoted films with high expectations and large initial theater bookings. Both were gone from the market within a few weeks and exhibitors who booked at the opening were permitted to double-bill before the term of their exhibition contract expired.

13 Schmalensee's (1978) model of preemption in the cereal market could produce this market fractioning and a geometric share distribution, or the looser log series distribution (see the paragraph following equation (2.11)).

14 This form of the Pareto law, which uses rank instead of the number of firms above a certain size, is known as Zipf's (1949) law.

15 The Gini coefficient is a measure of inequality of the distribution of revenues across movies and distributors. It ranges from 0 to 1, with a value of 0 indicating equality and higher values indicating less equality. It is the ratio of the area between the Lorenz curve and this line to the area above the line. It can be computed as $1 - 2\int_0^1 L(x)dx$, where $L(x)$ is the equation of the Lorenz curve. To calculate this, the Lorenz curve was first approximated by a 7th order polynomial using linear regression; the R^2 of the regression was 0.9992. The estimated Lorenz curve was integrated analytically to obtain the area beneath it.

16 On average, there were 4.43 films released per distributor. The top four distributors made 21 1/3 percent of the films in the sample and the top eight made 37 1/3 percent of the films in the sample.

17 We know film births and deaths are not constant (De Vany and Walls 1997), so we already know before we test that we are unlikely to find a power law.

18 Smith and Smith (1986) analyze film rental champions released before and through the 1950s, during the 1960s and in the 1970s for evidence that the characteristics of successful films have changed. The *Paramount Decree* of 1949 made the vertically integrated studios divest their theaters and altered contracting

in the 1950s and after, and could have changed film characteristics along the lines we describe in the text.

19 It is not uncommon for exhibitors to restrict passes and discounts in the opening weeks, or even for the exhibition license to require it; this raises prices before demand is revealed for what usually is the period of highest demand. But, if the film is playing poorly, passes and discounts or even double-billing are used to effectively lower the admission price.

20 Because film rentals are based on box-office revenues, distributors will usually earn less rental if a theater lowers its admissions price. Moreover, exhibitors can reduce the film rentals even while increasing their total earnings by manipulating admission prices. A stationary price overcomes some of these disadvantages. See De Vany and Eckert (1991) and De Vany and McMillan (1993) for a discussion of how the industry deals with information and incentive issues that arise in reporting revenues.

21 A similar mechanism is used in pricing Broadway plays. A large price increase after a successful opening chokes off the information dynamics and shortens the run, which undercompensates the artists who are paid by the week. The demand dynamics and compensation of artists makes a stationary price and adaptive run the most effective way to supply Broadway plays and compensate artists.

22 If there were a permanent and recognized increase in the demand for motion pictures, admission prices would rise generally at theaters throughout the industry. We are dealing here with the pricing of imperfect substitutes in the short run, a situation for which inflexible prices with some form of quantity rationing, like delivery lags or queues, are well suited. See, Gordon (1981) for a discussion of the general issues, Carlton (1978) for an analysis of delivery lags and De Vany (1976) for an analysis of queuing.

23 On the tenure of studio executives, we found that over 30 percent of letters in a survey of nearly 400 executives were returned with no forwarding address. On participation contracts as a means of lowering studio risk, Weinstein (1998) notes that nobody would make *Forrest Gump* until Tom Hanks agreed to forego his usual fee and take a percentage of profits instead.

24 *Heaven's Gate* ruined United Artists.

3 Quality evaluations and the breakdown of statistical herding in the dynamics of box-office revenue

1 The alternatives to the herd-type models include all types of agent-based models. See the work of Epstein and Axtell (1996) for examples of how all sorts of group behavior can emerge from the interaction of individual decision makers.

2 The early viewers of trailers and movies function as evaluators. See Avery *et al.* (1999) for a discussion of the market for evaluations, which movie studios attempt to influence both through their opening strategies and with "buzz" generated on their web sites.

3 Some models, such as the contagion model of Arthur (1994) or the simulated models of De Vany and Lee (2001), consider more than two choices.

4 In other models of sequential evaluation such as the Avery *et al.* (1999) model, the sequence is endogenous and both the agent and the position in the sequence must be denoted. Since we consider an exogenously determined sequence, denoting the agent by position is unambiguous.

5 De Vany and Walls (1996) model the process as a Bayesian decision problem in which the sequential choices reveal, with increasing accuracy, the true choice probabilities.

6 Ordinarily, the agent would choose the size of the sample to draw before the choice is made, as in sequential search models (Lippman and McCall 1976).

7 This property may seem paradoxical in a Gaussian world, where the rapid decay of the exponential tail means that conditional expectation declines rapidly in the upper tail. But, the movies are not a Gaussian business and the extreme events in the upper tail dominate the mean.

8 Recall that even these non-hit movies are exceptional movies in the lengths of their runs for most movies do not run more than four or five weeks. These movies were selected in order to have the sample contain movies with comparable runs, whether they be hits or non-hits.

9 Lehmann and Weinberg (2000) find that the optimal timing of video release is a function of how well a movie plays in its theatrical release, and that the common practice of uniformly releasing a movie on video 24 weeks after theatrical release is seldom optimal.

10 Our data only include weekly data for the first 10 weeks of release and they do not include the week in which the final revenues were earned. Thus, we can only calculate the correlations between opening and cumulative revenue as reported in Table 3.6.

11 A detailed econometric analysis of the distribution of cumulative revenues across movies is given in De Vany and Walls (2002). Using quantile regression analysis, they find that big budgets, movie stars and wide releases significantly increase the lower quantiles of the box-office revenue distribution, but that they do little to increase the upper quantiles—the ones that matter.

12 Swami *et al.* (1999) have developed a mathematical programming model of cinema screen management that presumes an exponential playoff rate.

13 This pattern of past revenue fostering future revenue is the Bose–Einstein dynamic of box-office revenues (De Vany and Walls 1996.)

14 A related finding is in Eliashberg and Shugan (1997) who found that film critic reviews do not correlate with opening week revenues but do correlate highly with revenues in later weeks and with total revenues.

4 Uncertainty in the movie industry: can star power reduce the terror of the box office?

1 See De Vany and Walls (1996) and De Vany (1997) for analyses of information dynamics in the context of the motion picture industry.

2 It may be possible to "steer" the information cascade, that is, to affect the conditional probabilities of branching to different paths. This is a subject of ongoing research.

3 Industry analyst Art Murphy states this bluntly, "Films succeed or fail on their own merits. ... Also, films of mass appeal are relatively impervious to 'critics'. Films that are going to be popular are popular." (Dale 1997).

4 Kenney and Klein (1983) and Blumenthal (1988) also examine block booking and blind bidding for motion pictures.

5 We focus on the domestic North American theatrical market because revenues in this market are an important determinant of revenues in foreign markets, video, pay television and non-pay television as well as ancillary revenues from soundtracks, books, video games, theme parks and other consumer products (Dale 1997, p. 22; Cones 1997, pp. 143–144).

6 The continuation or hazard function is defined as $f(x)/(1 - F(x))$.

7 Cassey Lee provided the list of stars. We also constructed an alternative star variable indicating if an actor had been in more than five films. This variable gave qualitatively similar results to *Premier*'s and James Ulmer's lists.

8 For example, *Pulp Fiction* initially opened on 1,338 screens, but the first week's gross per screen was so high that its release was expanded to 1,489 screens for the second week (Lukk 1997, p. 28).

9 Screen count data were only available for 1,500 of the 2,015 movies in our sample.

10 Although a film may play for 10 weeks, it may be earning a very small box-office gross in its final weeks of theatrical release. For example, in a study using data from *Variety*'s national *Top-50* chart, De Vany and Walls (1997) found that the median survival time on the chart was 4 weeks, and in a duration analysis they found that a movie had less than a 25 percent chance of lasting longer than 7 weeks or more and less than a 15 percent chance of lasting longer than 10 weeks or more. Thus, while over 50 percent of the films in our sample were still playing at the tenth week of release, only a small fraction of these would be earning enough revenue to be included in the *Top-50* chart.

11 The largest difference between the distribution functions was 0.4927. The marginal significance level was practically zero.

12 The largest difference between the distribution functions was 0.5084. The marginal significance level was practically zero.

13 We estimated gross profits as one half of box-office gross less the production budget. This overestimates profits from theatrical exhibition because it does not include promotional expenses.

14 This corresponds well to Vogel's (1990, p. 29) rule of thumb that about 70–80 percent of all major motion pictures either lose money or break even.

15 But, several small films could be made on the budget of a big-budget film, so one should compare the returns properly: $n \times r(Z)$ relative to $r(n \times Z)$, where $r(\cdot)$ is the returns function, Z is a small budget and n is the number of films. One should compare the distribution of returns of n films costing Z dollars each with the return distribution of one film with a budget of $n \times Z$.

16 Our simple notion of breakeven is that rental revenues are equal to the budget. In the industry there are numerous definitions of breakeven, and through "studio accounting" numerous hit films such as *Batman* will probably never break even (Cones 1997).

17 For example, *El Mariachi* and *The Brothers McMullen* had gross returns to budget of 292 and 417, respectively. But their absolute profits were small.

18 Ijiri and Simon (1971) model the firm size distribution using this form of the Pareto law.

19 This result has been replicated in other papers covering different time periods and countries. See De Vany and Walls (1996), Walls (1997) and Lee (1998).

20 The deeper reasons for the relationship have to do with information dynamics and are beyond the scope of this chapter. See Rosen (1981), De Vany and Walls (1996), De Vany (1997) and Bikhchandani *et al.* (1992).

21 The glare from the concession stand may blind their vision.

22 Ranked in terms of their association with a film becoming a hit, a long run is most important (marginal probability of 0.07), followed by a wide release (marginal probability of 0.055), and then star (marginal probability of 0.04).

23 Looking either at Column (1) or (3), one finds that Tom Cruise and Jim Carrey carry the biggest impact among the male stars, followed by Spielberg, Brad Pitt, Oliver Stone, Kevin Costner and Tom Hanks.

24 Bill Mechanic, Chairman of Twentieth Century Fox, lists no females among his top stars. His list: Tom Cruise, Harrison Ford, Mel Gibson, Tom Hanks, Arnold Schwarzenegger and John Travolta. Quoted in John Cassidy, "Chaos in Hollywood," *The New Yorker*, March 31, 1997.

25 Recall that profits are calculated for the theatrical market only and do not include foreign and other revenues. Cost is the estimated budget reported in the EDI data. The 0.5 figure is a rough estimate of the average rental rate. A high grossing film typically will earn a higher than average rental rate although a poor performing

film on which guarantees were paid may also earn a high rate. Consequently, this equation is a crude approximation to profits in the North American theatrical market.

26 We do not report quantile regressions because the mean absolute deviation estimator would not converge for the profit regressions. We are tackling this problem in another chapter.

27 In our current research we use probability models to estimate the "half lives" of stars and movie studios based on their movie portfolios of the past decade.

28 De Vany (1997) deals with this issue in more detail. We analyze how production decisions are related to past events in our current research.

29 Earthquakes also follow a power law. Trying to predict the next blockbuster is like trying to predict the next big earthquake.

30 Given the risk at the box office, it is not surprising that many movies are made on the basis of pre-committed foreign distribution funds and tie-ins to games and fast food promotions.

5 Does Hollywood make too many R-rated movies?: risk, stochastic dominance and the illusion of expectation

1 Ravid (1999) showed that G-rated movies earned higher average profit than R-rated movies.

2 Scherer and his coauthors (Scherer *et al.* 2000) show that the profit distribution from technological innovations is highly skewed with a Paretian tail. De Vany and Walls (1996) found the Bose–Einstein process to be a good model of motion picture revenue dynamics. Sornette and Zajdenweber (1999) explicitly model innovation as an α-stable Lévy motion and show that the innovation and international motion picture data follow the stable Paretian law.

3 The strategy of exploiting the asymmetric tails of the distribution to increase returns and lower large risks is developed by Anderson and Sornette (1999).

4 Of the 60 G-rated movies, 31 are animated films. However, the distributions of revenues, profits and returns do not differ between animated and non-animated G-rated movies: Kolmogorov–Smirnov tests for the equality of distribution functions resulted in marginal significance levels of 0.101, 0.258 and 0.224, respectively, which do not rise above conventional levels of significance. An anonymous referee pointed out that many people assume that animated G-rated movies are responsible for the high performance of G-rated movies. These tests show that the distributions do not differ; that is to say that both animated and live-action G-rated movies represent draws from the same probability distribution, even though the outcomes do sometimes differ.

5 A star may take "scale" to appear in a low-budget movie if they feel they can exhibit their craft in a strong role. These kinds of roles seem to occur in R-rated movies or are seen by the stars to occur there. This is not to be confused with gross or profit participation, which a star would only take in a high-budget movie with good prospects.

6 De Vany and Lee (2000) verify that the revenue increments are an α-Stable Lévy Motion whose increments scale as $\delta t^{1/\alpha}$. The process is dense with discontinuities, as McCulloch (1996) shows is true of an α-Stable Lévy Motion. The discontinuities are quite evident in the figure even though the series shown is a time-expanding average rather than the raw increments.

7 De Vany and Walls (1996) show that the statistical dynamics of box-office revenues converge asymptotically to a Pareto distribution, and in De Vany and Walls (1999) they demonstrate that the stable Paretian model is consistent with box-office revenue data. De Vany and Lee (2000) perform a battery of diagnostic tests on motion picture data which strongly support the stable Paretian model.

8 We are not asserting that studios use the normal distribution; the normal distribution model is a useful and familiar base case against which to contrast the stable Paretian model.

9 To be sure, if we were to include ancillary products—t-shirt sales, fast-food franchise tie-ins, toys etc.—the rate of return on G-rated films would dominate the other rating categories by an even greater amount.

10 See Huang and Litzenberger (1988) for a good exposition of stochastic dominance in portfolio theory.

11 A weaker form of stochastic dominance is second-order stochastic dominance. We will focus on first-order stochastic dominance as it corresponds to the empirical analysis that follows.

12 The diagnostic tests confirming the stable Paretian model for revenues are contained in De Vany and Lee (2000). See also Figures 5.1 and 5.2 which show the unstable mean and variance which Mandelbrot identified as salient characteristics of the stable distribution. The log–log linearity of the upper tail which Mandelbrot shows to be implied by the stable Paretian model is also visible in Figure 5.4.

13 Stuart *et al.* (1999, chapter 25) contains a discussion of the Kolmogorov–Smirnov test. De Vany and Walls (1999) and Sornette and Zajdenweber (1996) verify the Paretian tail of box-office revenues and both show it to be stable over time. These authors also reject log normality in favor of the Paretian distribution. De Vany and Walls (1996) formally reject log normality and 3-parameter log normality for motion picture revenues.

14 A personal communication from Louis Wilde indicates that the foreign and other market "windows" are as chaotic (meaning they probably follow an α-stable process) as our results indicate the domestic market to be.

15 Estimation of the upper tail is straightforward. For the lower tail, we estimated the exponent for losses, that is, the upper tail of the distribution of losses (negative profits). See De Vany and Walls (2000) for further diagnostics that show the stable Paretian model is an excellent fit to the data.

16 The model, therefore, predicts that extremely profitable movies are not unlikely, as they would be were the profit distribution normal. Even though the model is estimated before *Titanic*, it indicates that movies like it and even movies exceeding it occur with non-negligible probability. De Vany and Walls (2000) study the distribution of profits in more detail using the general stable distribution.

17 For high-budget, R-rated movies this is another confirmation of Ravid's (1999) conclusion that stars do not increase the profitability of films; in fact, stars do worse than that, they decrease the probability of positive profits and increase the probability of losses.

18 Mandelbrot (1967) calls this the scaling property meaning that the probability density function is invariant to linear transformations of the data.

19 Screenwriter William Goldman said so much in his Adventures in the Screen Trade when he said: "With all due respect, nobody knows anything."

6 Big budgets, big openings and legs: analysis of the blockbuster strategy

1 Distributors now also coordinate with video distributors on the video release (*Forbes Global Business & Finance*, June 15, 1998, pp. 26–28).

2 Revenues accrue to motion pictures from tie-ins with soundtracks, books, video games, theme parks, fast food franchises and other all sorts of other consumer products (Cones 1997, pp. 143–144).

3 "There are two kinds of talent that are critical to Hollywood—behind the screen and on-screen. The former depends on the ability to produce excellent products, but does not require public recognition. There are many executives, directors and writers who can command high talent fees because of their work and yet they remain unknown to the public. On-screen talent requires charisma and screen presence. The best-paid actors are those who have 'marquee value'—their appearance in a film will attract a loyal following and help the film to 'open'. This is the definition of a 'star' and usually means that the actor has qualities that ignite the public's imagination and their dreams" (Dale 1997, p. 42).

4 De Vany and Walls (1996) show several common release patterns.

5 Most movies die quickly, with the median survival time in *Variety's* Top-50 chart being only 4 weeks (De Vany and Walls 1997).

6 Art Murphy states this plainly, "Films succeed or fail on their own merits. No Ralph Naders, Moral Majorities etc. can impact the film business in the way that automobile tires, Corvairs or commercial advertisers are affected by consumer advocacy or pressures. Also, films of mass appeal are relatively impervious to 'critics'. Films that are going to be popular are popular" (Dale 1997, p. 4).

7 We would like to thank Cassey Lee for compiling the list of stars.

8 The marginal significance level for a two-sided runs test was 0.36.

9 The marginal significance level in a Wilcoxon matched-pairs signed-rank test is 0.0022.

10 Recall that all dollar figures are in constant 1982–84 dollars. We have labeled the four movies with the largest and smallest revenues and budgets. *El Mariachi* had the lowest budget at about \$4,800 but had revenues of \$1.4 million. *Waterworld* had the largest budget at \$114 million and revenues of only \$57.3 million. *Berlin Blues* has a budget of \$3.9 million and revenues of only \$1,304. *Jurassic Park* had revenues of \$245 million and a budget of \$43.2 million.

11 This basic regression equation has been used by previous researchers including Prag and Cassavant (1994), Litman and Ahn (1998) and Ravid (1999).

12 Recent research has discovered that motion picture box-office revenues follow power laws (De Vany and Walls 1999). In order to formally account for this, we reestimated equation (1) using McCulloch's (1997) symmetric stable regression. The results were qualitatively similar to the least-squares and robust regression estimates reported in table 8.

13 Greene (1997, p. 256) says that, "Coefficients of determination in cross sections of individual data as high as 0.2 are sometimes noteworthy."

14 Details of the estimation procedure and a Monte Carlo study are given in the STATA Reference Manual (1997, pp. 168–73) and Hamilton (1991), respectively. Briefly, the robust regression proceeds in three stages: (1) a least-squares regression is estimated and observations for which Cook's D >1 are deleted; (2) weighted regressions are then run using the weights proposed by Huber (1964) until convergence; and then (3) Beaton and Tukey (1974) biweights are used until convergence.

15 The robust standard errors are White's (1980) for least-squares and bootstrapped with 500 replications for the quantile regressions.

16 There probably are diminishing returns to a movie's entertainment value from adding more special effects, car crashes, explosions and cities wiped out after the numbers become large.

17 To examine if the revenue impact of stars varied with the budget, we included a slope dummy variable interacting log Budget × Star in the regressions. We found that the coefficient on the slope dummy was not statistically different from zero.

The evidence indicates that the effect of stars does not vary with the magnitude of the budget.

18 In the *Hollywood Reporter's* (1998) ranking of the "star power" of various actors, Tom Hanks and Jodie Foster scored 97 and 94, respectively, out of a possible 100 points.

19 Our results are consistent with Eliashberg and Shugan's (1997) finding that film critics do not correlate highly with opening week revenues but that they do correlate highly with later weeks revenues and cumulative revenues. They find that film critics are 'leading indicators' that can spot a high quality film but that the critics do not themselves create an information cascade.

7 Motion picture antitrust: the *Paramount* cases revisited

1 *United States* v. *Paramount Pictures, Inc.*, 66 F. Supp. 323(S.D.N.Y. 1946); 70 F. Supp. 53 (S.D.N.Y. 1947); 334 US 131 (1948); 85 F. Supp. 881 (S.D.N.Y. 1949); 339 US 974 (1950) (per curiam, no opinion); *United States* v. *Griffith Amusement Co.*, 68 F. Supp. 180(W.D. Okla. 1946); *United States* v. *Griffith*, 334 US 100 (1948); *United States* v. *Schine Chain Theaters, Inc.*, 63 F. Supp. 229 (W.D.N.Y. 1945); *Schine Chain Theaters, Inc.* v. *United States*, 334 US 110 (1948).

2 *United States* v. *Paramount Pictures et al.*, 1948–49 Trade Cas. 62,335 (S.D.N.Y. 1948) (the RKO decree); 1948–49 Trade Cas. 62,377 (S.D.N.Y. 1949) (the *Paramount* decree); *United States* v. *Loew's Inc., et al.*, 1950–51 Trade Cas. 62,765 (S.D.N.Y. 1951) (the Warner Bros. decree); 1950–51 Trade Cas. 62,861 (S.D.N.Y. 1951) (the Twentieth Century Fox decree); 1952–53 Trade Cas. 67,228 (S.D.N.Y. 1952) (the Loew's decree); and 1950–51 Trade Cas. 62,573, at pp. 63,680–63,682 (S.D.N.Y. 1950) (the Columbia, Universal and United Artists decree). These will be referred to collectively as the *Paramount* decrees.

3 Secs. I, 2; I5 U.S.C.A. secs. I, 2.

4 *United States* v. *Paramount Pictures, Inc.*, 66 F. Supp. 323, 357 (S.D.N.Y. 1946).

5 See Harris (July 27, 1986) and (January 3, 1988), and National Cable Television Association (November 1986, pp. 23–35, table 8).

6 *United States* v. *Griffith Amusement Co.*, 68 F. Supp. 180, 196 (W. D. Okla. (1946)).

7 See Conant (1960, pp. 1–10), Hellmuth (1962, pp. 395, 405, 411). See also Cassady (1958), *Comment, Yale Law Journal* (1965), Crandall (Spring 1975), Donahue (1987), and Whitney (1982). Two authors in the nonacademic literature who recognized the value of demand information are Knowlton (August 29, 1988), and Trustman (January, 1978). The exception in the academic economics literature is Kenney and Klein (October, 1983), which explains block booking as a means of making contracts more secure and also shows that previous explanations based on monopoly or price discrimination are wrong.

8 See Nelson (March/April 1970). William Goldman's rule for estimating demand is "Nobody knows anything." See Goldman (1983, p. 39).

9 The importance of timely demand information is attested to by the fact that the industry relies on real time (weekly and overnight on opening weekends) reporting of detailed, firm-by-firm, product-by-product information that would be considered private in most industries. The weekly listings in *Variety* give the pictures playing, prices and attendance for major theaters in metropolitan areas. Similarly, exhibitor reports of box-office gross receipts for most pictures are processed and published weekly in *Variety* and other publications.

10 Proceeding sequentially is an efficient Bayesian statistical decision procedure to estimate demand when the distribution is unknown and sampling is costly. See De Groot (1970).

11 Akst (April 11, 1988). Of every 10 pictures made, 3 or 4 make profits and 1 breaks even after including television, foreign, and home video sales. Only about 1 in 10 turns a profit from box-office sales alone. Knowlton (August 29, 1988, p. 73). In 1978, about 1 picture in 36 made "significant money." Trustman (January 1978, p. 73).

12 We estimate that the top-grossing picture in any year over the past two decades has claimed from just under 1–3.5 percent of total annual box-office gross. The top four pictures on *Variety's* "50 Top-Grossing Films" earn about one-half of the weekly box-office gross of those screens in *Variety's* sample.

13 See Conant (1960, p. 69).

14 The relation between transactions costs, long-term contracting, and vertical integration is emphasized by Coase (November, 1937), Williamson (May, 1971) and (October, 1979), and Klein *et al.* (October, 1978).

15 The royalty formula is $Royalty = r(pq - F)$, where r is the royalty rate, p is the admission price, q is admissions, and F is the allowance for the theater's fixed cost. Since F is the theater's private cost, in practice the distributor does not know it and the figure is reached through a negotiation in which the allowed figure is fixed for the coming year. F is the "house nut" in industry parlance. The royalty rate in a contemporary license might call for 90 percent of box-office revenue over the house nut. In addition, there will be floor rates that apply to the unadjusted box-office gross, so that the exhibitor pays the higher of 90 percent over fixed cost, or the minimum floor royalty applied to the total box-office gross (without deducting the fixed charge). The minimum royalty usually declines each week or each two weeks. For example, a typical contemporary license, for which we have information, might specify 70, 70, 60, 60 as the minimum royalty rate for each week of a four-week run and any additional weeks go at the industry standard 40 percent. The house nut is negotiated between the distributor and the exhibitor and usually applies to all the films of that distributor for the year unless it is renegotiated because of a change in the exhibitor's cost.

16 A theater manager could steal from the integrated firm as well as from the nonintegrated firm. The essential difference in the integrated and nonintegrated case is that privacy of information blocks the nonintegrated distributor from direct discovery of the theater's revenue.

17 Silver (1983) describes the difficulties the independent distributor faces in collecting royalties.

18 Cassady (1959, pp. 374–375, Conant (1960, pp. 16–20), and Hellmuth (1962, pp. 394–396).

19 Murphy (1983, pp. 251–252).

20 Cassady (1959, pp. 374–376).

21 (Cassady 1959, pp. 377–379).

22 (Cassady 1959, pp. 379–380). *Birth of a Nation* had played before 200 million Americans by 1946. Nash (October 16, 1988, p. 5).

23 Cassady (1959, pp. 381–384), and Conant (1960, pp. 21–23).

24 Cassady (1959, p. 382).

25 Cassady (1959, pp. 384–385).

26 Conant (1960, pp. 24–26).

27 66 F. Supp. 323 (S.D.N.Y. 1946).

28 66 F. Supp. 334 n. 2.

29 66 F. Supp. 339.

30 *United States* v. *Paramount Pictures, Inc.* 66 F. Supp. 323, 334–35 (S.D.N.Y. 1946). The distributors did not defend the practice as a device for eliciting exhibitor inputs, which is the explanation often given for retail price maintenance. See Goldberg (1984).

31 The distributor's profits are royalties less distribution cost. The market price will then depend on the total quantity of output, which is tq, where t is the number of theaters exhibiting the picture and q is admissions per theater. Let ϵ be the market price elasticity of demand (which depends on t and q since $Q = tq$). Assume that distribution cost is predominantly the cost of the number of prints and that print cost per admission is constant and equal to g. Given the contractual royalty rate r, the distributor's profit can be shown to be maximized where $p(1+1/\epsilon) = g/r$. On the other hand, the theater can be shown to maximize profit where $p(1 + 1/\gamma) = c_0'/(1 - r)$, where γ is the price elasticity of demand at each theater and c_0 is the marginal cost of admissions. Individual theater price elasticity of demand is greater than market price elasticity (because each theater is a substitute for others), so even if costs were somehow equal, the exhibitor would tend to charge a price that is less than the distributor's profit-maximizing one. The relevant marginal costs differ as well; $c'/(1 - r)$ in general does not equal the distributor's marginal cost g/r. In this case, the theater's optimal price is less than the distributor's optimal price. If only one theater were exhibiting the picture, this disparity in price elasticity would vanish and this may partly explain the exclusive first run. Nonetheless, cost differences would still create a disparity in optimal prices. Moreover, the distributor might lose admissions in larger markets by playing only one theater because filmgoers must travel too far to see it.

32 In the 1970s, concessions amounted to between 13 and 23 percent of total theater receipts according to Census reports. Conant (Autumn 1981, p. 99).

33 Twentieth Century Fox contracts in the early 1980s required that car heater charges be included in gross box-office receipts at drive-in theaters and expressly prohibited tying heaters to admissions. "Lead in" advertising, which the theater shows before showing the picture is another source of revenue diversion which distributors have tried to curb through the terms of the license agreement with incomplete success. See May (1983).

34 If the exhibitor charges a zero admission price, the film rental owed would be zero because it is a percentage of admission revenue. It is not impossible that an exhibitor could make more by giving out free passes to a Walt Disney film than by charging admission. The exhibitor would receive high profits on concession sales and would owe no film rental since admissions revenue would be zero. A minimum per capita royalty prevents this and similar abuses. SAC May 1983, and Harris (June 19, 1988). Such licenses do not constitute a vertical price-fixing scheme. *General Cinema Corporation* v. *Buena Vista Distribution Co., Inc.*, 681 F.2d. 594 (9th Cir. 1982).

35 *United States* v. *Paramount Pictures, Inc.* 334 U.S. 131, 142 (1948). Federal courts now insist on proof of conspiracy, not merely parallel conduct, to find collusion in motion picture antitrust cases. *United States* v. *Schine Chain Theaters, Inc.*, 575 F.2d 233, 236 (9th Cir. 1978); *Admiral Theatre Corp.* v. *Douglas Theatre Co.*, 585 F.2d. 877, 884, 887 (8th Cir. 1978).

36 *United States* v. *Paramount Pictures, Inc.*, 66 F. Supp. 324, 337 (S.D.N.Y. 1946).

37 *United States* v. *Paramount Pictures, Inc.*, 66 F. Supp. 336–337.

38 *United States* v. *Paramount Pictures, Inc.*, 66 F. Supp. 353–354.

39 The Herfindahl index, calculated as the sum of the squared production market shares, ranged from 0.160 to 0.213 with a mean of 0.167 during 1940–1950. Alicoate (1954, p. 125). This index is biased upward because we have treated

imports and independent production as though they were one firm each (individual firm data are not available for these groups). Given the volatility of market shares, static measures of concentration such as the Herfindahl index are defective measures of market power. The four-firm concentration ratio averaged 0.30 during 1940–50. The firms included in the set of the four largest firms by production changed almost every year.

40   *United States* v. *Paramount Pictures, Inc.*, 85 F. Supp. 881, 894 (S.D.N.Y. 1949), and Crandall (Spring, 1975, p. 60, table 3).

41   See De Vany and MacMillan (June, 1990).

42   *United States* v. *Paramount Pictures, Inc.*, 70 F. Supp. 53, 71 (S.D.N.Y. 1947).

43   *United States* v. *Paramount Pictures, Inc.*, 66 F. Supp. 323, 339–340 (S.D.N.Y. 1946).

44   See *Comment, Yale Law Journal* (1965, pp. 1084–1085), Crandall (Spring, 1975, p. 52), and Kenney and Klein (October 1983, p. 517).

45   Conant (1960, pp. 60–61, 65–70 and passim).

46   Conant (1960, pp. 155–156).

47   *United States* v. *Paramount Pictures, Inc.* 66 F. Supp. 323, 341 (S.D.N.Y. 1946).

48   See text accompanying note 30, supra.

49   *United States* v. *Paramount Pictures, Inc.*, 66 F. SUPP. 323, 341–342 (S.D.N.Y. 1946).

50   Alicoate (1940, p. 45) Runs at showcase and distributor-owned theaters were longer.

51   *United States* v. *Paramount Pictures, Inc.*, 66 F. Supp. 323, 343–345 (S.D.N.Y. 1946).

52   *United States* v. *Paramount Pictures, Inc.*, 334 U.S. 131, 146 (1948). Unreasonably long runs and clearances and favoritism shown to circuits were also found in *United States* v. *Schine Chain Theaters, Inc.* 63 F. Supp, 229. 238–234 (W.D.N.Y. 1945). These findings were reversed on grounds of ambiguous criteria in *Schine Chain Theaters, Inc. et al.* v. *United States*, 334 U.S. 110, 121–124 (1948).

53   *United States* v. *Paramount Pictures, Inc.* 70 F. Supp. 63, 68 (S.D.N.Y. 1947).

54   See text accompanying note 6, supra, and *United States* v. *Griffith*, 334 U.S. 100 (1948).

55   *United States* v. *Paramount Pictures, Inc.*, Supp. 333, 333 n. I, 346–347, 352 (S.D.N.Y. 1946). (*United States* v. *Schine Theaters Inc.* 63 F. Supp. 229, 239–240 (W.D.N.Y. 1945); *Schine Chain Theaters, Inc. et al.* v. *United States*, 334 U.S. 110, 116, 118–119, 120–124 (1948); *United States* v. *Griffith Amusement Co.*, 68 F. Supp. 180, 189–191 (W.D. Okla. 1946); *United States* v. *Griffith*, 334 U.S. 100 (1940).

56   *United States* v. *Paramount Pictures, Inc.*, 334 U.S. 131, 154 (1948). The Supreme Court instructed the District Court to reconsider its objections to franchises, but the District Court adhered to the view that distributors should be permitted to grant franchises to independents only. (*United States* v. *Paramount Pictures, Inc.*, 85 F. Supp. 881, 897 (S.D.N.Y. 1949).

57   Note that the exhibitor had a strong incentive to promote the picture, since film rentals did not depend on individual sums. Following *Paramount*, distributors could achieve the results of formula steals (and vertical integration) by leasing theaters and staffs for a fixed fee. They could then set their own admission price, and retain all of the box-office and concession revenues. This arrangement constituted short-run integration and it was called "four walling." Use of "four walling" by Warner Bros. was alleged to have violated its 1951 consent decree. In a subsequent decree in 1976, Warner Bros. agreed to discontinue the practice for 10 years. See Conant (Autumn 1981, p. 96 n. 197), citing *United States* v. *Warner Bros. Pictures, Inc.*, 1979–81 Trade Cas. 62,504 (S.D.N.Y. 1976).

58 The nonintegrated distributor could deal with an exhibitor suspected of misrepresenting box-office revenue by employing an independent observer, or by requiring the use of numbered tickets printed by a third party, or by dealing only with exhibitors with good reputations. That distributors still use these devices suggests they still suspect exhibitors of misrepresenting their box-office grosses. For example, five firms are listed under "Checking Theater Attendance" in *Motion Picture Almanac* (1985, p. 431). One firm offers "Blind (covert) theater checking, national coverage."

59 *United States* v. *Paramount Pictures, Inc.*, 66 F. Supp. 323, 348 (S.D.N.Y. 1946).

60 *United States* v. *Paramount Pictures, Inc.*, 334 U.S. 131, 156 (1948). United Artists did not employ block booking.

61 *United States* v. *Paramount Pictures, Inc.*, 334 U.S. 158.

62 68 F. Supp. 180 (W.D. Okla. 1946).

63 68 F. Supp. 185.

64 Lazarus (1983).

65 *United States* v. *Schine Chain Theaters, Inc.*, 63 F. Supp. 229, 232 (W.D.N.Y. 1945). The District Court in *Paramount* found that in 1943–44, total production by the 5 major and 3 minor studio-defendants was 260 pictures. Individual studio production ranged between 7 and 49 features with between about 5 and 15 percent of total US production excluding westerns, or between 4 and 12 percent including westerns. In 1943–44, RKO produced 38 features, but it would have been "financially impossible for RKO to operate its theaters on features distributed by RKO alone." *United States* v. *Paramount Pictures, Inc.*, 70 F. Supp. 53, 64–65 (S.D.N.Y. 1947).

66 In 1944, about 48 percent of film rentals paid by Loew's theaters were for Loew's pictures. In 1943–44, about 23 percent of the features shown in RKO theaters were distributed by RKO and 30 percent were distributed by other integrated distributors. In the four seasons of 1937–40, about 14 percent of Warner's domestic gross rentals were from its own theaters and about 24 percent from theaters in which other defendants had an interest. *United States* v. *Paramount Pictures, Inc.*, 70 F. Supp. 53, 69 (S.D.N.Y. 1947).

67 *Fortune* (August 1939, p. 106). Twentieth Century Fox block booked with less than 20 percent of theaters playing their films.

68 Approximately 7,000 theaters exhibited single features and 10,000 exhibited double features. Alicoate (1940, p. 45). A weekly program change would entail about 52 features per year as a requirement for single feature houses and about twice that for double-feature houses. Two changes per week in a double-feature house, which was common in sub-first-run theaters, would require about 200 films per year. The average length of run in 1940 was estimated at 2.25 days. (Alicoate 1940). Runs were longer, however, at the showcase theater owned by the majors and circuits. In the late 1940s, the average moviegoer attended 44 pictures a year, Trustman (January, 1978, p. 65).

69 See text accompanying note 25 supra.

70 Conant (1960, p. 145).

71 Hilton (1958, pp. 273–274) and Stigler (1968). The form of price discrimination usually assumed is for a monopolist to (1) identify two different groups of buyers, (2) charge different prices according to the intensity of their preferences, and (3) prevent resale between the two markets. The studios were largely incapable of doing (1) since they did not have the required information before a film played and did not charge fixed prices on first-run features, The royalty contract directly implemented (2) by charging a percentage of the box-office revenue rather than a fixed price. When fixed rental rates were used, they had less than complete success in (3) since they could not prevent resale among exhibitors. In the early days

prints could be shown at several theaters in a day by bicycle deliveries, enabling the exhibitors to break down any attempt at price discrimination and also to play the film at more locations than those contracted.

72 This theory was accepted by the courts in *United States* v. *Paramount Pictures, Inc.*, 66 F Supp 323, 350 (S.D.N.Y. 1946) and by Conant (1960, p. 79).

73 *United States* v. *Paramount Pictures, Inc.*, 66 F Supp. 323, 350 (S.D.N.Y. 1946).

74 *United States* v. *Paramount Pictures, Inc.*, 334 U.S. 131, 157 n. 11 (1948).

75 The inventory of first-run films in production valued at production cost is around 0.5 of sales measured as film rentals. See Vogel (1990, p. 43).

76 See Kenney and Klein (October 1983, pp. 520–522). Blind selling is a durable practice: in 1979, over 90 percent of the pictures of major distributors were sold blind. *Note, Harvard Law Review* (1979, p. 1131). In 1977, prints of the average independently produced picture were available only 11 days before release, and *Star Wars* was finished "only days before it started showing." *Note, Journal of Law and Commerce* (1985, p. 302). As of 1986, 22 states restricted blind bidding. Harmetz (October 23, 1986), and Comment, *Pace Law Review* (1982).

Blind selling or licensing may be done without bidding. The opposition of exhibitors to blind bidding stems at least in part from their opposition to bidding itself. On the difficulties of bidding as a licensing mechanism, see note 98 infra. The state legislation against blind licensing reflects complex motives. Exhibitors who want to book pictures for future dates will often license them before they have screened them—this practice is as old as the feature motion picture. The legislation seems redundant in light of the restrictions of the decrees on blind licensing. But, the state legislation was passed after *Paramount*. Our conjecture is that the law gives the exhibitor a release from the contract which contains a guarantee when there is evidence after the contract has been signed that the picture will not play well. This evidence may come from box-office reports to other states or from a screening. The presumption of an auction is that the contract is binding and the terms or conditions will not be changed later; otherwise, the auction is not authentic. The bidding methods employed after *Paramount*, in conjunction with the distributor strategy of setting a minimum guaranteed royalty (which is an optimal auction procedure), exposes the winning exhibitor to a severe form of a winner's curse which the antiblind bidding law provides some relief by making default easier.

77 *United States* v. *Paramount Pictures, Inc.* 66 F. Supp. 323, 350 (S.D.N.Y. 1946); 70 F. Supp. 53, 64 (S.D.N.Y. 1947).

78 *United States* v. *Paramount Pictures, Inc.*, 66 F Supp, 323, 350–351, 358–359 (S.D.N.Y. 1946); 70 F. Supp. 53, 66–67 (S.D.N.Y. 1947); 334 U.S. 131, 151 (1948).

79 *United States* v. *Griffith Amusement Co.*, 68 F. Supp. 180, 184, 192–193 (W.D. Okla. 1946).

80 *United States* v. *Griffith Amusement Co.*, 68 F. Supp. 185–186.

81 *United States* v. *Griffith Amusement Co.*, 68 F. Supp. 193.

82 *United States* v. *Griffith*, 334 U.S. 100 (1948).

83 *United States* v. *Paramount Pictures, Inc.*, 66 F. Supp. 323, 350–351, 358 (S.D.N.Y. 1946).

84 *United States* v. *Paramount Pictures, Inc.*, 334 U.S. 131, 149 (1948).

85 Theater admissions and concession sales are made in cash. Theaters must guard against employees retaining for themselves revenue that would otherwise go to the circuit. An on-site owner may he better able to monitor employees, and has stronger incentives to do so than other employees. There are many stories but not much hard evidence on this point. The Saturday night late show is sometimes referred to as "the manager's show" because there have been instances of

managers retaining ticket sales of this showing for their own pocket. One the-ater circuit owner uses binoculars to observe employees from across the street to detect cheating.

86 *United States* v. *Paramount Pictures, Inc.*, 334 U.S. 131, 169 (1948).
87 *United States* v. *Paramount Pictures, Inc.*, 66 F. Supp. 323, 354 (S.D.N.Y. 1946).
88 *United States* v. *Paramount Pictures, Inc.*, 66 F. Supp. 355.
89 *United States* v. *Paramount Pictures, Inc.*, 66 F. Supp. 353–354.
90 *Comment, Yale Law Journal* (1965).
91 *United States* v. *Paramount Pictures, Inc.*, 334 U.S. 131, 164 (1948).
92 Same as footnote 93
93 When they were required, screenings were poorly attended by exhibitors. See text accompanying note 78 supra. It is not obvious that a screening will improve an exhibitor's judgment. After he screened it, the head of one major theater circuit is said to have told George Lucas that nobody would want to see *Star Wars*. Generally, no theater circuit will bid for the right to exhibit a film or guarantee film rentals unless they screen it first. Bidding was not used to license films at the time of *Paramount*. It would have been impossible to license blocks and seasons of film production by auction.
94 *United States* v. *Paramount Pictures, Inc.* 334 U.S. 131, 175 (1948).
95 *United States* v. *Paramount Pictures, Inc.* 334 U.S. 166.
96 *United States* v. *Paramount Pictures, Inc.*, 85 F. Supp. 85 F Supp. 881 (S.D.N.Y. 1949) (requiring divestiture); 339 U.S. 974 (1950) (per curtain, upholding divestiture).
97 Under an auction licensing arrangement, distributors would have to com-pare complex exhibitor bids with differing ingredients almost on a daily basis. Exhibitors would have to decide which pictures to bid on, how much to bid, and formulate a contingency plan if their bid was not chosen. Strategic elements would come into play, such as the number and identities of the other bidders, the seller's reservation price and other pictures being offered by other distributors. Should the exhibitor underbid on this picture in the hope that other exhibitors will pass it over

   Should the seller's reserve price take the form of a guarantee, a minimum play time, a high floor percentage or all of them. How should pictures that fail to bring the reserve price be reoffered. Must they rebid or can their terms be negotiated. The winning bid must form a binding contract if the auction is to be credible, but then the flexibility that is required to deal with uncertainty may be lost (there can be no reviews or move-overs, for example, to adjust terms). Obstacles to the use of bidding to allocate motion pictures are not insuperable, but they do suggest that bidding is not clearly the superior way to license motion pictures and, in many ways, is unsuited to the purpose.
98 *United States* v. *Paramount Pictures, Inc.*, 334 U.S. 131, 172–173 (1948).
99 *United States* v. *Paramount Pictures, Inc.*, 85 F. Supp. 881, 894 (S.D.N.Y. 1949).
100 Alicoate (1954, p. 129).
101 In the ongoing research using the Top-50 motion picture charts in *Vari-ety*, David Brownstone and Arthur De Vany have found that a contemporary film's time on the charts is negatively related to the number of theaters showing it.
102 *United States* v. *Paramount Pictures, Inc.* 85 F. Supp. 881, 890 (S.D.N.Y. 1949).
103 *United States* v. *Griffith*, 334 U.S. 100, 107 (1948).
104 *United States* v. *Griffith Amusement Co.*, 68 F Supp. 180, 190 (W.D. Okla. 1946).
105 On the whole, the industry did not decline rapidly during the period of television's rapid growth from 1945 to 1955. Average weekly attendance at theatres fell from 80 million in 1940 to 60 million in 1950 and to 45.8 million in 1955. On the other hand, box-office gross remained somewhat level in nominal terms:

in 1940 it was $1,002,580,000; in 1950 it was $1,320,000,000; and in 1955 it was $1,185,100,000. See Alicoate (1957, p. 111). Admission price including tax was estimated to 24.1c in 1940, 52.8c in 1950, and about 60c in 1955. The total number of theaters in operation, including drive-ins, hardly changed: 19,032 theaters in 1940; 19,106 in 1950; and 19,003 in 1955. Walk-in theaters declined by about 4,500 and drive-ins increased by about the same amount. Alicoate (1957, p. 113). Nearly every motion picture company experienced a growth in gross income.

106 *United States* v. *Paramount Pictures, Inc.*, 66 F. Supp. 323 (S.D.N.Y. 1946).

107 *United States* v. *Paramount Pictures, Inc.* 334 U.S. 131 (1948). This was also the landmark case in *Paramount* according to the stock market evidence. See De Vany and MacMillan (June, 1990, p. 13).

108 See note 121 infra for details on the provisions with respect to the individual defendants.

109 Neither Columbia nor United Artists owned theaters. United Artists had no actors under contract; it was owned by some of the major stars of the period and did not use season contracts or block booking. Columbia was a producer-distributor only and there is no evidence that it used franchises or formula deals. RKO, Loew's, Twentieth Century Fox, Warner Bros., Universal and *Paramount* all produced at least 40 percent fewer films in 1950 than in 1940. Columbia produced more (59 in 1950 versus 51 in 1940) and United Artists produced about the same number (20 in 1940 versus 18 in 1950).

110 Crandall (Spring 1975, p. 66, figure 1 and Gini coefficients).

111 Film rentals as a percentage of admission revenue rose from 0.304 in 1948 to 0.331 in 1954, 0.421 in 1958, 0.456 in 1963, and 0.458 in 1967. Crandall (Spring 1975, p. 63, table 4). The average rental rate paid to the eight distributor defendants by their affiliated theaters was 27 percent, Brief of the Warner Defendants to the Supreme Court at 138, *United States* v. *Paramount Pictures, Inc.*, cited in Kenney and Klein (October 1983, p. 518 n. 53).

112 Crandall (Spring, 1975, p. 61) says. "Given that the decrees had required an end to the overtly collusive distribution policies and divestiture of many theaters from the divorced circuits of the major studios, this rapid increase in admission prices is startling." He explains these facts as indicating that the studios counteracted the provisions of the decrees by reducing output. This is contrary to the court's finding that production was open and competitive and raises the question as to why the majors had not exerted such control before the decision. His findings are contrary to his maintained hypothesis that the distribution policies were collusive. He attempts to reconcile the inconsistency of the facts with the collusion hypotheses by arguing that after the decrees the studios reduced production in order to raise prices.

113 The percentage of total annual box-office gross revenue earned by the top 20 grossing films increased steadily from 5 percent in 1949 to 24 percent in 1967. Motion picture Almanac (1987, pp. 233–288). Because total production declined, this figure somewhat overstates the increase in the share earned by the top grossing pictures.

114 Donahue (1987, p. 134).

115 A reserve price, first-price auction is revenue maximizing and efficient. The minimum guaranteed rental payment set by the distributor is the reserve price. The reserve price also affords the distributor some protection against exhibitor bidding rings. Whereas, it was common to renegotiate terms to prevent the exhibitor from large losses before *Paramount*, this practice stopped when bidding became prevalent. See note 98 supra.

116 Hellmuth (1962, p. 415 n. 35). See also Cassady and Cassady (1974, p. 4).

117 Industry managers estimated that the preferred number of screens per complex usually was 12, with 18 at some locations. See *Wall Street Journal* (March 17, 1988).

118 Harris (January 23, 1988) and National Cable Television Association (April 6, 1987, table I).

119 *Standard and Poor's* (March 26, 1987).

120 National Cable Television Association (April 6 1987, p. 3), and (November 1986, p. 27). The Paramount, RKO, Universal, United Artists, and Columbia decrees do not prohibit them from entering the exhibition business. RKO in 1948 and Paramount in 1949 were the first of the five major defendants to sign decrees, negotiated before the District Court's final decision on remand from the Supreme Court. The decrees as to Universal, United Artists, and Columbia contained no prohibition on theater ownership because these firms did not own theaters at the time of the case. Conant (1960, p. 105) and Helmuth (1962, p. 412). Twentieth Century Fox, Loew's (which owned MGM), and Warner Bros. did not negotiate decrees until their appeal from the District Court's final decision had been rejected by the Supreme Court. *United States* v. *Paramount Pictures, Inc.*, 85 F. Supp. 881 (S.D.N.Y. 1949); 339 U.S. 974 (1950). Their decrees, signed in 1951 and 1952, prohibit them from reentering exhibition. See note 2 supra, the Warner Bros. decree, sec. VI, para. B, at p. 64,273; the Twentieth Century Fox decree, sec. VI, at p. 64,550; and the Loew's decree, sec. VI, para. H, at p. 67,331.

121 Delugach (October 22, 1981), Harris (July 27, 1986), Cieply and Barnes (August 21, 1986), and National Cable Television Association (November 1986, p. 28).

122 Laura Landro (September 17, 1986) and (October 20, 1986), and National Cable Television Association (November 1986, pp. 4–5, Table 1).

123 *Standard and Poor's* (March 26, 1987) and National Cable Television Association (April 6, 1987, pp. 4–5), citing Lawrence Cohn, "New Economy of Scale in Hollywood, Majors Buys Wide-Reaching," *Variety*, January 14, 1987, p. 7.

124 *United States* v. *Griffith Amusement Co.*, 68 F. Supp. 180, 189 (1946).

125 *United States* v. *Griffith Amusement Co.*, 68 F. Supp. 180, 189 (1946).

126 Coase (1972, p. 68).

127 See text accompanying note 4 supra.

128 *Comment, Yale Law Journal* (1965, pp. 1102–1108).

8 Was the antitrust action that broke up the movie studios good for the movies?: evidence from the stock market

1 *United States* v. *Paramount Pictures, Inc.*, 661. Supp. 323 (S.D.N.Y. 1946).

2 See Conant (1960) for the arguments made in favor of the hypothesis that the integrated studios and their exhibitor contracts were the basis of a price fixing conspiracy.

3 Justice Vaught of the US District Court, industry observers and scholars have expressed similar conclusions; see Schickel (1984) and De Vany and Eckert (1991).

4 These and other common practices in the industry at this time are explained in De Vany and Eckert, 1991 who argue that they were efficient responses to uncertainty and incentive problems. The stark uncertainty of the motion picture business is characterized in Goldman's (1983) "nobody knows" principle which is implied by the stable Paretian distribution theory of box-office revenues developed by De Vany and Walls, 1997, 1999, and De Vany and Lee, 2001. De Vany, 2003 shows that the flexibility inherent in these contracts was essential to the efficient pricing and timing of bookings in the complex and volatile dynamics of motion picture box-office revenues.

5   If a stock did not trade on a given day, the price on the previous day was used for the price on that day.

6   Crandall (1975) at p. 60, table I.

7   De Vany and Lee (2000) show that concentration measures are extremely unstable and that the expected value of the HHI does not exist mathematically (it is infinite).

8   In fact, the price of exhibiting a "hit" is always high given the design of the exhibition license (De Vany 2003).

9   See note 6.

10 The Department of Justice regulated entry into exhibition using a test of whether a theater was "needed" in a city that was virtually identical to the test of public convenience and necessity applied by the ICC and the CAB when they controlled entry into trucking and airlines. For a discussion of entry control into exhibition, see the reference cited as "Comment."

11 Some members of the public, primarily church groups, also objected to block booking as it obligated the exhibitor to show films the group sometimes found to be offensive. In these instances, the objection to block booking was a form of censorship.

12 It is interesting to note that, in later years and under a different administration, the Department of Justice began to bring suits against exhibitors, claiming that motion picture "splits" were a violation of the Sherman Antitrust Act. In an exhibitor split, motion pictures are allocated among cooperating exhibitors. Justice claims these cooperative allocations are made to reduce film rental payments to distributors, though that seems unlikely to be true. One irony of splits is that each exhibitor takes the releases of a given distributor "down the line" and that creates a franchise-like arrangement rather like those prevailing before *Paramount*. It is hard to suppress the thought that a long-run relationship between distributor and theater is so beneficial that they will find some way to establish it.

13 Distributors began to face the "end period problem" as exhibitors began to go out of business in the decline of the middle 1950s. They could not prevent a marginal exhibitor from placing a high bid for their film, showing it, and then going out of business without payment of owed film rentals. The distributor could not even recover the print leased to the exhibitor if he went out of business (De Vany and Eckert 1991).

14 Film Daily 1950, at p. 138.

9   Stochastic market structure: concentration measures and motion picture antitrust

1   Lerner (1934).

2   See Tirole (1988, p. 66, pp. 218–219).

3   See Saving (1970) and Curry and George (1983).

4   Marginal cost is assumed to be identical for all firms in the dominant group. See Martin (1993).

5   Martin (1993, pp. 166–167 and pp. 169–170). The model can be extended for case of differentiated products. In this case, market power would be affected by the degree of product differentiation in addition to conjectured reactions of rivals, price elasticity of demand and market share.

6   Schmalensee (1987).

7   Salop (1987, p. 10).

8   Fisher (1987, pp. 30–31).

9   *United States* v. *Paramount Pictures, Inc.*, 85F. Supp. 881, 894 (S.D.N.Y. 1949). See Conant (1960, pp. 44–45) and De Vany and Eckert (1991, p. 90). During

the 1943–44 season, the eight defendants had received 95 percent of domestic film rentals (Westerns excluded).

10 Recall that the original distributors-defendants were: Loews, Paramount, RKO, Twentieth Century Fox, Warner Bros, Columbia, Universal and United Artists. Loews and United Artists are now part of MGM while Columbia is now part of Sony Pictures. The new entry into this upper echelon is the Disney-owned Buena Vista.

11 Note that in the *Paramount* case, concentration was also measured in terms of the number of films produced and distributed. For example see *United States* v. *Paramount Pictures, Inc.*, 66F. Supp. 323, 894 (S.D.N.Y. 1946). See Conant (1960, p. 36). Another case which looked at market share in terms of features distributions is *United States* v. *Loew's Inc.* See Conant (1960, p. 45). This approach is questionable because releasing a large number of films does not necessarily translate into greater box-office success.

12 Values of CR5 and CR8 for 1943–44 are computed from *United States* v. *Loew's Inc.*, 334 US 131. Final Finding of Fact 100, February 8, 1950. See Conant (1960, table 13, p. 46).

13 *United States* v. *Paramount Pictures, Inc.*, 85F. Supp. 881, 894 (S.D.N.Y. 1949).

14 US Department of Justice and Federal Trade Commission (1997, p. 15).

15 These additional factors relate to the presence/absence of conditions conducive to collusion entry–exit conditions and efficiency gains from merger. See US Department of Justice and Federal Trade Commission (1997, pp. 18–34).

16 US Department of Justice and Federal Trade Commission (1997, p. 16).

17 US Department of Justice and Federal Trade Commission (1997, p. 16, fn.18).

18 US Department of Justice and Federal Trade Commission (1997, p. 19): "It is likely that market conditions are conducive to coordinated interaction when the firms in the market previously have engaged in express collusion and when the salient characteristics of the market have not changed appreciably since the most recent such incident."

19 See for example Scherer and Ross (1990). This is also explicitly acknowledged in US Department of Justice and Federal Trade Commission (1997).

20 The correlation between the HHI and CR6 is significantly lower than those reported for some manufacturing industries (larger than 0.9). See Scherer and Ross (1990). Our results support House's (1977, p. 75) finding that concentration indices are not highly correlated especially at high levels of concentration.

21 Note that a comparison between Figures 9.2 and 9.3 reveals many instances where they move in opposite directions. For example, the HHI declined between week 3 and week 4, and between week 5 and week 6. During the same two periods, the CR6 increased. This could happen if the increase in the market shares of the top six distributors is accompanied by increases in market shares of some lower ranked distributors.

22 De Vany and Walls (1997).

23 We can extend this to include more ranks, for example, rank-1, rank-2 and rank-3. The conclusions we can draw are essentially the same. However, as we include more ranks, each distributor is likely to have more films in a given week. Clearly, a more sophisticated model would be one in which distributors maintain a portfolio of films.

24 Global transmission of information affects all agents simultaneously while local information is transmitted locally from one agent to other neighboring agents (e.g. network of immediate friends).

25 De Vany and Eckert (1991, p. 110).

26 De Vany and Eckert (1991, p. 102, fn. 16).

27 See Cones (1997, pp. 31–32).
28 De Vany and Walls (1996, 1999), Lee (1999), and De Vany and Lee (2001).

10 Motion picture profit, the stable Paretian hypothesis and the curse of the superstar

1 The distribution is also known as the stable distribution or the Lévy stable distribution. It was famously called the stable-Paretian distribution by Mandelbrot and Fama. For consistency and conformity with current practice, we shall call it the stable distribution here.

2 Other distributions, such as the student-t can be used to model heavy tails, but they are ad hoc in that their use cannot appeal to the generalized central limit theorem.

3 See Samorodnitsky and Taqqu (1994).

4 The Gini coefficient of profit, a measure of inequality, is 0.608.

5 The Gini coefficient of inequality of losses, 0.461, is less than the coefficient of profit.

6 This is the estimate from our sample of 2,015 movies released in the North American theatrical market from 1984 through 1996.

7 We analyze a sample of 2,015 movies exhibited in North America between 1984 and 1996 inclusive. The data were obtained from AC Nielson EDI. The composition of the sample—ratings, genres, revenues, budgets, star presence, etc.—is given in detail in De Vany and Walls (1999).

8 Skewness-kurtosis and Shapiro–Francia tests both reject the null hypothesis of Normality at significance levels of practically zero.

9 Parameter estimation and computation of the theoretical density function were performed using John Nolan's (1998) STABLE software package.

10 The empirical density function is calculated using the Gaussian kernel with the width parameter equal to 2(inter-quartile range)$n^{-1/3}$.

11 Another diagnostic for detecting variation near the extreme values is the variance-stabilized PP-plot, where a transformation is applied to make the variance in the PP-plot uniform. The variance-stabilized PP-plot is nearly the same as the PP-plot shown in Figure 10.2 so we have not included it here.

12 See Nolan (1999) for the formulas of the pointwise confidence bands.

13 The $\chi^2_{df=1}$ test statistic is 3.04 and the upper 5th percentage point of the χ^2 distribution with one degree of freedom is 3.84.

14 That is to say, owing to their asymmetry, movies with stars stochastically dominate movies that do not have stars. Both have a theoretical variance that is infinite, so variance does not exist mathematically and cannot be a measure of risk. The sample variance is finite for both kinds of movies, but it is not an estimate of the expected variance. The sample variance grows with the number of observations.

15 Using the range from −$100 to $100 million gives a negative estimate of expected profit. One must calculate over the whole range of outcomes because of the heavy probability mass that lies in the extreme tails. For our numerical integration, we used a range from −$5 billion to $5 billion.

16 Movies in this range correspond to the overpopulated category of money losing R-rated movies featuring stars that De Vany and Walls (2001) discovered.

17 This calculation is biased because the cost of movies in our data already include the star's pay. It serves only to illustrate the point. More detailed data on movie budgets would be required to do the calculation precisely and they simply are not available.

18 John Brodie and Anita Busch (1996) noted the heavy losses studios incurred in superstar "megapix."

19 In a collection of 22 superstar contracts that we reviewed we find that all of them have profit participation. More evidence comes from the number of superstars who have formed their own production companies to take back-end participation through their company in addition to collecting a fee. The growing incidence of stars who take producer credits, and thereby hold a profit interest, is further evidence of the rise of back-end participation.

20 Since cost is roughly linear in time, the expected time to completion will, like cost, be proportional to the time already expended in production (on this "self-similarity" property of power laws see Schroeder 1990).

21 Caves (2000) makes the observation that handing off part of the production sequence to another agent at each step in the production sequence helps to obviate moral hazard as a source of the angel's nightmare. The angel's nightmare may exist even if there is no moral hazard so long as the production process has the Paretian property.

22 Many other iterative, complex production processes are likely to be self-similar and experience this form of cost overrun.

23 See De Vany and Eckert 1991, Caves *et al.* 2000 for an analysis of the "nobody knows" principle and its basis in the stable distribution.

24 See De Vany and Eckert (1991) on the exhibition license and Filson *et al.* (2001) for an analysis of the way exhibition licenses allocate returns and risk.

11 Contracting with stars when "nobody knows anything"

1 Cumulative gross is a proxy measure of influence or productivity, but I am not implying that these artists are causally responsible for the grosses of their movies.

2 This is variant of Mandelbrot's dead poet metaphor follows from the linearity of the conditional expectation in n, see Schroeder, 1990.

3 This holds for movie audiences as well, as shown in Chapter 2.

4 See De Groot (1970) on optimal stopping and Dixit and Pindyck (1994) on the value of waiting for new information.

12 How extreme uncertainty shapes the movie business

1 Economists, who are familiar with homogeneous production functions, may see this as homogeneity of degree H in the density function.

Bibliography

Adler, Moshe. Stardom and talent. *American Economic Review*, 75(1): 208–212, 1985.

Alchian, Armen A. Uncertainty, evolution and economic theory. *Journal of Political Economy*, 58(3): 211–222, 1950.

Alicoate, Jack, ed. *1950 Film Daily Year Book of Motion Pictures*. Wid's Films and Film Folk, Inc., New York, 1950.

Anderson, J. V. and D. Sornette. Have your cake and eat it too: increasing returns while lowering large risks. Working paper, Institute for Geophysics and Planetary Physics, UCLA, 1999.

Anderson, L. R. and C. A. Holt. Informational cascades in the laboratory. *American Economic Review*, 87(5): 847–862, 1997.

Arthur, Brian. The economy and complexity. In Daniel Stein, editor, *Lectures in the Sciences of Complexity: Santa Fe Institute Studies in the Sciences of Complexity*, pages 713–740. Addison-Wesley, Redwood City, CA, 1989.

Arthur, W. Brian. Complexity in economics and financial markets. *Complexity*, 1: 20–25, 1995.

Atkinson, A. B. and A. J. Harrison. *Distribution of Personal Wealth in Britain*. Cambridge University Press, Cambridge, 1978.

Banerjee, Abhijit V. A simple model of herd behavior. *Quarterly Journal of Economics*, 107(3): 797–817, August 1992.

Beaton, A. E. and J. W. Tukey. The fitting of power series, meaning polynomials, illustrated on band-spectroscopic data. *Technometrics*, 16: 146–185, 1974.

Bhagat, Sanjai and Roberta Romano. Event studies and the law: part ii: technique and corporate litigation. *American Law and Economics Review*, 4: 141–167, 2002.

Bikhchandani, Sushil, David Hirshleifer and Iwo Welch. A theory of fads, fashion, custom, and cultural change as informational cascades. *Journal of Political Economy*, 100(5): 992–1026, 1992.

Binder, John. Measuring the effect of regulation with stock price data, *Rand Journal of Economics*, 8: 167–183, 1985.

Bittlingmayer, George and Thomas W. Hazlett. Dos kapital: has antitrust action against microsoft created value in the computer industry? *Journal of Financial Economics*, 55: 329–359, 2000.

Blumenthal, Marsha A. Auctions with constrained information: blind bidding for motion pictures. *Review of Economics and Statistics*, 70(2): 191–198, 1988.

Box, G. E. P. and D. R. Cox. An analysis of transformations. *Journal of the Royal Statistical Society B*, 26: 211–243, 1964.

Brock, William A. Scaling laws for economists. Technical report, University of Wisconsin, Madison, 1993.

Brodie, John and Anita Busch. Stars breaking bank while megapix tank. *Variety*, 90: 1, 1996.

Carlton, Dennis. Market behavior with demand uncertainty and price inflexibility. *American Economic Review*, 68: 571–587, 1978.

Cassady, Ralph, Jr., and Ralph Cassady. *The Private Antitrust Suit in American Business Competition*. Bureau of Business and Economic Research, University of California, Los Angeles, 1974.

Caves, Richard E. *Creative Industries: Contracts between Art and Commerce*. Harvard University Press, Cambridge, 2000.

Chisholm, Darlene C. Continuous degrees of residual claimancy: some contractual evidence. *Applied Economics Letters*, 3(11): 739–741, November 1996.

Chisholm, Darlene C. Profit-sharing versus fixed-payment contracts: evidence from the motion pictures industry. *Journal of Law, Economics and Organization*, 13(1): 169–201, April 1997.

Chung, Kee H. and Raymond A. K. Cox. A stochastic model of superstardom: an application of the Yule distribution. *Review of Economics and Statistics*, 76(4): 771–775, 1994.

Collins, Allen, Chris Hand and Martin C. Snell. What makes a blockbuster? Economic analysis of film success in the United Kingdom. *Managerial and Decision Analysis*, 23(6), 2002.

Comment. An experiment in preventive anti-trust: judicial regulation of the Motion Picture Exhibition Market under the Paramount decrees. *Yale Law Journal*, 74: 1041–1112, 1965.

Conant, Michael. *Antitrust in the Motion Picture Industry: Economic and Legal Analysis*. University of California Press, Berkeley, 1960.

Cones, John W. *The Feature Film Distribution Deal*. Southern Illinois University Press, Carbondale, 1997.

Cox, D.R. and D. Oates. *Analysis of Survival Data*. Chapman and Hall, New York, 1984.

Crandall, Robert W. The postwar performance of the motion-picture industry. *20 Antitrust Bulletin*, 49, 1975.

Curry, B. and K. D. George. Industrial concentration: a survey. *Journal of Industrial Economics*, 31(3): 203–255, 1983.

Dale, Martin. *The Movie Game: The Film Business in Britain, Europe and America*. Cassell, London, 1997.

Day, Richard. Adaptive processes and economic theory. In Ricard H. Day and Theodore Groves, editors, *Adaptive Economic Models*, pages 1–38. Academic Press, New York, 1975.

De Groot, Morris. *Optimal Decision Theory*. Wiley, New York, 1968.

De Vany, Arthur. Uncertainty, waiting time and capacity utilization—a stochastic theory of product quality. *Journal of Political Economy*, 84: 523–541, 1976.

De Vany, Arthur. Contracting in the movies when "nobody knows anything". In Victor Ginsburg, editor, *Proceedings of the Rotterdam Conference on Cultural Economics*, North-Holland (forthcoming 2003).

De Vany, Arthur S. and Cassey Lee. Motion pictures and the stable Paretian hypothesis. IMBS working paper, University of California at Irvine, 2000.

De Vany, Arthur S. and Cassey Lee. Quality signals in information cascades and the dynamics of the distribution of motion picture box office revenues. *Journal of Economic Dynamics and Control*, 25: 593–614, 2001.

De Vany, Arthur S. and Ross Eckert. Motion picture antitrust: the Paramount cases revisited. *Research in Law and Economics*, 14: 51–112, 1991.

De Vany, Arthur and Hank McMillan. Markets, hierarchies, and antitrust in the motion picture industry. Working paper, Economics Department, University of California, Irvine, 1993.

De Vany, Arthur S. and W. David Walls. Innovation tournaments: an analysis of rank, diffusion and survival in motion pictures. Paper presented, NBER Conference on New Products, Cambridge, MA, 1993.

De Vany, Arthur S. and W. David Walls. Bose–Einstein dynamics and adaptive contracting in the motion picture industry. *The Economic Journal*, 439(106): 1493–1514, 1996.

De Vany, Arthur S. and W. David Walls. The market for motion pictures: Rank, revenue and survival. *Economic Inquiry*, 4(35): 783–797, November 1997.

De Vany, Arthur S. and W. David Walls. Private information, demand cascades and the blockbuster strategy. IMBS working paper, University of California at Irvine, 2000.

De Vany, Arthur and W. David Walls. Does Hollywood make too many R-rated movies? Risk, stochastic dominance, and the illusion of expectation. *Journal of Business*, 75(3): 425–451, April 2002.

De Vany, Arthur and W. David Walls. Motion picture profit, the stable Paretian hypothesis, and the curse of the superstar. *Journal of Economic Dynamics and Control*, 2002.

De Vany, Arthur and W. David Walls. Quality evaluations and the breakdown of statistical herding in the dynamics of box office revenues. IMBS working paper, University of California at Irvine, 2002.

De Vany, Arthur S. and W. David Walls. Uncertainty in the movie industry: does star power reduce the terror of the box office? *Journal of Cultural Economics*, 23(4): 285–318, November 1999.

De Vany, Arthur S. and W. David Walls. Motion picture profit, the stable Paretian hypothesis, and the curse of the superstar. *Journal of Economic Dynamics and Control*, in press.

Dixit, Avinash K. and Robert S. Pindyck. *Investment under Uncertainty*. Princeton University Press, Princeton, 1994.

Eliashberg, Jehoshua and Steven M. Shugan. Film critics: influencers or predictors? *Journal of Marketing*, 61(2): 68–78, April 1997.

Fama, Eugene. Mandelbrot and the stable Paretian hypothesis. *Journal of Business*, 36(4): 420–429, 1963.

Fama, Eugene. The behavior of stock market prices. *Journal of Business*, 38(1): 34–105, 1965.

Feller, William. *An Introduction to Probability Theory and Its Applications*. John Wiley & Sons, Inc., New York, 1957.

Filson, Darren, Fernando Fabre, Alfredo Nava and Paola Rodriguez. At the movies: risk sharing and the economics of exhibition contracts. Working paper, Claremont Graduate University, 2001.

Fisher, Franklin M. Horizontal mergers: triage and treatment. *Economic Perspectives*, 1(2): 23–40, 1987.

Ghosh, A. The size distribution of international box office revenues and the stable Paretian hypothesis? Working paper, Economics department, University of California at Irvine, 2000.

Goldman, William. *Adventures in the Screen Trade*. Warner Books, New York, 1983.

Gordon, Robert. Output fluctuations and gradual price adjustment. *Journal of Economic Literature*, XIX: 493–530, 1981.

Greene, William H. *Econometric Analysis*. Prentice-Hall, New York, third edition, 1997.

Hamilton, L. C. How robust is robust regression? *Stata Technical Bulletin*, 2: 21–26, 1991.

Harmetz, Aljean. Film studios buying up theaters in major cities. *New York Times*, part. C, p. 20, October 23, 1986.

Harris, Kathryn. Box office potential entices large chains, movie makers. *Los Angeles Times*, part. IV, p. 5, July 27, 1986.

Heckman, James and Burton Singer. A method for minimizing the impact of distributional assumptions in econometric models for duration data. *Econometrica*, 52(2): 271–320, 1984.

Hill, Bruce and Michael Woodrofe. Stronger forms of Zipf's law. *Journal of the American Statistical Association*, 70(349): 212–219, March 1975.

House, John C. The measurement of concentrated industrial structure and the size distribution of firms. *Annals of Economic and Social Measurement*, 6(1): 73–107, 1977.

Huang, C. F. and R. H. Litzenberger. *Foundations for Financial Economics*. North-Holland, New York, 1988.

Huber, P. J. Robust estimation of a location parameter. *Annals of Mathematical Statistics*, 35: 73–101, 1964.

Ijiri, Yuji and Herbert A. Simon. Effects of mergers and acquisitions on business firm concentration. *Journal of Political Economy*, 79(2): 314–322, Mar/Apr 1971.

Ijiri, Yuji and Herbert A. Simon. Interpretations of departures from the Pareto curve firm-size distributions. *Journal of Political Economy*, 82(2): 315–332, Mar/Apr 1974.

Ijiri, Yuji and Herbert A. Simon. *Skew Distributions and the Sizes of Business Firms*. North-Holland, New York, 1977.

Jovanovic, Boyan. Micro shocks and aggregate risk. *Quarterly Journal of Economics*, 17:395–409, 1987.

Kalbfleisch, J. D. and R. L. Prentice. *The Statistical Analysis of Failure Time Data*. Wiley, New York, 1980.

Kaplan, E. L. and P. Meier. Nonparametric estimation from incomplete observations. *Journal of the American Statistical Association*, 53: 457–481, 1958.

Kenney, Roy W. and Benjamin Klein. The economics of block booking. *Journal of Law and Economics*, 26: 497–540, 1983.

Kwoka, John. The herfindahl index in theory and practice. *Antitrust Bulletin*, Winter: 915–947, 1985.

Langer, E. J. The illusion of control. *Journal of Personality and Social Psychology*, 32: 311–338, 1975.

Lee, Cassey. Bose–Einstein dynamics in the motion picture industry revisited. Mimeo, Economics Department, University of California at Irvine, 1998.

Lehmann, D. R. and C. B. Weinberg. Sale through sequential distribution channels: an application to movies and videos. *Journal of Marketing*, 64(3): 18–33, 2000.

Levinthal, Daniel A. and James G. March. The myopia of learning. *Strategic Management Journal*, 14: 95–112, 1993.

Levy, M. and S. Soloman. Of wealth power and law: the origin of scaling in economics. *Racah Institute of Physics, Hebrew University*, 1998.

Lippman, Steven A. and John J. McCall. Job search in a dynamic economy. *Journal of Economic Theory*, 12(3): 365–390, June 1976.

Litman, Barry. The motion picture entertainment industry. In Walter Adams, editor, *The Structure of American Industry*, pages 183–216. Macmillan, New York, 1990.

Litman, B. and H. Ahn. Predicting financial success of motion pictures: the early 90s experience. In *The Motion Picture Mega-Industry*, chapter 10, pages 172–197. Allyn and Bacon, Needham Heights, Massachusetts, 1998.

Litwak, Mark, *Reel Power: The Struggle for Influence and Success in the New Hollywood*. William Morrow, New York, 1986.

Lukk, Tiiu. *Movie Marketing: Opening the Picture and Giving it Legs*. Silman-James Press, Los Angeles, 1997.

McCulloch, J. Huston. Continuous time processes with stable increments. *Journal of Business*, 51(4): 601–619, October 1978.

McCulloch, J. Huston. Financial applications of stable distributions. In G. S. Maddala and C. R. Rao, editors, *Statistical Methods in Finance*, volume 14 of *Handbook of Statistics*, pages 393–425. North-Holland, New York, 1996.

McCulloch, J. Huston. Measuring tail thickness to estimate the stable index alpha: a critique. *Journal of Business and Economic Statistics*, 15(1): 74–81, January 1997.

McKensie, Jordi. Box office revenue distributions, Paretian tails, and survival analysis. *Draft Thesis*, 2003.

Mandelbrot, B. New methods in statistical economics. *Journal of Political Economy*, 71: 421–440, 1963a.

Mandelbrot, B. The variation of certain speculative prices. *Journal of Business*, 36: 394–419, 1963b.

Mandelbrot, Benoit. *Fractals and Scaling in Finance*. Springer Verlag, New York, 1997.

Mantegna, R. N. and H. E. Stanley. Scaling in financial markets. *Nature*, 376: 46–49, 1995.

Martin, Stephen. *Advanced Industrial Economics*. Blackwell, Oxford, 1993.

May, R. M. Patterns of species abundance and diversity. In R. Patrick, editor, *Diversity: Benchmark Papers in Ecology*, pages 340–379. Hutchinson Ross, Stroudsburg, PA, 1983.

Medved, Michael. *Hollywood vs. America: Popular Culture and the War on Traditional Values*. Harper Collins, New York, 1992.

National Cable Television Association. The "compulsory cartel": a survey of the motion picture studios' drive for dominance over program supply and exhibition, Washington, DC. November 1986.

Nelson, Randy A., Michael R. Donihue, Donald Waldman and Calbraith Wheaton. What's an Oscar worth? Unpublished manuscript, Department of Economics, Colby College, Maine, undated.

Nelson, Richard R. Firm and industry response to changed market conditions: an evolutionary approach. *Economic Inquiry*, 18: 179–202, 1977.

Nolan, John P. Univariate stable distributions: parameterizations and software. In Robert J. Adler, Raisa E. Feldman and Murad S. Taqqu, editors, *A Practical Guide* to *Heavy Tails: Statistical Techniques and Applications*, pages 527–533. Birkhauser, Berlin, 1998.

Nolan, John P. Fitting data and assessing goodness-of-fit with stable distributions. Mimeograph, Mathematics Department, American University, Washington, DC, June 1999.

Pagano, M. and K. Gauvreau. *Principles of Biostatistics*. Brooks Cole Publishing, Pacific Grove, CA, 1993.

Pareto, V. *Cours d'Economique Politique*. MacMillan, Paris, 1897.

Prag, Jay and James Cassavant. An empirical study of determinants of revenues and marketing expenditures in the motion picture industry. *Journal of Cultural Economics*, 18(3): 217–35, 1994.

Presson, Paul K. and Victor A. Benassi. Illusion of control: a meta-analytic review. *Journal of Social Behavior and Personality*, 11(3): 493–510, 1996.

Ravid, S. A. Information, blockbusters and stars: a study of the film industry. *Journal of Business*, 72: 463–486, 1999.

Rosen, Sherwin. The economics of superstars. *American Economic Review*, 71: 167–183, 1981.

Ross, S. *Introduction to Probability Models*. Academic Press, San Diego, CA, fifth edition, 1993.

Rothschild, Michael. Searching for the lowest price when the distribution of prices is unknown. *Journal of Political Economy*, 82(4): 689–711, July/August 1974.

Salop, Steven. Symposium on mergers and antitrust. *Economic Perspectives*, 1(2): 3–12, 1987.

Samorodnitsky, G. and M. S. Taqqu. *Stable Non-Gaussian Random Processes*. Chapman and Hall, New York, 1994.

Scherer, F. M. and Ross, David. *Industrial Market Structure and Economic Performance*, Third edition. Houghton Mifflin, Boston, MA, 1990.

Scherer, F. M., D. Harhoff and J. Kukies. Uncertainty and the size distribution of rewards from innovation. *Evolutionary Economics*, 10: 175–200, 2000.

Schickel, Richard. *D. W. Griffith: An American Life*. Free Press, New York, 1984.

Schmalensee, Richard. Entry deterrence in the ready-to-eat breakfast cereal industry. *Bell Journal of Economics*, 9: 305–327, 1987.

Schroeder, Manfred. *Fractals, Chaos, Power Laws*. W. H. Freeman and Co., New York, 1990.

Simonoff, J. S. and I. R. Sparrow. Predicting movie grosses: Winners and losers, blockbusters and sleepers. *Chance* 13 (Summer): 15–24, 2000.

Smith, Rodney, Michael Bradley and Greg Jarrell. Studying firm-specific effects of regulation with stock market data: an application to oil price regulation. *Rand Journal of Economics*, 4(4): 467–489, 1986.

Smith, Sharon P. and V. Kerry Smith. Successful movies: a preliminary empirical analysis. *Applied Economics*, 18(5): 501–507, 1986.

Sornette, Didier and Daniel Zajdenweber. The economic return of research: the Pareto law and its implications. *European Physical Journal B*, 8(4): 653–664, 1999.

Stata Corporation. *Stata Reference Manual*, release 5, volume 3. Stata Press, College Station, TX, 1997.

Steindl, J. *Random Processes and the Growth of Firms: A Study of the Pareto Law*. Hafner, New York, 1965.

Swami, Sanjeev, Jehoshua Eliashberg and Charles B. Weinberg. SilverScreener: a modeling approach to movie screens management. *Marketing Science (Special Issue on Managerial Decision Making)*, 18(3): 352–372, 1999.

The Hollywood Reporter. Star power '98, 1998.

Thompson, Rex. Conditioning the return-generating process on firm specific events: a discussion of event study methods. *Journal of Financial and Quantitative Analysis,* 20(2): 151–168, 1985.

Uchaikin, Vladimir V. and Vladimir M. Zolotarev. *Chance and Stability: Stable Distributions and their Applications.* VSP, Utrecht, 1999.

United States v. Paramount Pictures, Inc., 66 F. Supp. 323, 354 (S.D.N.Y. 1946).

US Department of Justice and Federal Trade Commission. *Horizontal Merger Guidelines.* Revised April 8, 1997.

Vogel, Harold L. *Entertainment Industry Economics: A Guide for Financial Analysis.* Cambridge University Press, New York, second edition, 1990.

Wallace, W. Timothy, Alan Seigerman and Morris B. Holbrook. The role of actors and actresses in the success of films: how much is a movie star worth? *Journal of Cultural Economics,* 17(1): 1–24, 1993.

Walls, W. David. Increasing returns to information: evidence from the Hong Kong movie market. *Applied Economics Letters,* 4(5): 187–190, May 1997.

Walls, W. David. Product survival at the cinema: evidence from Hong Kong. *Applied Economics Letters,* 5(4): 215–219, April 1998.

Weinstein, Mark. Profit-sharing contracts in Hollywood: evolution and analysis. *Journal of Legal Studies,* 27(1): 67–112, 1998.

White, Halbert. A heteroskedasticity-consistent covariance matrix estimator and direct test for heteroskedasticity. *Econometrica,* 48: 817–838, 1980.

Williamson, James R. Appendix A: U. S. Department of Justice, merger guidelines. Press release, May 30. In *Federal Antitrust Policy During the Kennedy–Johnson Years.* Greenwood Press, London, 1968.

Wyatt, Justin. From roadshowing to saturation release: majors, independents, and marketing/distribution innovations. In Jon Lewis, editor, *The New American Cinema,* pages 64–86. Duke University Press, Durham, 1998.

Zipf, G. K. *Human Behavior and the Principle of Least Effort.* Addison-Wesley, Reading, Massachusetts, 1949.

Zolotarev, V. M. *One-dimensional Stable Distributions,* volume 65 of *American Mathematical Society Translations of Mathematics Monographs.* American Mathematical Society, Providence, 1986 (Translation of the original 1983 Russian edition.).

Index

Printed in Great Britain
by Amazon